Pierre N. V. Tu

Introductory Optimization Dynamics

Optimal Control with Economics and Management Science Applications

With 85 Figures

Springer-Verlag
Berlin Heidelberg New York Tokyo
1984

Dr. Pierre Ninh Van Tu
Associate Professor
Department of Economics
The University of Calgary
2500 University Drive, N. W.
Calgary, Alberta T2N 1N4
Canada

ISBN 3-540-13305-4 Springer-Verlag Berlin Heidelberg New York Tokyo
ISBN 0-387-13305-4 Springer-Verlag New York Heidelberg Berlin Tokyo

Printing and binding: Weihert-Druck GmbH, Darmstadt

2142/3140-543210

PREFACE

Optimal Control theory has been increasingly used in Economics and Management Science in the last fifteen years or so. It is now commonplace, even at textbook level. It has been applied to a great many areas of Economics and Management Science, such as Optimal Growth, Optimal Population, Pollution control, Natural Resources, Bioeconomics, Education, International Trade, Monopoly, Oligopoly and Duopoly, Urban and Regional Economics, Arms Race control, Business Finance, Inventory Planning, Marketing, Maintenance and Replacement policy and many others. It is a powerful tool of dynamic optimization.

There is no doubt social sciences students should be familiar with this tool, if not for their own research, at least for reading the literature. These Lecture Notes attempt to provide a plain exposition of Optimal Control Theory, with a number of economic examples and applications designed mainly to illustrate the various techniques and point out the wide range of possible applications rather than to treat exhaustively any area of economic theory or policy. Chapters 2,3 and 4 are devoted to the Calculus of Variations, Chapter 5 develops Optimal Control theory from the Variational approach, Chapter 6 deals with the problems of constrained state and control variables , Chapter 7, with Linear Control models and Chapter 8, with stabilization models. Discrete systems are discussed in Chapter 9 and Sensitivity analysis in Chapter 10. Chapter 11 presents a wide range of Economics and Management Science applications. Only deterministic Control Theory will be dealt with: Stochastic Control, Differential Games and other related topics are beyond the scope of these introductory Lecture Notes.

My motivation to write this introductory text originates from a realisation, some six years ago, of the Economics, Business and other Social Sciences students' need of a simple text on Optimal Control Theory tailored to their own requirements and illustrated with familiar examples on the one hand and the scarcity at the time of such books, on the other. Students faced two extreme choices: they could either acquire a superficial understanding of the Theory by reading a chapter or two in Mathematical Economics textbooks, or wade their way through Mathematics and Engineering texts at the risk of getting lost by the degree of abstraction on the one hand and the unfamiliar Engineering examples on the other. Instructors had to write their own texts. These Lecture Notes cover a major part of a half course given at the University of Calgary to junior graduate and senior undergraduate Economics students with no more than a general knowledge of Linear Algebra and Calculus, including Differential and Difference Equations. The Mathematical Appendix is intended to remedy any further deficiencies students might have in Differential and Difference Equations.

No originality is normally claimed for Lecture Notes. These are no exception. The author owes an immense debt to the literature (listed in the References) which is so rich and varied that it is almost impossible to acknowledge all sources individually. No attempts will be made in this direction. The students who have taken the course have undoubtedly contributed much to the improvement of the exposition and reduction of fallacies. My colleagues have also made their contribution through intellectual conversations. Professors M.C. Kemp, R. Pindyck and R. Dobell have also made some suggestions which are much appreciated.

My thanks are also extended to the Department of A & M Economics of the University of Southampton where I spent my sabbatical leave in 1980-1981 for the provision of an ideal environment and research facilities needed to bring the first draft of this manuscript to completion. Mary Blount is to be commended for her exemplary patience and technical competence in the typing of most the manuscript. Finally, this publication has been made possible, in part, by a grant from the Endowment Fund of the University of Calgary, for which I am grateful. Needless to say that the remaining errors and deficiencies are mine alone.

CONTENTS

		Page
Preface		V
Chapter 1	INTRODUCTION	1
1.1	The Dynamic Optimization Problem	1
1.2	The Control Problem	1
1.3	The State of the Dynamic System	2
1.4	The Control Variables	3
1.5	Reachability, Controllability and Observability	3
1.6	The Objective Functional	4
1.7	Some Examples	5
1.8	The Calculus of Variations and Optimal Control Problems	6
Chapter 2	THE CALCULUS OF VARIATIONS	8
2.1	Functionals and their Variations	9
2.2	A Necessary Condition: The Euler Equation . .	11
2.3	Generalizations of Euler's Equation	16
	2.3.1 Case of Several Variables	16
	2.3.2 Case where f involves derivatives of nth order	17
2.4	Particular Cases of the Euler Equation	19
	2.4.1 Absence of x	19
	2.4.2 Absence of t	21
	2.4.3 Absence of $\dot{x}$	22
	2.4.4 $f(x, \dot{x}, t)$ is linear in $\dot{x}$	24
2.5	Variational Problems with Constraints	26
	2.5.1 Point and Differential Equation Constraints	26
	2.5.2 Isoperimetric Constraint	28
2.6	Some Economic Applications	32
	2.6.1 Dynamic Pure Competition	32
	2.6.2 Dynamic Utility and Capital Accumulation .	33
	2.6.3 Capital Theory	34
	2.6.4 Time Optimal Problem in Economic Planning .	36
	2.6.5 Optimal Education and Balanced Growth . .	37
	2.6.6 Micro Foundations of Macro Models	40
	2.6.7 Constrained Optimal Consumption Plan . . .	41
	2.6.8 Optimal Waste Disposal	42
	2.6.9 The Perimetric Problem of Non-Renewable Resources	44
2.7	Summary and Conclusion	45

Page

Chapter 3 BOUNDARY CONDITIONS IN VARIATIONAL PROBLEMS 48

3.1 Two fixed End Point and Natural Boundary Problems 48
3.2 Variable End Points 50
3.3 Broken Extremals and the Erdman-Weierstrass Corner Conditions 64
3.4 Canonical Form of the Euler Equation 72
3.5 Some Economic Applications 75

3.5.1 Dynamic Monopoly 75
3.5.2 Optimal Economic Growth 77
3.5.3 Capital Theory with Exhaustible Resources . 81
3.5.4 Optimal Mining with Incomplete Exhaustion . 85

3.6 Summary 88

Chapter 4 SECOND VARIATIONS AND SUFFICIENCY CONDITIONS 90

4.1 Introduction 90
4.2 Variations of Functionals 91
4.3 The Legendre Condition 92
4.4 The Jacobi Condition 93
4.5 The Weierstrass Condition for Strong Extrema . . 95
4.6 The Legendre-Clebsch Condition 98
4.7 Sufficient Conditions: An Important Special Case 102
4.8 Summary and Conclusion 105

Appendix to Chapter 4 109

Chapter 5 OPTIMAL CONTROL: THE VARIATIONAL APPROACH 110

5.1 Introduction 110
5.2 From the Calculus of Variations to Optimal Control 110
5.3 Pontryagin's Maximum Principle 113
5.4 Transversality Conditions 122

5.4.1 Problems with Fixed Final Time T 123
5.4.2 Problems with Free Final Time T 124
5.4.3 Transversality Conditions in Infinite Horizon Problems 130

5.5 Second Variations and Sufficient Conditions . . . 135
5.6 Some Economic Applications 140

5.6.1 Dynamic Monopoly 140
5.6.2 Optimal Growth 141
5.6.3 Non-Renewable Resources 144
5.6.4 Optimal Population 146
5.6.5 Optimal Phasing of Deregulation 148

5.7 Summary and Conclusion 151

		Page
Chapter 6	CONSTRAINED OPTIMAL CONTROL PROBLEMS	153
6.1	Introduction	153
6.2	Optimal Control with Equality Constraints	153
6.3	Optimal Control with Inequality Constraints . .	158
	6.3.1 Bounded Control Variables	158
	Application: Permanent Capital in the Resource Industries	169
	6.3.2 Bounded State Variables	171
	Application: Optimal Investment in Physical and Human Capital	176
6.4	Dynamic Programming, Hamilton-Jacobi Equation and the Euler Equation	183
6.5	Summary and Conclusion	188
	Appendix to Chapter 6	190
Chapter 7	LINEAR OPTIMAL CONTROL	193
7.1	Introduction	193
7.2	Bang Bang Control and Time Minimum Problem . . .	195
	Economic Application: Optimal Monetary Policy .	202
7.3	Singular Control	205
7.4	Singular Control and the Calculus of Variations .	213
7.5	Singularity and Controllability	214
7.6	Some Economic Applications	216
	7.6.1 Optimal Economic Growth	216
	7.6.2 Resource Economics	219
	Reproducible Resources	220
	Non-Renewable Resources	224
	7.6.3 Optimal Domestic and Foreign Investment . .	226
7.7	Summary and Conclusion	233
Chapter 8	STABILIZATION CONTROL MODELS	234
8.1	Introduction	234
8.2	Linear Regulator Problems	235
8.3	Linear Tracking Problems	240
8.4	Controllability	245
8.5	Observability	247
8.6	Some Economic Applications	248
	8.6.1 The Multiplier-Accelerator Model	248
	8.6.2 Production and Inventory Stabilization Model	251
	8.6.3 Economic Stabilization: The Austrian Case	253
8.7	Conclusion	255

Page

Chapter 9 DISCRETE CONTROL SYSTEMS 256

9.1 Introduction 256
9.2 Discrete Calculus of Variations 256
Example 9.1 Discrete Optimal Growth Model . . . 258
9.3 Discrete Maximum Principle 261
Example 9.2 Optimal Management of Renewable Resources 264
Example 9.3 Discrete Linear Regulator 266
Example 9.4 Linear Tracking and Economic Stabilization 268
Example 9.5 Optimal Wage Price Control 273

Chapter 10 SENSITIVITY ANALYSIS 277

10.1 Introduction 277
10.2 Sensitivity Theory 278
10.3 Cross Sensitivity 281
10.4 Objective Functional Sensitivity 282
10.5 Stability and Sensitivity 283
10.6 Some Economic Applications 285

10.6.1 Tax and Taste Sensitivity and Cross Sensitivity 285
10.6.2 Optimal Growth Model: Sensitivity and Comparative Dynamics 287

10.7 Conclusion 291

Chapter 11 SOME ECONOMIC AND MANAGEMENT APPLICATIONS 293

11.1 Introduction 293
11.2 Some Economic Applications 294

11.2.1 Optimal Economic Growth 294

11.2.1.1 The One-Sector Optimal Growth Model . 295
11.2.1.2 Technical Progress in the Aggregate Model 301
11.2.1.3 Two-Sector Models 301
11.2.1.4 The Multisectoral Optimal Growth Models 308
11.2.1.5 Numerical Methods for Optimal Growth Models 310

11.2.2 Economic Stabilization Models 312
11.2.3 Dynamic Theory of the Firms 312
11.2.4 International Trade 313
11.2.5 Regional Economics 316
11.2.6 Optimal Urban Economics 319
11.2.7 Education, Labour Training and Human Capital 322
11.2.8 Natural Resources 323
11.2.9 Optimal Control of Pollution 323
11.2.10 Optimal Population Control 325
11.2.11 Optimal Control of the Armament Build-up 326

Page

11.3 Some Management Science Applications: A Dynamic Theory of the Managerial Firm . . . 331

11.3.1 Optimal Financing Model 331
11.3.2 Optimal Production and Inventory Models . 334
11.3.3 Marketing Models 334
11.3.4 Maintenance Models 338

11.4 Conclusion 340

MATHEMATICAL APPENDIX: REVIEW OF DIFFERENTIAL AND DIFFERENCE EQUATIONS

A.1 Introduction 341
A.2 Differential Equations 341

A.2.1 First Order Linear Differential Equation Systems . 343
A.2.2 Fundamental Matrix 345
A.2.3 The nth Order Differential Equation 348
A.2.4 Non-Homogeneous First Order Differential Equation Systems 348

A.3 Difference Equations 351
A.4 Stability of Differential and Difference Equations . . 355
A.5 Phase Diagrams and Non-Linear Differential Equations . . 357
A.6 Isoclines 361
A.7 Non-Linear Differential Equations 364
A.8 Phase Diagrams of Difference Equations 367
A.9 Liapunov's Second (or Direct) Method 369

REFERENCES 373

CHAPTER I

INTRODUCTION

1.1 The Dynamic Optimization Problem

The central economic problem of optimal allocation of scarce resources having alternative uses, at a point of time and over time, involves both static and dynamic optimization. The choice between less consumption now and more later--between short-term pain and long-term gain--is a dynamic optimization problem. So is the allocation of investment among the various sectors over the whole economic planning period.

It might be thought that the fulfillment of the static optimality conditions would automatically ensure dynamic optimality. But this is not the case (see Dorfman, Samuelson and Solow, 1958 Ch. 12): different statically optimal programmes trace out different intertemporal paths not all of which are dynamically efficient.

An important tool of dynamic optimization is the Calculus of Variations (CoV). This has been applied in Economics since 1924 (see Evans 1924, 1930, Ramsey 1928, and Hotelling 1931). The CoV, however, has its limitations which were successfully overcome by Pontryagin's (1962) Maximum Principle, which is also called the Optimal Control theory (O.C.).

1.2 The Control Problem

Driving a car, setting a thermostat, filling a bathtub, launching a missile, allocating scarce resources among competing needs, regulating the economy, setting a limit to wages and prices, choosing an appropriate monetary and fiscal policy to maintain full employment without inflation, are all examples of control. OC consists of choosing among all admissible control variables $u(t)$, the one which brings a dynamic system from some initial state $x(t_0)$ at time t_0 to some terminal state $x(T)$ at some terminal time T, in such a way as to impart a maximum or minimum to a certain objective functional which is also called performance index.

1.3 The State of the Dynamic System

The state of a dynamic system is a collection of numbers $x(t) \equiv (x_1(t), x_2(t), \ldots, x_n(t))$ which, once specified at time $t = t_0$, are completely determined for all times $t \geq t_0$ by the choice of the control vector $u(t) = (u_1(t), \ldots, u_r(t))$. The numbers $x_i(t)$ $(1 \leq i \leq n,\ t_0 \leq t \leq T)$ are called state variables and a state space is an n-dimensional space of which the state variables $x_i(t)$ $(1 \leq i \leq n)$ are co-ordinates. Similarly, $u(t)$ is an r-control vector. For example, $x(t)$ may represent the economic variables such as GNP, consumption, investment, and so on and u may stand for control variables or policy instruments such as Government expenditure or money supply.

The state at time t of a system is represented by a system of differential or difference equations which is called the dynamic system. For example

$$\dot{x}(t) = f[x(t), u(t), t]$$

or (1)

$$x(k+1) = f[x(k), u(k), k]$$

These dynamic systems could be linear or non linear, autonomous (without t or k in (1)) or non-autonomous, constant or variable coefficient differential or difference equations.

1.4 The Control Variables

The above dynamic systems are controlled by some appropriately chosen instruments. Only admissible controls or instruments, i.e., those which obey the various restrictions imposed by the physical conditions of the problem, need be considered. For example, if $u_i(t)$ represents the proportions of national income saved for capital formation in sector $i (i = 1, \ldots, r)$ then clearly $0 \leq u_i(t) \leq 1 \; \forall \; i, t$ and $0 \leq \sum_1^r u_i(t) \leq 1$. In general, we describe this physical constraint by the requirement that control variables must be chosen from the set of admissible controls $\Omega(u(t))$ i.e., $u(t) \in \Omega \, (u(t))$. In the above example, $\Omega \, (u(t)) \equiv \left\{ u_i(t) \; : \; 0 \leq u_i(t) \leq 1, \; 0 \leq \sum_1^r u_i(t) \leq 1 \right\}$.

If $u(t)$ is a function of time only, we have an open-loop control. An example is the setting of a washing machine or a dishwasher in which each cycle is of a given duration. If $u(t)$ is also a function of some state variables, i.e., if $u(t) = u[x(t), t]$ we have a closed loop control. An example is the thermostat control which is a function of the room temperature $x(t)$. Another example of closed loop control is the government expenditure $u(t)$ which is a function of GNP $x(t)$ and election time t : the state variable, GNP or actual room temperature will activate the control as necessary, i.e. $u(t) = u \; [x(t) \; t]$.

1.5 Reachability, Controllability and Observability

A state x^1 is said to be reachable from some arbitrary state x^0 at t_0 if a control $u_1 (T) \in \Omega \; (u(t))$ could be found such that $x(u_1, x^0, t_1) = x^1$ for some time $t_1 \geq t_0$. The collection of all such x^1 is called the set of reachable states at t (from x^0 at t_0) with respect to the admissible control set $\Omega(u)$.

Controllability refers to the fact that some final state x^1, usually located at the origin (i.e., $x^1 = 0$) for convenience, could be reached from some initial state x^0 at t_0 by some appropriate choice of $u(t) \varepsilon \Omega$ (u). Thus controllability is a necessary condition for the existence of a solution.

By Observability is meant the ability to determine the initial state x^0 from observation of current data and output. Output denotes some relation between the state and control vectors, i.e., $y(t) = g[x(t), u(t), t]$. The observability problem arises only when output cannot be measured explicitly.

1.6 The Objective Functional

Control variables must be chosen such as to maximize or minimize some functional (called performance index)

$$J = \int_{t_0}^{T} f_0 \ (x, u, t) \ dt \tag{2}$$

where f_0 is some real-valued well-behaved scalar function. If $f_0 \ (x, u, t) = \pi(x, p)e^{-rt}$ or $f_0 = u(c)^{-rt}$ then J is the present value of future profit (π) or consumption utility $u(c)$ discounted at rate r. This is a familiar economic concept.

There are three alternative ways to formulate the objective functional (2). They are equivalent and reducible to each other. They are:

$$J = \int_{t_0}^{T} f(x,u,t) \ dt + S \ (s(T), T) \tag{3}$$

where f and S are both assumed continuously differentiable. $S(x(T), T)$ is the "scrap" value function, giving the worth of the program at the terminal time period T, for example the scrap value of a machine or a car at the end of the planning period. This is called the Bolza problem.

The special case of (3) in which $S = o$, i.e.,

$$J = \int_{t_0}^{T} f(x, u, t)\, dt \tag{4}$$

is called the Lagrange problem.

Finally, the special case of (3) in which $f(x, u, t) = 0$, i.e.,

$$J = S(x(T), T) \tag{5}$$

is called the Mayer problem.

By a re-definition of variables, all three problems are equivalent. For example, the Bolza problem can be converted into a Mayer problem by defining the added variable $x_{n+1}(t)$ as

$$x_{n+1}(t) = \int_{t_0}^{t} f(x, u, t')\, dt'$$

$$x_{n+1}(t_0) = 0$$

in which case (3) becomes

$$J = x_{n+1}(T) + S(x(T), T)$$

1.7 Some Examples

Example 1

A driver uses throttle acceleration $u_1(t)$ and breaking deceleration $u_2(t)$ to control a car. Let the distance of the car from the original parking position by $y(t)$, and the total distance to be covered be d, a specified constant. Then, starting at time $t_0 = 0$ from a position of rest at $y(0) = 0$, the car obeys the dynamic system

$$d^2y/dt^2 \equiv \ddot{y} = u_1(t) + u_2(t)$$

or, defining $x_1(t) \equiv y(t)$ and $x_2(t) \equiv \dot{y}(t)$

$$\begin{bmatrix} \dot{x}_1(t) \\ \dot{x}_2(t) \end{bmatrix} = \begin{bmatrix} 0 & 1 \\ 0 & 0 \end{bmatrix} \begin{bmatrix} x_1(t) \\ x_2(t) \end{bmatrix} + \begin{bmatrix} 0 & 0 \\ 1 & 1 \end{bmatrix} \begin{bmatrix} u_1(t) \\ u_2(t) \end{bmatrix}$$

of the form

$$\dot{x}(t) = Ax(t) + Bu(t) \qquad (1)$$

obviously $\left(x_1(0),\ x_2(0)\right) = (0,0)$ and $\left(x_1(T),\ x_2(T)\right) = (d,0)$.

If the driver wants to arrive at destination in minimal time,

$$J = \int_0^T dt$$

<u>Example 2</u>

In a neo-classical model of economic growth described by

$$\dot{k}(t) = sf(k) - nk$$

where k is capital per head of worker, and $k(0) = k$, $k(T) = k_T$,

$f(k)$ is per capita production function,

s is the marginal propensity to save

n is the given constant rate of population growth,

the objective functional most commonly used is to maximize the discounted stream of utility (w) throughout the planning period, i.e.,

$$J = \int_0^T w(c)e^{-rt}\, dt$$

where $c = (1-s)\ f(k)$, per-capita consumption

r = constant rate of discount ($t_0 = 0$)

1.8 The Calculus of Variations and Optimal Control Problems

In the standard Calculus of Variations problem, the objective is to maximize or minimize some functional

$$J = \int_0^T f_0\ (x,\ \dot{x},\ t)\ \ dt$$

subject or not to some differential or algebraic equation of constraint

$$\dot{x} = f\ (x,\ \dot{x},\ t$$

In the standard Optimal Control the objective is to maximize or minimize some functional

$$J = \int_0^T f_0\ (x,\ u,\ t)\ \ dt$$

subject to

$$\dot{x} = f(x,\ u,\ t)$$

When $\dot{x} = u$, the similarity between the two becomes obvious. In fact, OC problems could be solved by the CoV techniques (Euler equation) and the Maximum Principle could be treated as a natural development of the CoV. The next three chapters are now devoted to the CoV and the following four chapters, to OC theory.

CHAPTER 2

THE CALCULUS OF VARIATIONS

The Calculus of Variations is a branch of Mathematics dealing with optimization of functionals. The variational problem goes back to the antiquity. The first solution seems to have been that of queen Dido of Carthage in about 850 B.C. Virgil reported that, having been promised all the land lying within the boundaries of a bull's hide, the clever queen cut the hide into many thin strips, tied them together in such a way as to secure as much land as possible within this boundary.[1] The solution is of course a circle. This is a typical isoperimetric problem of the Calculus of Variations. However, it was not until the late seventeenth century that substantial progress was made when a rigorous solution of the brachistochrone problem was provided by Newton, de l'Hospital, John and Jacob Bernouilli in 1696. This problem consists of determining the shape of a curve joining A to B such that a frictionless particle sliding along it under the influence of gravity alone moves from A to B in the shortest time. The solution is a cycloid. This played an important part in the development of the Calculus of Variations.[2]

In Economics, the use of the Calculus of Variations goes back to the 1920's with the works of Evans (1924, 1930), Ramsey (1928) and Hotelling (1931). Evans and Roos attempted to find the optimal price path for the whole planning period such as to maximize the profit functional of the monopolist. This is a typical problem of the Calculus of

Variations. Ramsey (1928) is another pioneer. He wanted to find a saving programme which would minimize the discrepancy between a certain bliss level of utility and the net actual utility (utility of consumption net of the disutility of labour) for all periods up to infinity. This optimal saving problem, which has constituted a source of inspiration in optimal economic growth theory, was solved by the tool of the Calculus of Variations. Hotelling's (1931) problem of optimal mining which has remained a classical pioneering work in the Economics of Natural Resources was again an exercise in the Calculus of Variations. Today, the Calculus of Variations and Optimal Control are commonplace in Economics and rightly so: Economics being defined as the Science of optimal allocation of scarce resources having alternative uses both at a point of time and overtime.

2.1 Functionals and Their Variations

Functionals play an important part in the Calculus of Variations. A functional, say norm $x = \|x\|$ or $J(x) = \int_a^b x(t)dt$, is a rule of correspondence that assigns to each function x in R a unique real number $\|x\|$ or $J(x)$. An analogy exists between functions and functionals. While the arguments of a function are variables, e.g. $x = x(t)$, the arguments of a functional are themselves functions, e.g. $J(x(t)) = \int_a^b x\,dt$. Just as a function is completely determined when its variables are given specific values, a functional is completely determined by the choice of one particular variable function from the set of all admissible functions. The increment of the argument of a function is the difference $dt = t^*-t$ (see fig. 2.1) while the variation (δx) of a functional is the difference $\delta x(t) = x^*(t)-x(t)$ between two functions of the same class. In the study of a function, one is concerned with finding points which make a function an extremum while in the case of a functional, one is interested in finding functions which impart an extremum to the functional (see fig. 2.1 and 2.2).

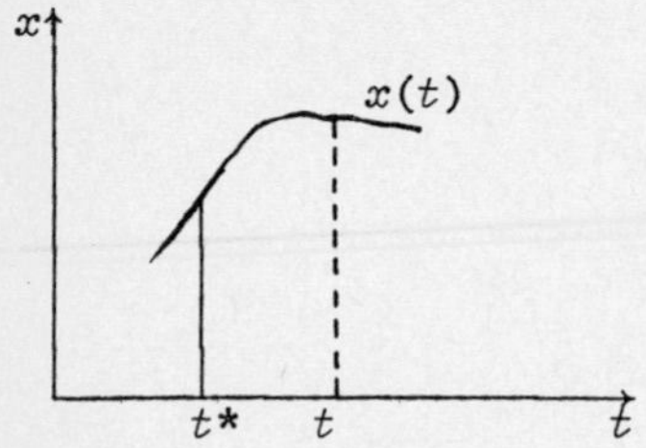

Fig. 2.1 Function $x = x(t)$

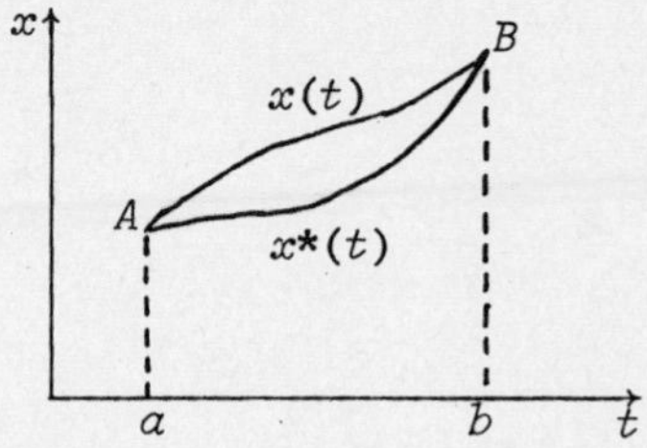

Fig. 2.2 Functional $J(x) = \int_a^b x\,dt$

Just as in the case of functions, the increment or variation of a functional $J(x)$ is

$$\Delta J(x) = J(x+\delta x) - J(x)\ . \tag{1}$$

Putting arbitrary $\delta x = h$, and using Taylor expansion, this could be written as

$$\begin{aligned} \Delta J(h) &= J(x+h) - J(x) \\ &= \phi(h) + Q(h) + 0\|h\|^2 \\ &= \delta J(h) + \delta^2 J(h) + 0\|h\|^2 \end{aligned} \tag{2}$$

where $\phi(h)$ is the linear term in the Taylor series referred to as first variations $\delta J(h)$ and $Q(h)$ is the quadratic term indicating second variations $\delta^2 J(h)$ and $0\|h\|^2 \to 0$ as $h \to 0$. Just as a function $f(t)$

achieves a relative maximum (minimum) at t^* when $f(t^*) \geq f(t)$ $(f(t^*) \leq f(t))$ for all t sufficiently close[3] to t* and global maximum (minimum) for all other $t \neq t^*$, a functional $J(x)$ is said to obtain a local or relative maximum (minimum) along $x^*(t)$ when $\Delta J(x^*) \geq 0$ (≤ 0), i.e. $J(x^*) \geq J(x)$ $(J(x^*) \leq J(x))$ for all neighbouring functions sufficiently close to x^* and global maximum (minimum) when $J(x^*) \geq J(x)$ $(J(x^*) \leq J(x))$ for all other functions $x(t) \neq x^*(t)$. The order of closeness (see footnote 3) is immaterial for the study of the first necessary condition but becomes important in the study of weak and strong variations in the examination of sufficient conditions.

2.2 A Necessary Condition: The Euler Equation

Let $C[a,b]$ represent the class of all continuous functions defined on a closed interval $[a,b]$ and $C^i[a,b]$ represent all functions defined on $[a,b]$ and having continuous ith derivative $(1 \leq i \leq n)$. By convention, $C^0[a,b] \equiv C[a,b]$. There is no loss of generality in setting $a = 0$ and $b = T$ so that $C[a,b] = C[0,T]$.

Consider a variational problem of the simplest form

$$J(x) = \int_0^T f(x,\dot{x},t)\, dt \tag{3}$$

where the end points $A(0,x(0))$ and $B(T,x(T))$ are fixed, $f(x,\dot{x},t) \in C^2[0,T]$, $x(t) \in C^2[0,T]$ and $\dot{x} \equiv dx/dt$ and x is a scalar function (see fig. 2.3, 2.4).

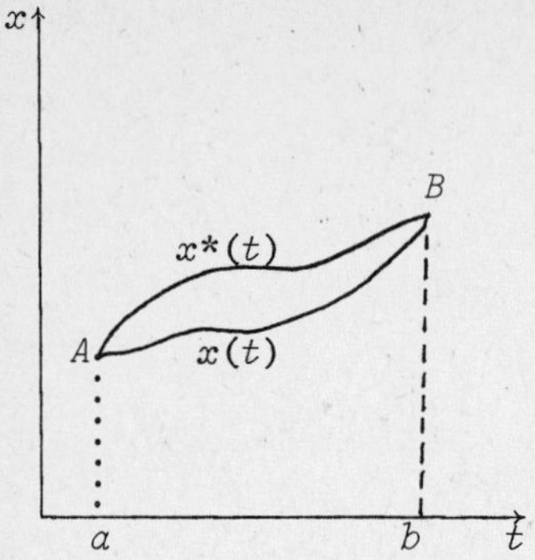

Fig. 2.3

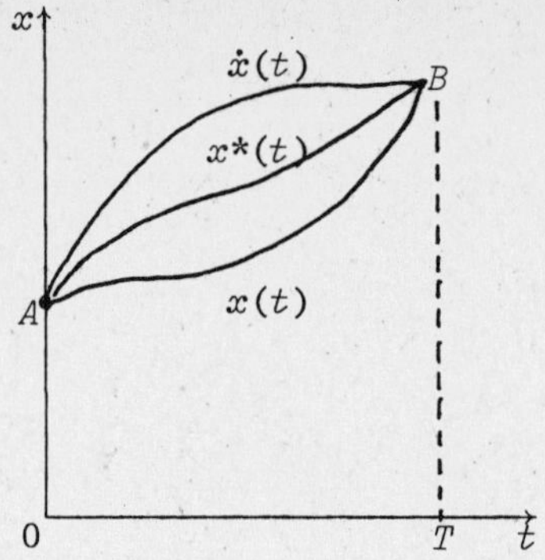

Fig. 2.4

The problem is to choose among all admissible functions, i.e. all functions $x(t) \in C^2[0,T]$ originating at A and ending at B, the one, say $x^*(t)$, which imparts a maximum or minimum to the functional $J(x)$. We know that a necessary condition for an extremum is $\delta J(x) = 0$. Suppose

$$\delta J(x) = \int_0^T g(t)\ h(t)\ dt = 0 \tag{4}$$

where $g(t) \in C[0,T]$ and $h(t)$ is some arbitrary function obeying $h(0) = 0 = h(T)$ in fig. 3.4. What can be said about $g(t)$? The following fundamental Lemma says that $g(t) = 0$. More precisely the Fundamental Lemma is as follows:

Lemma 2.1 Fundamental Lemma

Let $g(t)$ *be any continuous function in* $[0,T]$ *and* S *be the set of all arbitrary continuous and differentiable functions* $h(t)$ *in* $[0,T]$ *such that* $h(0) = 0 = h(T)$ *where* T *is fixed. If*

$$\int_0^T g(t)\ h(t)\ dt = 0 \tag{5}$$

for all $h \in S$ *then* $g(t) = 0$ *for all* $t \in [0,T]$.

Proof See Footnote 4.

This Lemma plays a fundamental part in the proof of the Euler's equation in the following theorem.

Theorem 2.1

Let $J(x) = \int_0^T f(x,\dot{x},t)\ dt$ *(where* $\dot{x} = dx/dt$*) be defined on* $C'[0,T]$ *and satisfy the boundary conditions* $x(0) = x_0$, $x(T) = x_T$. *Then a necessary condition for* $J(x)$ *to have an extremum is that* $x(t)$ *satisfy the following Euler's equation*

$$f_x - \frac{d}{dt} f_{\dot{x}} = 0\ . \tag{6}$$

Proof. A necessary condition for $J(x)$ to be an extremum is $\delta J(x) = 0$, i.e.

$$\delta J = \int_0^T \Big[f_x(x,\dot{x},t)h + f_{\dot{x}}(x,\dot{x},t)\dot{h}\Big]\ dt = 0 \tag{7}$$

where $f_x \equiv \partial f(x,\dot{x},t)/\partial x$, $f_{\dot{x}} \equiv \partial f(x,\dot{x},t)/\partial \dot{x}$ and $h(t)$ is a continuous and arbitrary function called displacement function such that $h(0) = 0 = h(T)$. Integration by parts gives

$$\begin{aligned} \int_0^T \dot{h}\ f_{\dot{x}}\ dt &= h\ f_{\dot{x}}\Big|_0^T - \int_0^T \Big(\frac{d}{dt} f_{\dot{x}}\Big)h\ dt \\ &= 0 - \int_0^T \Big(\frac{d}{dt} f_{\dot{x}}\Big)h\ dt \end{aligned} \tag{8}$$

since $h(0) = 0 = h(T)$. Note that arguments of functions are omitted (to alleviate notations). Substitution into (7) gives

$$\delta J = \int_0^T \left(f_x - \frac{d}{dt} f_{\dot{x}}\right) h \, dt = 0 \tag{9}$$

which by the Fundamental Lemma gives the Euler equation

$$f_x - \frac{d}{dt} f_{\dot{x}} = 0 \; . \tag{6}$$

QED

Written out in full, with $f_{x\dot{x}} \equiv \dfrac{\partial^2 f}{\partial x \partial \dot{x}}$ etc., (6) is

$$f_x - f_{\dot{x}t} - f_{x\dot{x}} \, \dot{x} - f_{\dot{x}\dot{x}} \, \ddot{x} = 0 \tag{10}$$

(6) or (10) is called the Euler, or Euler-Lagrange equation whose integral curves are the extremals of the functional.

Note that (10) implies that $f(x,\dot{x},t)$ has continuous first and second derivatives with respect to all its arguments, at all points (x,t) where $f_{\dot{x}\dot{x}} \neq 0$. If $f_{\dot{x}\dot{x}} = 0$, $J(x)$ is said to be a degenerate functional.

Note finally that the Euler equation only gives necessary, not sufficient conditions. The problem of sufficient conditions in the Calculus of Variations is very complicated. It will be discussed later.

Example 2.1

Find the extremal of $\int_0^1 (a\dot{x}^2 + bt)dt$, given $x(0) = 0$, $x(1) = 2$, $a \neq 0$.

The integrand is $f(\dot{x}) = a\dot{x}^2 + bt$. Euler equation gives

$$f_x - \frac{d}{dt} f_{\dot{x}} = 0 - \frac{d}{dt} 2a\dot{x} = 0$$

i.e., $2a\ddot{x} = 0$ or $\ddot{x} = 0$ since $a \neq 0$. Integrating this we obtain $\dot{x}(t) = c$ and $x(t) = ct + d$ where c and d are the constants to be determined by the boundary conditions $x(0) = 0$ and $x(1) = 2$.

$x(0) = ct + d = 0$ implies $d = 0$ and $x(1) = 2 = c(1) + 0$. The solution is thus a straight line going through the origin

$$x(t) = 2t$$

Example 2.2

Find the extremum of $\int_0^{10} f(x, \dot{x}, t)\, dt$ where $f(x, \dot{x}, t) \equiv a\dot{x}^2 + bx + ct$, with $x(0) = 1$, $x(10) = 5$ the Euler equation gives

$$f_x - \frac{d}{dt} f_{\dot{x}} = b - \frac{d}{dt}\; 2a\dot{x} = b - 2a\ddot{x} = 0$$

or $\ddot{x} = b/2a$ whose solution obtained by integrating twice, is $\dot{x} = (b/2a)t + \beta$ and $x(t) = \frac{bt^2}{4a} + \beta t + \gamma$. From $x(0) = 1$, and $x(10) = 5$, we have

$$x(t) = (b/4a)\; t^2 + (.4 - 2.5\; b/a)t + 1$$

where a, b are the known coefficients in $f = a\dot{x}^2 + bx + ct$. The solution is thus a polynomial function, convex if a and b are of the same sign and concave if they are of opposite signs.

Example 2.3

Find the optimal output $x(t)$ level of a producer facing a constant market price $p = \$4$ and a total cost function of $\dot{x}^2 + x^2$ discounted at the market rate of interest of $r = 0.2$ from an initial output level of $x(0) = 0$ to a terminal output $x(1) = 10$. The problem is one of maximizing

$$\int_0^1 (4x - \dot{x}^2 - x^2)\; e^{-.2t}\; dt\;,\; x(0) = 0;\; x(1) = 10$$

The discounted profit function is $f(\dot{x}, x, t) \equiv (4x - \dot{x}^2 - x^2)\; e^{-.2t}$ Euler equation gives

$$f_x - \frac{d}{dt} f_{\dot{x}} = e^{-.2t}\;(4 - 2x\;) - \frac{d}{dt}\, e^{-.2t}\;(-2\dot{x}) = 0$$

i.e., $\ddot{x} - .2\dot{x} - x = -\,2$

whose solution is

$$x(t) = c_1 e^{\lambda_1 t} + c_2 e^{\lambda_2 t} + 2$$

where $\lambda_1, \lambda_2 = .1 \pm \sqrt{.01 + 1}$ =, (1.105, - .905) the roots of the auxiliary equation of the complementary function, 2 is the particular integral, and c_1, c_2 are arbitrary constants to be determined by $x(0) = 0$ and $x(1) = 10$,

i.e., $$x(0) = c_1 + c_2 + 2 = 0$$

$$x(1) = c_1 e^{\lambda_1} + c_2 e^{\lambda_2} + 2 = 10$$

giving c_1, c_2 = (3.358, -5,358), i.e., the solution is

$$x(t) = 3.358\, e^{1.105t} - 5.358\, e^{-.905t} + 2$$

2.3 Generalizations of Euler's Equation

2.3.1 Case of several variables

If $J[x] = \int_0^T f(x,\dot{x},t)\, dt$ where $x = (x_1, x_2, \ldots, x_n)$ and $\dot{x} = (\dot{x}_1, \dot{x}_2, \ldots, \dot{x}_n)$ where $\dot{}$ indicates time derivatives,

$$\delta J = \int_0^T \left(\sum_1^n h_i f_{x_i} + \sum_1^n \dot{h}_i f_{\dot{x}_i} \right) dt = 0 \,. \tag{11}$$

$$= \int_0^T \left(f_{x_i} - \frac{d}{dt} f_{\dot{x}_i} \right) h_i \, dt = 0 \quad \forall\, h_i(t)$$

by integration by parts, remembering that $h_i(0) = h_i(T) = 0$ for all i.

This gives, by the Fundamental Lemma, and since i is arbitrary,

$$f_{x_i} - \frac{d}{dt} f_{\dot{x}_i} = 0 \;\forall i \tag{12}$$

or, in full

$$f_{x_i} - f_{\dot{x}_i t} - f_{x_i \dot{x}_i} \dot{x}_i - f_{\dot{x}_i \dot{x}_i} \ddot{x}_i = 0 \; (1 \le i \le n). \tag{13}$$

Example 2.3.1

Find the extremal of $\int_0^{10} (\dot{x}_1^2 + \dot{x}_2^2 + e^t)\, dt$ for

$[x_1(0), x_1(10)] = (1, 11)$ and $[x_2(0), x_2(10)] = (2, 6)$

The integrand is $f(\dot{x}_1, \dot{x}_2, t) \equiv \dot{x}_1^2 + \dot{x}_2^2 + e^t$

The Euler equation gives

$$\frac{\partial f}{\partial x_i} - \frac{d}{dt}\frac{\partial f}{\partial \dot{x}_i} = 0 - \frac{d}{dt} 2\, \dot{x}_i \quad (i = 1, 2)$$

i.e., $\ddot{x}_i = 0$, $\dot{x}_i = a_i$; and the solution gives the linear functions

$x_i(t) = a_i t + b_i$ $(i = 1, 2)$ where a_i and b_i, for the given boundary conditions, are $(a_1, a_2, b_1, b_2) = (1, .4, 1, 2)$

i.e., $x_1(t) = t + 1$

$x_2(t) = 0.4t + 2$

2.3.2 Case where f involves derivatives of nth order $(n \geq 1)$

$$J(x) = \int_0^T f(t, x, \dot{x}, \ddot{x}, \ldots, \overset{n}{x})\, dt \tag{14}$$

with fixed end points $\overset{i}{x}(0) = \overset{i}{x}_0$, giving $h(0) = \dot{h}(0) = \ldots = \overset{n}{h}(0) = 0$

$\overset{i}{x}(T) = \overset{i}{x}_T$ giving $h(T) = 0 = \dot{h}(T) = \ldots = \overset{n}{h}(T)$.

A necessary condition for an extremum of $J(x)$,

is

$$0 = \delta J = \int_0^T \left(f_x h + f_{\dot{x}} \dot{h} + \ldots + f_{\overset{n}{x}} \overset{n}{h}\right) dt \tag{15}$$

Integration by parts gives

$$\int_0^T f_{\dot{x}}\, \dot{h}\, dt = f_x\, h \Big|_0^T - \int_0^T h \frac{d}{dt} f_{\dot{x}}\, dt$$

$$= 0 - \int_0^T h \frac{d}{dt} f_{\dot{x}}\, dt \tag{16}$$

$h(0) = 0 = h(T)$; and repeated integrations by parts give

$$\int_0^T f_{\ddot{x}} \ddot{h}\, dt = f_{\ddot{x}} \dot{h}\Big|_0^T - \int_0^T \dot{h} \frac{d}{dt} f_{\ddot{x}}\, dt$$

$$= 0 - \frac{d}{dt} f_{\ddot{x}} h\Big|_0^T + \int_0^T h \frac{d^2}{dt^2} f_{\ddot{x}} h\, dt$$

$$= 0 - 0 + \int_0^T \frac{d^2}{dt^2} f_{\ddot{x}} h\, dt . \tag{17}$$

Similarly for higher order derivatives. Repeated application of integration by parts, making use of $h(0) = \dot{h}(0) = \ldots = h^n(0) = 0$ and $h(T) = 0 = \dot{h}(T) = \ldots = h^n(T)$, gives

$$\delta J = \int_0^T \Big(f_x - \frac{d}{dt} f_{\dot{x}} + \frac{d}{dt} f_{\ddot{x}} + \ldots + (-1)^n \frac{d^n}{dt^n} f_{x^n}\Big) h\, dt = 0 . \tag{18}$$

By the Fundamental Lemma, (18) gives

$$f_x - \frac{d}{dt} f_{\dot{x}} + \frac{d^2}{dt^2} f_{\ddot{x}} + \ldots + (-1)^n \frac{d^n}{dt^n} f_{x^n} = 0 . \tag{19}$$

(19) is called Euler-Poisson equation.

It can be seen that the Euler equation (6) emerges as a special case of (19) where $n = 1$, i.e.

$$f_x + (-1) \frac{d}{dt} f_{\dot{x}} = 0 .$$

Example 2.3.2

Find the extremal of $\int_0^1 (\dddot{x}^2 + \dot{x} + at^2)\, dt$ for $x(0) = 0$, $\dot{x}(0) = 1$, $x(1) = 1$ and $\dot{x}(1) = 1$.

The objective function is $f(\dddot{x}, \dot{x}, t) \equiv \ddot{x}^2 + \dot{x} + at^2$

the Euler-Poisson equation gives

$$f_x - \frac{d}{dt} f_{\dot{x}} + \frac{d^2}{dt^2} f_{\ddot{x}} = 0 - \frac{d}{dt} 1 + \frac{d^2}{dt^2} 2\ddot{x} = 0$$

i.e., $\ddddot{x} = 0$. Integrating, $\dddot{x} = c_1$, $\ddot{x} = c_1 t + c_2$

$$\dot{x} = \frac{c_1 t^2}{2} + c_2 t + c_3; \text{ and}$$

$$x(t) = \frac{c_1 t^3}{6} + \frac{c_2 t^2}{2} + c_3 t + c_4$$

where c_1, c_2, c_3, c_4 are to be determined by $(x(0), \dot{x}(0), \dot{x}(1)) = (0, 1, 1, 1)$.

This gives $(c_1, c_2, c_3, c_4) = (0, 0, 1, 0)$ i.e., the solution $x(t)$ is a 45° line going through the origin, i.e., $x(t) = t$

2.4 Particular Cases of the Euler Equation

2.4.1 Absence of x

The functional $J[x]$ is

$$J[x] = \int_0^T f(\dot{x}, t)\, dt \tag{20}$$

where f does not contain x explicitly. Euler's equation is reduced to

$$\frac{d}{dt} f_{\dot{x}} = 0 \tag{21}$$

i.e. $f_{\dot{x}}$ is a constant. This is a first order differential equation which does not contain x. Solution of (21) gives

$$\dot{x} = g(t, c) \tag{22}$$

where c is an arbitrary constant. The solution for (22) is found by integrating $\dot{x}$.

If f depends only on $\dot{x}$, i.e. $J = \int_0^T f(\dot{x})\, dt$, the Euler's equation is

$$\frac{d}{dt} f_{\dot{x}} = f_{\dot{x}\dot{x}}\, \ddot{x} = 0 \tag{23}$$

i.e. either $\ddot{x} = 0$, in which case $\dot{x} = c$ (c is some constant) and $x(t) = c_1 t + c_2$, or $f_{\dot{x}\dot{x}} = 0$. If $f_{\dot{x}\dot{x}}$ has a real root, i.e. $\dot{x}(t) = c$, the solution is

$$x(t) = c_3 t + c_4\ . \tag{24}$$

In either case, the solution to (23) is $x(t) = at + b$ where a, b, c, c_1, c_2, c_3, c_4 are constants, i.e., the solution to the Euler equation is a family of straight lines.

Example 2.4.1a

Find the extremal of $\int_0^T (\dot{x}^2 t/2 + t^2)\, dt$. This integrand is independent of x, i.e., $f(\dot{x}, t) = \dot{x}^2 t/2 + t^2$

Euler equation gives $f_{\dot{x}} = c$ (some constant) $= t\dot{x}$, i.e., $\dot{x} = c/t$ whose solution given by integration is $x(t) = c \ln t + b$ where c and b are arbitrary constant, to be determined by initial conditions. Note that integrating the Euler equation $0 - \frac{d}{dt}(t\dot{x}) = -(\dot{x} + t\ddot{x}) = 0$ one obtains the same result, but in a more laborious way.

Example 2.4.1b The Shortest Distance Problem

Find the shortest curve joining two fixed points A and B. Noting that the distance along each segment of an arc or path $x(t)$

is $ds = [(dx)^2 + (dt)^2]^{\frac{1}{2}} = [1 + \dot{x}^2]^{\frac{1}{2}}\, dt$ one can see that this is reduced to minimizing the functional $J(x)$

$$J(x) = \int_0^T (1 + \dot{x}^2)^{\frac{1}{2}}\, dt\ .$$

This is the special case 2.4.1 in which the integrand does not involve either x or t explicitly. Euler equation gives $f_{\dot{x}}$ = constant, i.e.

$$\ddot{x}[\dot{x}^2(1 + \dot{x}^2)^{-3/2} + (1 + \dot{x}^2)^{-\frac{1}{2}}] = 0$$

the solution of which gives $\ddot{x} = 0$ or $x(t) = at + b$ where a and b are constants, i.e. a family of straight lines. Clearly only one such line goes through the two fixed points $A[0,x_0]$ and $B[T,x_T]$. This is, of course, obvious from the start.

2.4.2 Absence of t

$$J(x) = \int_0^T f(x,\dot{x})\, dt \tag{25}$$

Euler's equation gives

$$f_x - \frac{d}{dt} f_{\dot{x}} = f_x - f_{\dot{x}x}\dot{x} - f_{\dot{x}\dot{x}}\ddot{x} \tag{26}$$

which gives, on multiplication by $\dot{x}$

$$f_x\dot{x} - f_{\dot{x}x}\dot{x}^2 - f_{\dot{x}\dot{x}}\dot{x}\ddot{x} \equiv \frac{d}{dt}(f - \dot{x}f_{\dot{x}}) = 0 \tag{27}$$

the first integral of which is

$$f - \dot{x}f_{\dot{x}} = c \tag{28}$$

where c is some constant.

Example 2.4.2

Ramsey's (1928) optimal saving problem may be simplified as

$$\text{Max. } J(k) = \int_0^T u(c)\, dt$$

where $u(c)$ is the utility of per capita consumption c where $c = f(k) - \dot{k}$, with the production $f(k)$ as a concave increasing function of capital per worker k and $\dot{k}$ is per capita investment. This is case 2.4.2 where the integrand $u(c) \equiv u[f(k) - \dot{k}]$ does not involve t explicitly. Euler equation gives immediately

$$u(c) - \dot{k}(\partial u/\partial \dot{k}) = a \text{ (constant)}$$

$$\dot{k} = [a - u(c)]/u'(c)$$

where $\partial u/\partial \dot{k} = u'(c)(-1)$.

See Allen (1938, p. 540).

2.4.3 Absence of $\dot{x}$

$$J(x) = \int_0^T f(x,t)\, dt \tag{29}$$

Euler's equation gives

$$f_x = 0 . \tag{30}$$

This is not a differential equation but a non-linear (in general) Algebraic equation. In general, boundary conditions cannot be satisfied, as there are no constants of integration. In other words, the solution exists only if the curve $x = x(t)$ happens to pass through the specified boundary points.

Example 2.4.3a

Consider a businessman who wants to maximize his total revenue functional $\int_0^T R(x)\,dt$ where $R(x) \equiv xp(x)$ and $p(x) = -ax + b$, his linear downward sloping demand curve. The Euler equation gives

$$R_x - \frac{d}{dt} R_{\dot{x}} = -2ax + b = 0$$

i.e., $x^* = b/2a$: his optimal output would be kept at the constant level $b/2a$. Clearly this could not satisfy boundary conditions unless these are $x(0) = x(T) = b/2a$.

Example 2.4.3b

As an illustration of the case where the integrand does not involve $\dot{x}$ (case2 .4.3), consider the profit functional $J = \int \pi(x, t)\,dt$ where $\pi(x) = px - a_1x^2 - b_1x - c_1$. Euler equation gives

$$\pi_x = p - 2a_1x - b_1 = 0$$

i.e. optimal output is the one at which marginal cost is equal to price, a static optimisation rule:

$$x^*(t) = (p-b_1)/\,2a_1 .$$

This is not a differential equation and clearly boundary conditions cannot be satisfied in general. It can be seen that $x^* = 0$ for $p = b_1$, i.e. when price reaches the level of average variable cost, the firm will shut down: zero production is the optimal policy.

Example 2.4.3c

Consider Gordon's problem (1967 p. 277) of a mining firm trying to maximize the net present value J of the net profit (Π), discounted at rate r, from the output $q(t)$ of his mining resource i.e.,

$$J = \int_0^T \Pi[q(t),\ t]e^{-rt} - \lambda\ q(t)\ \ dt + \lambda S$$

where λ is the Lagrangean multiplier and S the stock of his non-renewable resource. The Euler equation $F_q - \frac{d}{dt} F\ \dot{q} = 0$ where $F \equiv \Pi(.)e^{-rt} - \lambda q(t)$ above, with $\dot{q}$ absent, gives

$$F_q - \frac{d}{dt} F_{\dot{q}} = e^{-rt}\Pi_q - \lambda - 0 = 0$$

i.e., $\partial\Pi/\partial q = \lambda e^{rt}$ i.e., marginal profit equal to user cost. It is clear that this is not a differential equation and his boundary conditions (Gordon 1967, equations (6) and (7)) $\Pi(q(0),\ 0) - \lambda q(0)$ and $\Pi\ (q(T),\ T) = \lambda q(T)e^{rT}$ cannot be satisfied in general. For example if $\Pi = pq - aq^2 - bq - c$ then $\Pi q = 0$ gives $q^*(t) = (p-b)\ /2a$ which could only be satisfied if $q(0) = q(T) = (p-b)/2a$, a very special case indeed.

2.4.4 <u>$f(x,\dot{x},t)$ is linear in $\dot{x}$</u>

f could be written as

$$f(x,\dot{x},t) = \alpha[x(t),t] + [\beta(x(t),t]\dot{x}(t)\ . \tag{31}$$

Euler's equation gives

$$\frac{\partial\alpha}{\partial x} + \frac{\partial\beta}{\partial x}\dot{x} - \frac{d\beta}{dt} = 0 \tag{32}$$

or

$$\frac{\partial\alpha}{\partial x} - \frac{\partial\beta}{\partial t} = 0 \tag{33}$$

which, as in 2.4.3, is not a differential equation but an algebraic equation. In general, this functional is independent of the path of integration: the objective functional $J(x)$ has the same value for every curve which satisfies the boundary conditions. The Variational problem is then of no interest.

Note that this is a case of degenerate functionals mentioned earlier where $f_{\dot{x}\dot{x}}$ is zero identically: this happens whenever $f(x,\dot{x},t)$ depends on $\dot{x}$ linearly.

Example 2.4.4a

Consider a producer trying to minimize $J = \int_0^T f(x, \dot{x}, t)\, dt$ where $f(x, \dot{x}, t) = x^2 + 2xt\dot{x}$ is the production cost x^2 of output and $2xt\dot{x}$ is the cost associated with expansion or contraction of his product x. This is a case where f is linear in $\dot{x}$ and $\partial\alpha/\partial x = \partial\beta/\partial t = 2x$ i.e., J is independent of the path of integration. The Euler equation gives

$$f_x - \frac{d}{dt} f_{\dot{x}} = 2x + 2t\dot{x} - 2x - 2t\dot{x} \equiv 0$$

which vanishes identically : as expected, the Calculus of Variations provides no useful information.

Example 2.4.4b

As an example of case 2.4.4 where $\dot{x}$ enters the integrand linearly, we can consider the problem of maximizing $\int_0^T \Pi(x, \dot{x})dt$ where

$$\pi(x,\dot{x}) = px - a_1x^2 - b_1x - c_1 - b_2\dot{x} .$$

Euler equation then gives

$$\pi_x = p - 2a_1x - b_1 = \frac{d}{dt}(-b_2) = 0$$

i.e.

$$x^*(t) = (p-b_1)/2a_1 .$$

Clearly this is not a differential equation and the solution does not generally satisfy boundary conditions. In this example, only the special case where $x_0 = x_T = (p-b_1)/2a_1$ could be satisfied.

2.5 Variational Problems With Constraints

In the Calculus of Variations, subsidiary conditions are sometimes imposed by the physical nature of the problem. The extremum of a functional defined under such constraints is called conditional or constrained extremum. An important implication of constraints is the variations $\delta x_i (1 \le i \le n)$ are not arbitrary and consequently the Fundamental Lemma cannot be applied.

This problem is solved by using the well-known substitution method or the Lagrange multiplier approach. As this method is commonplace in Economics, it does not justify lengthy treatment. We shall discuss the cases of point constraints, differential equation constraints and isoperimetric constraints.

2.5.1 Point and Differential Equation Constraints

Consider the problem of extremizing the functional

$$J = \int_0^T f(x,\dot{x},t)\, dt \qquad (34)$$

subject to

$$g_i(x,\dot{x},t) = 0 \quad (1 \le i \le r < n) \qquad (35)$$

where x is an n-vector and $\dot{x}$ its time derivative

$f(x,\dot{x},t)$ is a scalar function and t is time

$g_i(x,\dot{x},t)$ is called a differential equation constraint and when $\dot{x}$ is absent,

$g_i(x,t)$ is called a point constraint.

The cases of point and differential equation constraints are treated together since the same method applies to both.

Define the Lagrange function L as

$$L \equiv f(x,\dot{x},t) + p\cdot g(x,\dot{x},t) \tag{36}$$

or, in scalar notation

$$L \equiv f(x,\dot{x},t) + \sum_{i=1}^{r} p_i g_i(x,\dot{x},t) \tag{37}$$

with $x(0) = x_0$ and $x(T) = x_T$ and the augmented functional J_a as

$$J_a \equiv \int_0^T L(x,\dot{x},p,t)\ dt\ . \tag{38}$$

The variations of δJ_a are

$$\begin{aligned} \delta J_a &= \int_0^T (L_x\ \delta x + L_{\dot{x}}\ \delta\dot{x} + L_p\ \delta p)\ dt \\ &= \int_0^T [(L_x - \frac{d}{dt} L_{\dot{x}})\delta x + L_p\ \delta p]\ dt \end{aligned} \tag{39}$$

by integration by parts. Along an extremum, $\delta J_a = 0$ and the constraints must also be satisfied, i.e. by Euler equations

$$L_x - \frac{d}{dt} L_{\dot{x}} = 0 \tag{40}$$

$$L_p - \frac{d}{dt} L_{\dot{p}} = 0 \tag{41}$$

where

$$L_x \equiv f_x + g_x p \equiv \begin{bmatrix} \frac{\partial f}{\partial x_1} \\ \cdot \\ \cdot \\ \cdot \\ \frac{\partial f}{\partial x_n} \end{bmatrix} + \begin{bmatrix} \frac{\partial g_1}{\partial x_1} & \frac{\partial g_2}{\partial x_1} & \cdots & \frac{\partial g_r}{\partial x_1} \\ \cdot & \cdot & & \cdot \\ \cdot & \cdot & & \cdot \\ \cdot & \cdot & & \cdot \\ \frac{\partial g_1}{\partial x_n} & \frac{\partial g_2}{\partial x_n} & \cdots & \frac{\partial g_r}{\partial x_n} \end{bmatrix} \begin{bmatrix} p_1 \\ p_2 \\ \cdot \\ \cdot \\ \cdot \\ p_r \end{bmatrix}$$

$$L_{\dot{x}} \equiv \begin{bmatrix} \frac{\partial f}{\partial \dot{x}_1} \\ \cdot \\ \cdot \\ \cdot \\ \frac{\partial f}{\partial \dot{x}_n} \end{bmatrix} + \begin{bmatrix} \frac{\partial g_1}{\partial \dot{x}_1} & \frac{\partial g_2}{\partial \dot{x}_1} & \cdots & \frac{\partial g_r}{\partial \dot{x}_1} \\ \cdot & \cdot & & \cdot \\ \cdot & \cdot & & \cdot \\ \cdot & \cdot & & \cdot \\ \frac{\partial g_1}{\partial \dot{x}_n} & \frac{\partial g_2}{\partial \dot{x}_n} & \cdots & \frac{\partial g_r}{\partial \dot{x}_n} \end{bmatrix} \begin{bmatrix} p_1 \\ p_2 \\ \cdot \\ \cdot \\ \cdot \\ p_r \end{bmatrix}$$

for the differential equation constraints case, and

$$L_{\dot{x}} \equiv \begin{bmatrix} \frac{\partial f}{\partial \dot{x}_1} \\ \cdot \\ \cdot \\ \cdot \\ \frac{\partial f}{\partial \dot{x}_n} \end{bmatrix}$$

for the point constraints case where $g_i(x,t) = 0$ (i.e. g_i does not contain the vector $\dot{x}$).

Note that $L_{\dot{p}} = 0$ i.e. (41) just says that the constraints are satisfied, i.e. $L_p = g = 0$.

2.5.2 Isoperimetric Constraint

Originally, the isoperimetric problem consists in finding among all closed curves of a given length l, the one which encloses the largest area. This is attributed to queen Dido's land transaction problem of finding among all the perimeters of a fixed length, the one which encloses the largest land area. In a wider sense, the isoperimetric problem is a variational problem with a constraint expressed by a certain integral taking on a prescribed value. A sailor with a fixed stock of provisions to last for the whole journey has to solve an isoperimetric problem when he decides how much to consume each day. A society having a fixed stock of some non-renewable resource faces an isoperimetric problem in its decision concerning the optimal rate of extraction and consumption of the resource.

Mathematically, the problem is one of extremising a functional

$$J = \int_0^T f(x,\dot{x},t)\ dt \qquad (42)$$

subject to

$$x_i(0) = x_{i0}\ ;\ x(T) = x_{iT} \qquad (1 \le i \le n) \qquad (43)$$

and

$$\int_0^T g_i(x,\dot{x},t)\ dt = l_i \qquad (1 \le i \le r < n) \qquad (44)$$

where l_i are constants.

The constraints (44) are called isoperimetric conditions. Define new functions

$$y_i(t) \equiv \int_0^t g_i(x,\dot{x},t)\ dt \qquad (45)$$

with

$$y_i(0) = 0$$

$$y_i(T) = l_i \qquad (1 \le i \le r)\ . \qquad (46)$$

Differentiation of (45) gives, for all i

$$\dot{y}_i(t) = g_i(x,\dot{x},t)$$

or

$$g_i - \dot{y}_i = 0 \ .$$

With this, the isoperimetric constraints have been reduced to the differential constraints discussed earlier. Application of the Lagrangian multipliers method gives

$$\begin{aligned} J_a &\equiv \int_0^T F(x,\dot{x},t)\; dt \\ &\equiv \int_0^T [f(x,\dot{x},t) + \sum_1^r p_i(t)(g_i-\dot{y}_i)]\; dt \end{aligned} \qquad (47)$$

where $F(x,\dot{x},t) \equiv f(x,\dot{x},t) + \sum_1^r p_i(g_i-\dot{y}_i)$ or in vector notation, $F \equiv f + p.(g-\dot{y})$.

Euler equations give

$$\frac{\partial F}{\partial x_j} - \frac{d}{dt}\left(\frac{\partial F}{\partial \dot{x}_j}\right) = 0 \qquad (1 \le j \le n) \qquad (48)$$

$$\frac{\partial F}{\partial y_i} - \frac{d}{dt}\left(\frac{\partial F}{\partial \dot{y}_i}\right) = 0 \qquad (1 \le i \le r) \ . \qquad (49)$$

Since F does not depend on y explicitly, (49) gives

$$0 - \frac{d}{dt}(-p_i) = 0 \qquad (50)$$

i.e. $p_i(t)$ are constant for all i.

The solution of the Euler equations contains $2n$ integration constants and r constant Lagrangian multipliers p_i. These are determined by $x_i(0) = x_{i0}$, $x_i(T) = x_{iT}$ and the r isoperimetric conditions (44)

Example $$\text{Max. } J = \int_0^T \dot{x}^2 \, dt$$

subject to

$$x(0) = x_0 \; ; \quad x(T) = x_T \quad \text{and } T \text{ specified}$$

and

$$\int_0^T (1+x)dt = l \qquad (l \text{ constant}) .$$

Define $y(t) \equiv \int_0^t (1+x) \, dt$ with $y(0) = 0$ and $y(T) = l$

$$\dot{y}(t) = 1 + x(t) .$$

The Lagrange equation is $F \equiv \dot{x}^2 + p(1+x-\dot{y})$.

Euler equation gives $2\ddot{x} = p$

the solution of which is

$$x(t) = \frac{p}{4} t^2 + at + b .$$

Boundary conditions give

$$x(0) = x_0 = b$$
$$x(T) = x_T = \frac{p}{4} T^2 + aT + b$$

$$\int_0^T (1+x)\, dt = \int_0^T (1 + \frac{p}{4} t^2 + at + b)\, dt = l$$

i.e. $$p = \frac{12}{T^3} [l - \frac{a}{2} T^2 - (1+b)T] .$$

The last 3 equations determine the three unknowns a, b and p.

2.6 Some Economic Applications

Example 2.6.1 Dynamic Pure Competition

A producer in a perfectly competitive market wants to find the optimal production path $x(t)$ ($0 \le t \le T$) such that, starting from a historical level of production x_0 at $t = 0$ and reaching a production target x_T at a specified final period T, his profit functional $J(x)$ is maximized. His costs include production costs $C_1 = a_1x^2 + b_1x + c_1$ and costs associated with production increase ($\dot{x}$) such as building extra capacity in anticipation of rising output, labour hiring and training, management recruiting, inflation and the like. These are $C_2 = a_2\dot{x}^2 + b_2\dot{x} + c_2t$. His total revenue is px where p is constant in this market. His profit is

$$\pi(x,\dot{x},t) = px - a_1x^2 - b_1x - c_1 - a_2\dot{x}^2 - b_2\dot{x} - c_2t$$

and his profit functional $J(x)$ is

$$J(x) = \int_0^T \pi(x,\dot{x},t)\, dt .$$

Euler equation gives

$$\pi_x \equiv p - 2a_1x - b_1 = \frac{d}{dt} (-2a_2\dot{x} - b_2) \equiv \frac{d}{dt} \pi_{\dot{x}}$$

i.e.

$$2a_2\ddot{x} - 2a_1 x + p - b_1 = 0$$

or putting $b^2 \equiv a_1/a_2$, $c \equiv (p-b_1)/2a_2$ $(a_2 \neq 0)$

$$\ddot{x} - b^2 x + c = 0$$

Solving this gives his optimal output $x^*(t)$ (assuming $a_1 \neq 0$), as

$$x^*(t) = Ae^{bt} + Be^{-bt} + c/b^2$$

where A and B are determined by $x(0) = x_0$ and $x(T) = x_T$ for $t = 0$ and $t = T$ (T is fixed).

Example 2.6.2 Dynamic Utility and Capital Accumulation

Consider the optimal growth problem in which utility is a function of consumption level C and its growth rate $\dot{C}$ i.e., $U(C, \dot{C})$ (ChaKravarty 1968). This is a fairly realistic case for many developing countries where the only consolation for a low level of current consumption is the high rate of economic growth and hence of consumption. The objective is to maximize

$$J = \int_0^\infty U(C, \dot{C})\, dt$$

where $C \equiv Y - \dot{K} = Y(K) - \dot{K} \equiv f(K, \dot{K})$

= Gross national product Y left over after capital accumulation $\dot{K}$ is met

$$\dot{C} = \dot{Y} - \ddot{K} = Y'(K)\, \dot{K} - \ddot{K} \equiv g(K, \dot{K}, \ddot{K})$$

The Euler-Poisson equation gives

$$\frac{\partial U}{\partial K} - \frac{d}{dt}\frac{\partial U}{\partial \dot{K}} + \frac{d}{dt^2}\frac{\partial U}{\partial \ddot{K}} = 0$$

$$U_1 f_1 + U_2 g_1 - \frac{d}{dt}\left(U_1 f_2 + U_2 g_2\right) + \frac{d^2}{dt^2}\left(U_2 g_3\right) = 0$$

where f_i, g_i are the partial derivatives of f and g e.g., $f_2 = \partial f/\partial \dot{K}$, $g_3 = \partial g/\partial \ddot{K}$ etc...

For example if $Y = bK$ and $U(C, \dot{C}) = C^a + \gamma \dot{C}$ with $0 < a, \gamma < 1$. The Euler-Poisson equation gives

$$0 = U_K - \frac{d}{dt} U_{\dot{K}} + \frac{d^2}{dt^2} U_{\ddot{K}} = ab(bK - \dot{K})^{a-1} - a(1-a)(bK - \dot{K})^{a-2}(b\dot{K} - \ddot{K})$$

Dividing through by $a(bK - \dot{K})^{a-2}$ ($\neq 0$) gives

$$(1-a)\ddot{K} + b(a-2)\dot{K} - b^2 K = 0$$

whose solution is

$$K(t) = A_1 e^{bt} + A_2 e^{bt/(1-a)}$$

where A_1 and A_2 are arbitrary constants to be determined from boundary conditions, i.e. by solving the last equation with $K(0) = K_0$ and $K(T) = K_T$, both given, for the two unknown A_1 and A_2.

Example 2.6.3 Capital theory

As an illustration of the case of several variables, consider the problem of optimal capital accumulation in a multisectoral economy (Samuelson 1960, Grandville 1980). The economy has n capital goods $K(t) = [K_1(t), \ldots, K_n(t)]$, n consumption goods $C(t) = [C_1(t), \ldots, C_n(t)]$. Output is used for consumption $C(t)$ and capital formation $\dot{K}(t) = [\dot{K}_1(t), \ldots, \dot{K}_n(t)]$ purposes. The production function f, assumed homogeneous of degree one, with any one commodity, say the first one, singled out as numéraire, gives

$$\dot{K}_1(t) = f\,[K_1(t), \ldots, K_n(t); \dot{K}_2(t) + C_2(t), \ldots] - C_1(t)$$

where $\partial f/\partial K_i > 0$ for $i = 1, 2, \ldots, n$ and $\partial f/\partial \dot{K}_i < 0$ ($i = 2, \ldots, n$)

With all K_i fixed at $t = 0$ and $t = T$ except K_1 the accumulation of which is to be maximized, the problem is to maximize, subject to $\dot{K}_1(t) = f(\cdot) - C_1(t)$ above,

$$K_1(T) - K_1(0) = \int_{0.}^{T} \dot{K}_1(t)\, dt$$

i.e.

$$\text{Max. } J = \int_0^T F(K, \dot{K}, \lambda)dt$$

where $F(K, \dot{K}, \lambda) \equiv \dot{K}_1(t) + \lambda\left\{f\ [K_1(t), \ldots, \dot{K}_2(t) + C_2(t), \ldots] - \dot{K}\ (t) - C_1(t)\right\}$

Euler equations are

$$\frac{\partial F}{\partial \lambda} - \frac{d}{dt}\frac{\partial F}{\partial \dot{\lambda}} = \frac{\partial f}{\partial \lambda} - 0 = 0 \quad \text{which gives}$$

$$\dot{K}_1(t) = f\left[K_1, \ldots, K_n; \dot{K}_2 + C_2, \ldots, \dot{K}_n + C_n\right] - C_1(t)$$

and

$\dfrac{\partial F}{\partial K_i} - \dfrac{d}{dt}\dfrac{\partial F}{\partial \dot{K}_i} = 0$ $(i = 1, 2, \ldots, n)$ which give, for the first commodity,

$$\lambda\frac{\partial f}{\partial K_1} - \frac{d}{dt}(1 - \lambda) = 0 \quad \text{or} \quad -\frac{\dot{\lambda}}{\lambda} = \frac{\partial f}{\partial K_1}$$

and for the remaining n-1 commodities i $(i = 2, 3, \ldots, n)$

$$\frac{d}{dt}\frac{\partial f}{\partial \dot{K}_i} = \frac{\partial f}{\partial K_i} + \frac{\partial f}{\partial K_1}\frac{\partial f}{\partial \dot{K}_i} \qquad (i = 2, 3, \ldots,)n$$

$$\text{or } \frac{d}{dt}\frac{\partial f}{\partial \dot{K}_i}\Big/\frac{\partial f}{\partial \dot{K}_i} = \frac{\partial f}{\partial K_i}\Big/\frac{\partial f}{\partial \dot{K}_i} + \frac{\partial f}{\partial K_1}$$

But $\dfrac{\partial f}{\partial \dot{K}_i} = \dfrac{\partial \dot{K}_1}{\partial \dot{K}_i} = p_i$ = the price ratio of the ith flow commodity expressed in terms of the first, and $\dfrac{\partial f}{\partial K_1} = \dfrac{\partial \dot{K}_1}{\partial K_1} = r_1$ the own rate of interest in terms of the first commodity and $r_i = \dfrac{\partial \dot{K}_i}{\partial K_i} = -\dfrac{\partial f}{\partial K_i}\Big/\dfrac{\partial f}{\partial \dot{K}_i}$ = own rate of interest of the ith commodity. The above results given by the Euler equations are thus

$$r_1 = r_i + \dot{p}_i/p_i \qquad (i = 2, \ldots, n)$$

a fundamental relationship in Capital theory which says that the equilibrium own rate of return of any commodity is equal to the own rate of return of the numéraire commodity r_1 net of capital gain $\dot{p}_i/p_i$. If prices are falling, the own-rate r_i must exceed the interest rate r_1 to induce people to hold the commodity i in question.*

<u>Example 2.6.4</u> Time Optimal Problem in Economic Planning

The problem in the last example can be formulated as one of minimizing the time taken for a country to build up the capital stock from some historical initial level to a certain prescribed target as quickly as possible. Re-writing the investment function $\dot{K}_1(t) = f(\cdot) - C_1(t)$ of the last problem in Example 2.6.3 as

$$dt = L\left(K_1, K_2, \ldots, \frac{dK_2}{dK_1}, \ldots\right) dK_1 \quad \left(\frac{\partial L}{\partial K_i} > 0\right)$$

we have the time optimal problem (of minimizing the time T taken by the program)

$$\text{Min.} \quad J = \int_0^T dt = \int_{K_1^0}^{K_1^T} L\left[K_1, K_2(K_1), \ldots, \frac{dK_2}{dK_1}(K_1), \ldots\right] dK_1$$

The problem emerges in a new light: it is the problem of a developing country which, starting from a certain historical capital stock, wants to achieve a certain standard of living and industrialization enjoyed by some developed country taken as a model, as quickly as possible.

This is a standard Calculus of Variations problem. The Euler equation gives

---

*This fundamental relationship first discovered by Samuelson (1937) then Solow (1956) has maintained its importance in Capital Theory over the years. See Grandville (1980).

$$\frac{\partial L}{\partial K_i} - \frac{d}{dK_1}\frac{\partial L}{\partial \dot{K}_i} = 0 \qquad (i = 2, \ldots, n)$$

Here K_1 plays the role of t used up to now.

Under normal conditions, the Euler equations above can be solved to give the optimal capital structure

$$K_i\ (K_1) = \psi^i\ (K_1;\ K_1^\circ, \ldots, K_n^\circ\ ;\ K_1^T, \ldots, K_n^T\ )$$

$$(i = 2, 3, \ldots, n)$$

and using this to integrate $dt = LdK_1$ gives

$$t(K_1) = \psi^\circ\ (K_1;\ K_1^\circ, \ldots, K_n^\circ\ ;\ K_1^T, \ldots, K_n^T\ )$$

The results are completely equivalent to the previous model in Example 2.6.3

Example 2.6.5 Optimal Education and Balanced Growth

As another illustration, let us consider the problem of optimal investment in physical (K) and human (L_1) capital in the context of balanced growth (Tu 1969). Given an exogenous birth rate (b), the problem is to choose the optimal number of people to be trained (L_3), bearing in mind that too few students would cause a shortage of skilled manpower and hence a bottleneck for economic growth and too many would trigger a brain drain, a waste for the training country. The population is thus composed of highly trained manpower $L_1(t)$, untrained labour $L_2(t)$ and students $L_3(t)$ assumed to be a constant proportion ρ of $L_1(t)$ in a balanced growth economy, i.e., for the initial number of births B_o, the population is

$$L(t) = \int_{t-a_1}^{t-a_2} B_o e^{b\tau}\, d\tau = L_1(t) + L_2(t) + L_3(t)$$
$$= (1+\rho)L_1(t) + L_2(t)$$

The utility function with constant elasticity $(1-v)$, is

$$U(C) = (1-v)^{-1} C^{1-v}$$

where

$$C = f(L_1, L_2, K) - \dot{K} - E$$

i.e., consumption C is output $Y = f(L_1, L_2, K)$ not spent on investment $(\dot{K})$ and education (E), where the production function f is assumed homogeneous of degree one and factors received their marginal product, i.e.,

$$Y = f(L_1, L_2, K) = f_1 L_1 + f_2 L_2 + f_3 K = w_1 L_1 + w_2 L_2 + rK$$

where w_1, w_2, r = factors' rental, a_2 = school starting age and a_1 = retirement age. Thus the objective is to maximize

$$J = \int_0^T U(C) e^{-it}\, dt \quad \text{where}$$

$$C(t) = -\dot{K}(t) + rK(t) + wL_1(t) + w_2 L(t)$$

where $w \equiv w_1 - w_2(1+\rho)$

Putting $\quad F \equiv U(C)e^{-it} \equiv (1-v)^{-1} C^{1-v} e^{-it}$

we obtain from Euler equations

$$\frac{\partial F}{\partial L_1} - \frac{d}{dt}\frac{\partial F}{\partial \dot{L}_1} = (\omega - c\rho)\, C^{-v} e^{-it} = 0$$
$$= \frac{df}{dC}\frac{\partial C}{\partial L_1} - \frac{d}{dt}\left(\frac{dF}{dC}\frac{\partial C}{\partial \dot{L}}\right) = 0$$

i.e., $\omega \equiv w_1 - w_2(1+\rho) = c\rho$

and
$$\frac{\partial F}{\partial K} - \frac{d}{dt}\frac{\partial F}{\partial \dot{K}} = e^{-it} C^{-v} (r - i - v\dot{C}/C) = 0$$
$$= \frac{dF}{dC}\frac{\partial C}{\partial K} - \frac{d}{dt}\frac{dF}{dC}\frac{\partial C}{\partial \dot{K}} = 0$$

The first Euler equation written as

$$w_1 - w_2 = \rho(c + w_2)$$

says that in an optimal program, the incremental gain $w_1 - w_2$ in terms of earning differential accruing to the educated is exactly equal to the incremental cost $c + w_2$ in supplying them, c being the actual training cost, w_2 being earnings foregone and ρ being the shadow price or imputed value involved in keeping a fraction of the work force in educational institutions.

The second equation gives $\dot{C}/C = (r-i)/v$ or in terms of K,

$$-\ddot{K} + v^{-1}\,(rv + r - i)\dot{K} - v^{-1}r(r - i)K = w_2\,[\dot{L} - v^{-1}(r - i)L]$$

solving

$$K^*(t) = A_{11}e^{v-1(r-i)t} + A_{12}\,e^{rt} + A_{13}\,e^{bt}$$

The optimal number of students to be trained is

$$L_3^*(t) = A_{21}\,e^{v-1(r-i)t} + A_{22}\,e^{rt} + A_{23}\,e^{bt}$$

and the educational budget involved is

$$E^*(t) = A_{31}\,e^{v-1(r-i)t} + A_{32}\,e^{rt} + A_{33}\,e^{bt}$$

where A_{ij} $(i, j = 1, 2, 2)$ are arbitrary constants to be determined by transversality conditions. (For detailed calculations, see Tu 1969).

A generalisation to the case of m types of trained manpower $(L_1, \ldots, L_m)$ one type of untrained labour (L_u) is straightforward:

$$L(t) = \sum_1^m (1+\rho_i)\,L_i + L_u$$

and the Euler equations are now

$$\frac{\partial F}{\partial L_i} - \frac{d}{dt}\frac{\partial F}{\partial \dot{L}_i} = 0 = (\omega_i - c_i\rho_i)\,C_{-v}\,e^{-it} \qquad (i = 1, \ldots m)$$

where $\omega_i \equiv w_i - w_u\,(1+\rho_i)$ $\quad (i = 1, 2, \ldots, m)$ where w_u is unskilled wage rate

and

$$\frac{\partial F}{\partial K} - \frac{d}{dt}\frac{\partial F}{\partial \dot{K}} = 0 \text{ as before.}$$

Example 2.6 .6   Micro Foundations of Macro Models

The Calculus of Variations provides a tool in searching the Micro foundations for the Keynesian Model (see for example, Sargent (1979), Jorgenson (1967), Nerlove (1972), Takayama (1974) and for a survey, Söderström (1976). One such area of search lies in the dichotomy between static profit maximisation based on the conventional classification of factors into fixed and variable categories and the dynamic counterpart characterised by the recognition that inputs are neither completely fixed nor variable but adjustable at a cost. Static profit maximization is replaced by the maximization of the discounted present value of the net return $R(t)$ in all future periods, which is the discounted profit functional. The adjustable factor is the change in capital stock $K(t)$ net of depreciation, i.e., $\dot{K}(t) = I(t) - \delta K(t)$ where $I(t)$ is gross investment and δ a constant depreciation rate. Given the production function $Q(t) = F[K(t), L(t)]$ where $L(t)$ is labour, and total revenue $p(t)Q(t)$ where $p(t)$ is price, the objective function is

$$f(K, L, I, t) \equiv e^{-rt} \; [p(t)Q(t) - w(t)L(t) - C(I)]$$

where $w(t)$ is wage rate and $I(t) = \dot{K}(t) + \delta K(t)$ defined above. This is a typical problem in the Calculus of Variations. The objective functional is the Present value

$$J = \int_0^\infty f(K, L, I, t) \; dt$$

The two Euler equations, with derivatives written as f_L, f_K etc., are

$$f_L - \frac{d}{dt} f_{\dot{L}} = f_L \equiv \frac{\partial f}{\partial L} = e^{-rt} \; (pF_L - w) = 0$$

$$f_K - \frac{d}{dt} f_{\dot{K}} = e^{-rt} \quad pF_K - (\delta+r) \; C'(I) + C''(I)(\ddot{K} + \; \dot{K}) = 0$$

The first Euler equation gives the static result of wage (w) labour marginal product (pF_L) equality: instantaneous adjustment takes place at each period. The second Euler equation gives the optimal rate of investment. Here emerges the crucial difference brought about by the cost adjustment function $C(I)$ which could be linear, convex or concave. For the linear case, $C'(I)$ is constant, $C''(I) = 0$ and the result is similar to the static case. For the convex or concave case, the solution $K^*(t)$ of the second Euler equation can only be defined by reference to $C(I)$. For further details, see Treadway (1969), Rothschild (1971) and Söderström (1976). Thus the Macro aggregations based on the traditional classification of factors between fixed and variable factors are oversimplified: the degree of "fixedness" or "mobility" of factors crucially depends on the cost of adjustment and the degree of substitution between current production and adjustment activities. A re-examination along these lines sheds new lights on the specification of the aggregate production and investment function in Macro economics.

Example 2.6.7 Constrained Optimal Consumption Plan

As an illustration of the problem of point and differential equation constraints, consider the problem of maximizing the present value (discounted at rate r) of consumption utility, i.e., $u(c)e^{-rt}$ subject to the differential equation constraint

$$\dot{k}(t) = i(t) - \delta k(t)$$

and point constraint

$$f(k) - c(t) - i(t) = 0$$

i.e., net per capita investment $\dot{k}$ is gross per capita investment $i(t)$ less depreciation $\delta k(t)$ and per capita output $f(k)$ is allocated between per capita consumption $c(t)$ and investment $i(t)$.

Define

$$F(c, i, k, \dot{k}, t, p, \lambda) \equiv e^{-rt} [u(c) + p(i - \delta k - \dot{k}) + \lambda(f - c - i)]$$

the problem is to maximize

$$J = \int_0^\infty F(c, i, k, \dot{k}, t, p, \lambda)\, dt$$

Euler equations give

$$F_c - \frac{d}{dt} F_{\dot{c}} = F_c - 0 = 0 \qquad \text{or} \qquad u'(c) = \lambda(t)$$

$$F_i - 0 = 0 \qquad \text{or} \qquad p(t) = \lambda(t)$$

$$F_p - \frac{d}{dt} F_{\dot{p}} = F_p - 0 = 0 \qquad \text{or} \qquad \dot{k}(t) = i(t) - \delta k(t)$$

$$F_\lambda - \frac{d}{dt} F_{\dot{\lambda}} = F_\lambda - 0 = 0 \qquad \text{or} \qquad f(k) = c(t) + i(t)$$

$$F_k - \frac{d}{dt} F_{\dot{k}} = -p\delta + \lambda f'(k) + \dot{p} = 0$$

which, in view of $p(t) = \lambda(t)$, gives

$$\dot{p}(t) = [\delta - f'(k)]\, p(t)$$

This, together with $F_p = 0 \Rightarrow \dot{k}(t) = i(t) - \delta k(t)$

gives the solution to the system. This pair of autonomous non-linear differential equations could be solved by phase diagram, as will be shown in a later chapter.

<u>Example 2.6 .8</u> Optimal Waste Disposal

As a further application of point and differential equation constraints, consider Plourde's (1972) model of waste disposal. The utility function is $U(C_1, C_2) = u(C_1) + v(C_2)$ where $u(C_1)$ is consumption utility, $v(C_2)$ is waste disutility function. Consumption good C_1 is produced with labour L_1 $C_1(t) = f(L_1)$ and pollution C_2 is proportional to the waste stock G, i.e., $C_2(t) = bG(t)$. Waste accumulates according to $\dot{G} = \lambda\ f(L_1) - g(L_2) - \alpha G(t)$ i.e., in any period, a constant proportion γ of output $f(L_1)$ accumulates as waste and a constant proportion α of waste bio-decomposes and a variable quantity $g(L_2)$ is removed by labour L_2. Note that total labour force $L(t) = L_1(t) + L_2(t)$.

The objective is to maximize, subject to the above equations, the functional

$$J = \int_0^\infty U(C_1, C_2)\ e^{-\delta t}\ dt \qquad (\delta > 0)$$

This amounts to maximizing

$$\int_0^\infty F(\cdot)\ dt$$

where, omitting arguments, e.g., $f(L_1)$ is written as f etc...,

$$F(\cdot) \equiv e^{-\delta t}\ [u(C_1) + v(C_2)] + p(\gamma f - g - \alpha G - \dot{G}) + \lambda_1(f - C_1)$$
$$+ \lambda_2(G - C_2) + \lambda_3(L - L_1 - L_2)$$

Euler equations give

$$F_{c_1} - \frac{d}{dt} F_{\dot{c}} = e^{-\delta t}\ u'(C_1) - \lambda_1 = 0$$

$$F_{c_2} - \frac{d}{dt} F_{\dot{c}_2} = e^{-\delta t}\ v'(C_2) - \lambda_2 = 0$$

$$F_G - \frac{d}{dt} F_{\dot{G}} = -\alpha p + \lambda_2 - \dot{p} = 0$$

$$F_{L_1} - \frac{d}{dt} F_{\dot{L}_1} = (p\gamma + \lambda_1)\ f'\ (L_1) - \lambda_3 = 0$$

$$F_{L_2} - \frac{d}{dt} F_{\dot{L}_2} = pg'(L_2) - \lambda_3 = 0$$

Basically these results say that in an optimal program, resources should be allocated such that the discounted marginal utility of consumption is equal to the marginal cost of supplying it, the discounted marginal disutility of pollution is equal to the net marginal cost of waste removal. Similarly the remaining equations imply the fulfillment of marginal conditions. For further details, see Plourde (1972).[5]

Example 2.6.9 The perimetric problem of Non Renewable Resources

As an illustration of Queen Dido's perimetric problem, let us consider the optimal extraction policy for non-renewable resources.

Consider a society endowed with a known fruit stock (s) of some non-renewable resource which is essential to the economy, i.e.,

$$\int_0^T q(t)\, dt = s$$

where $q(t)$ is the quantity of the resource extracted for consumption at time t. The objective is to maximize the utility of consumption $u(q)$ with $u''(q) < 0 < u'(q)$, discounted at rate r, i.e.

$$\text{Max.} \quad \int_0^T u(q)\, e^{-rt}\, dt$$

Define the remaining stock at t $(0 \le t \le 0)$ as

$$x(t) = s - \int_0^t q(\tau)\, d\tau$$

i.e. $\dot{x}(t) = -q(t)$

with $x(0) = s$ and $x(T) = 0$

The objective functional is $\int_0^T F(\dot{x}, q, p(t)\, dt$ where

$$F \equiv u(q)\, e^{-rt} - p(q + \dot{x})$$

The Euler equations give

$$F_q - \frac{d}{dt} F_{\dot{q}} = u'(q)\, e^{-rt} - p = 0$$

$$F_x - \frac{d}{dt} F_{\dot{x}} = \qquad \dot{p} \qquad = 0$$

i.e., $p(t)$ is a constant function and optimal extraction rate $q^*(t)$ should be such that

$$u'(q^*) = p\, e^{rt}$$

where $p(t) = p =$ constant for all $t \in [0, T]$, i.e., such that the marginal utility of consuming non-renewable resource $u'(q^*)$ should increase exponentially at the discount rate r, which, in view of the concavity of $u(q)$, implies that later generations should consume less than earlier generations.

2.7 Summary and Conclusion

In this chapter, the basic ideas of the Calculus of Variations have been introduced. The fundamental lemma was presented and proved since it is important to understand the main theorem giving the Euler equations which must be satisfied in any extremum problem regardless boundary conditions.

Euler equations were discussed in detail and generalised to the cases of several variables as well as higher order derivatives. The results could be summarised as

$$f_{x_i} - \frac{d}{dt} f_{\dot{x}_i} = 0 \quad \forall\, i$$

where $J = \int_0^T f(x,\dot{x},t)\, dt$, the objective functional and

$$f_x - \frac{d}{dt} f_{\dot{x}} + \frac{d}{dt} f_{\ddot{x}} -\dots\ .. + (-1)^n \frac{d^n}{dt^n} f_{x^n} = 0$$

for the case where $J = \int_0^T f(x,\dot{x},\ddot{x},\dots,x^n,t)\, dt$.

The particular cases where f does not contain t or x or $\dot{x}$ explicitly were also discussed and examples provided.

The problem of constrained extremisation however is of utmost importance in Economics and other social sciences as well as in Physics and Engineering. This was discussed in some detail and the results could be summarised as follows: Set up the Lagrange equation $L = f + pg$ where f is the objective function, g is the constraint equation (or system of equations) and p is the Lagrange multiplier scalar (or vector) and treat L as f in the unconstrained case. This leads to the solution of the Euler equations

$$L_x - \frac{d}{dt} L_{\dot{x}} = 0 \ .$$

$$L_p = 0$$

Euler equations are generally non-linear differential equations the solution of which must satisfy boundary conditions. These are of several types: fixed and free end points, fixed and free initial and terminal times and the various combinations of these. These will be discussed in Chapter 3, where some economic applications will be presented.

Chapter 2: FOOTNOTES

1. See, for example, Menger (1956), Kline (1962), Smith (1974) or Virgil's Aeneid, Book I versus 367 seq.

2. See Kline (1962).

3. Two continuous and differentiable functions $x(t)$ and $y(t)$ defined on a closed interval are called close to each other or order o if they do not deviate from each other by a distance greater than $\varepsilon > o$, i.e., $|x(t) - y(t)| < \varepsilon$. They are said to be close of order 1 if their first derivatives are also close to each other, i.e., if $|x(t) - y(t) < \varepsilon$ and $|x'(t) - y'(t)| < \varepsilon$. The order of closeness is immaterial in the study of first variations but it assumes its importance in the examination of second variations and strong extrema.

4. Proof of the Fundamental Lemma.
Suppose $g(t) \neq o$, say $g(t) > o$, in $[0, T]$. Then, by continuity, $g(t) \neq o$ for some positive interval $[a, b]$ in $[0, T]$ where $o < a < b < T$. Let $h(t) \equiv (t-a)(b-t)$ for $t \in [a, b]$ and $h(t) = o \quad \forall\, t \notin [a, b]$ (See fig. 2.5) Clearly $h(t)$ satisfies all the conditions of the Lemma. But then $\int_o^T g(t)\ (t-a)(b-t)\ dt \neq o$. This contradiction proves the Lemma.

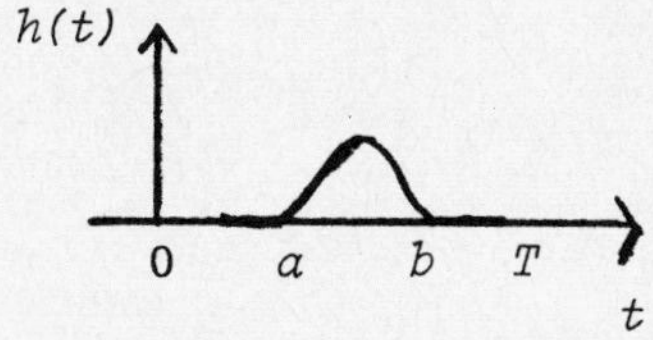

Fig. 2.5

5. Note that we have changed Plourde's Control approach to the Variational problem and inequality into equality constraints in order to illustrate the various cases of the Euler equations discussed in this chapter.

CHAPTER 3

BOUNDARY CONDITIONS IN VARIATIONAL PROBLEMS

As we have seen in chapter 2, the solution of the problem of finding an extremum of the functional

$$J(x) = \int_0^T f(x,\dot{x},t)\, dt \tag{1}$$

amounts to solving the Euler equation $f_x - \frac{d}{dt} f_{\dot{x}} = 0$. Since this is generally a second order differential equation, its solution involves two arbitrary constants which are determined by boundary conditions. These differ from problem to problem. They will now be discussed, starting from the simplest case of two fixed end points.

3.1 Two Fixed End Point and Natural Boundary Problems

It will be recalled that a necessary condition for the functional $J(x)$ in (1) to have an extremum is $\delta J = 0$, where

$$\begin{aligned} \delta J &= \int_0^T (f_x h + f_{\dot{x}} \dot{h})\, dt \\ &= \int_0^T (f_x - \frac{d}{dt} f_{\dot{x}}) h\, dt + h f_{\dot{x}} \Big|_0^T = 0 \ . \end{aligned} \tag{2}$$

Since the Euler equation must be satisfied, the first term on the RHS of (2) must vanish, leaving

$$h f_{\dot{x}} \Big|_0^T = 0 \tag{3}$$

If both the initial $A(0,x_0)$ and terminal point $B(T,x_T)$ are fixed, the displacement function $h(t)$ vanishes at these points i.e. $h(0) = 0 = h(T)$ or $x(0) = x_0$ and $x(T) = x_T$ and (3) is satisfied. This case, encountered repeatedly in chapter 2, is called the two-fixed-end-point problem: the extremal $x^*(t)$ goes through these fixed points and $x(0) = x_0$, $x(T) = x_T$ determine the two arbitrary constants of integration.

If, however, the end points $x(0)$ and $x(T)$ are not specified, $h(t)$ does not vanish at these points then in order for $\delta J = f_{\dot{x}} h \Big|_0^T = 0$, $h(0)$ and $h(T)$ being non zero, $f_{\dot{x}}$ must vanish at these points i.e.

$$f_{\dot{x}} = 0 \quad \text{at} \quad t = 0 \quad \text{and} \quad t = T \tag{4}$$

this is the natural boundary case and (4) is called the natural boundary conditions. They serve to determine the two constants of integration.

Example 3.1.1

Consider the Brachistochrone problem, examined in chapter 2, of minimizing $\int_0^T (1 + \dot{x}^2)^{\frac{1}{2}} \, dt$. Euler equation gives $\ddot{x} = 0$ or $x(t) = at + b$. If $x(0) = 4$ and $T = 10$, but $x(T) = x(10)$ is unspecified, we have $f_{\dot{x}}\Big|_{t=10} = 0$, i.e.

$$f_{\dot{x}}\Big|_{t=10} = \frac{\dot{x}(10)}{[1 + \dot{x}^2(10)]^{\frac{1}{2}}} = 0$$

giving $\dot{x}(10) = a = 0$. This, together with $x(t) = at + b = 0 + b = 4$ at $t = 10$ gives the extremal $x^*(t)$ as $x^*(t) = 4$ which, as expected, is a straight line, joining $A(0,4)$ to $B(10,4)$, parallel to the t-axis.

Example 3.1.2

Consider the problem of minimizing $J(x)$ where

$$J(x) = \int_0^2 (\dot{x}^2 + x\dot{x} + 2\dot{x} + 4x)\, dt$$

where $x(0)$ and $x(2)$ are unspecified. This is the special case (1) where $t_0 = 0$, $T = 2$ but $x(t_0)$ and $x(T)$ are free (see fig. 3.4a).

Euler equation is $\ddot{x} = 2$, the solution of which is

$$x(t) = t^2 + c_1 t + c_2$$

where c_1 and c_2 are determined by $f_{\dot{x}} = 2\dot{x} + x + 2 = 0$ for $t = 0$ and 2 i.e. $2(2t + c_1) + t^2 + c_1 t + c_2 + 2 = 0$, for $t = 0$ and $t = 2$, giving $2c_1 + c_2 + 2 = 0$ for $t = 0$ and $2(4 + c_1) + 4 + 2c_1 + c_2 + 2 = 0$ for $t = 2$, i.e., $c_1 = -6$, $c_2 = 10$.

Substitution into the above gives

$$x(t) = t^2 - 6t + 10$$

3.2 Variable End Points

The discussion of the Two-Fixed-End-Point and Natural Boundary Problems may be considered an introduction to the more general case in which either the initial point $A(t_0, x(t_0))$ or the terminal point $B(T, x(T))$ or both may be free; more specifically either the initial time t_0 or the final time T or both and also either $x(0)$ or $x(T)$ or both may be unspecified. These will be discussed in this section.

Consider the problem of finding an extremum of the functional

$$J(x) = \int_{t_0}^{T} f(x, \dot{x}, t)\, dt$$

where t_0, T, $x(t_0)$ and $x(T)$ are all free. These four unknowns must be determined.

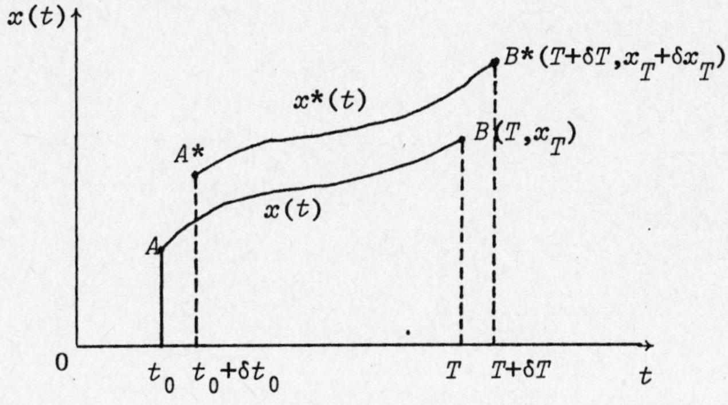

Fig. 3.1

Consider two neighbouring curves $x(t)$ and $x^*(t)$ in fig. 3.1. Let $h(t) = x^*(t) - x(t)$, i.e. $x^*(t) = x(t) + h(t)$. Since the variations $h(T)$ at the terminal point B do not affect the variations $h(t_0)$ at the initial point A, without loss of generality, we may consider point A as fixed, i.e. $x(0) = x_0$, $t(0) = t_0 = 0$ both fixed and B as the variable end point with both $x(T)$ and T unspecified. The above diagram is now simplified as in fig. 3.2.

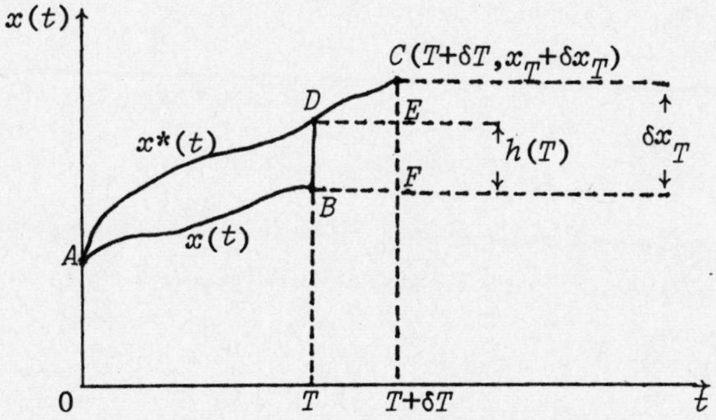

Fig. 3.2

It can be seen from fig. 3.2 that $BD = h(T)$, $FC = \delta x_T$ $EC \approx \dot{x}(T)\,\delta T$, i.e. $BD = FC - EC$ or

$$h(T) = \delta x_T - \dot{x}(T)\delta T \tag{5}$$

where δx_T is the difference between the terminal points of curves $x^*(t)$ and $x(t)$ (see fig. 3.2).

The variations of the functional $J(x)$ are given by $\Delta J(x)$, i.e.

$$\begin{aligned}\Delta J &= \int_0^{T+\delta T} f(t,x+h,\dot{x}+\dot{h})\,dt - \int_0^{T} f(t,x,\dot{x})\,dt \\ &= \int_0^{T} [f(t,x+h,\dot{x}+\dot{h}) - f(t,x,\dot{x})]\,dt \\ &\quad + \int_T^{T+\delta T} f(t,x+h,\dot{x}+\dot{h})\,dt\,.\end{aligned} \tag{6}$$

The last term on the RHS of (6) could be written as

$$\int_T^{T+\delta T} f(t,x+h,\dot{x}+\dot{h})\,dt \approx f(.)\Big|_{t=T+\theta\delta T}\delta T \tag{7}$$

by using the mean value theorem, $(0<\theta<1)$ and

$$f(.)\Big|_{t=T+\theta\delta T} = f(.)\Big|_{t=T} + \epsilon$$

where $\epsilon \to 0$ as $\delta T \to 0$ and $\delta x_T = 0$. The last term of (6) would thus be roughly equal to $f(.)\big|_T\,\delta T$.

The first two terms on the RHS of (6) can be approximated by use of Taylor series

$$\int_0^T (f_x h + f_{\dot{x}} \dot{h})\, dt + R$$

where R is the remainder involving higher order terms.

Thus the first variations are, neglecting R and ϵ,

$$\delta J = \int_0^T (f_x h + f_{\dot{x}} \dot{h})\, dt + f(.)\big|_T \delta T$$

$$= \int_0^T (f_x - \frac{d}{dt} f_{\dot{x}})h\, dt + hf_{\dot{x}}\big|_T + f(.)\big|_T \delta T \qquad (8)$$

by integration by parts.

Since variations at the end do not affect variations in the interior interval $(0,T)$, it is necessary for $\delta J = 0$ that Euler equation be satisfied, i.e.

$$f_x - \frac{d}{dt} f_{\dot{x}} = 0 \ . \qquad (9)$$

Hence (8) is reduced to

$$\delta J = f_{\dot{x}} h\big|_T + f(.)\big|_T \delta T = 0 \ . \qquad (10)$$

Substituting $h(T)$ $(\equiv h\big|_T)$ from (5) gives

$$\boxed{[f(.)\big|_T - \dot{x} f_{\dot{x}}\big|_T]\delta T + f_{\dot{x}}\big|_T \delta x_T = 0} \qquad (11)$$

which are called Transversality Conditions. They will determine the two unknowns x_T and T.

Two cases must be distinguished:

Case 1: If the variations δx_T and δT are independent of each other, their coefficients both vanish and (11) gives the transversality conditions

$$f(.)\big|_T - \dot{x}f_{\dot{x}}\big|_T = 0 \tag{12}$$

$$f_{\dot{x}}\big|_T = 0 \tag{13}$$

which together mean simply that

$$f(.)\big|_T = 0 = f_{\dot{x}}\big|_T \,. \tag{14}$$

Case 2: If the end point B must move along a certain curve $x(t) = g(t)$ (see fig. 3.3), then we have, in this case,

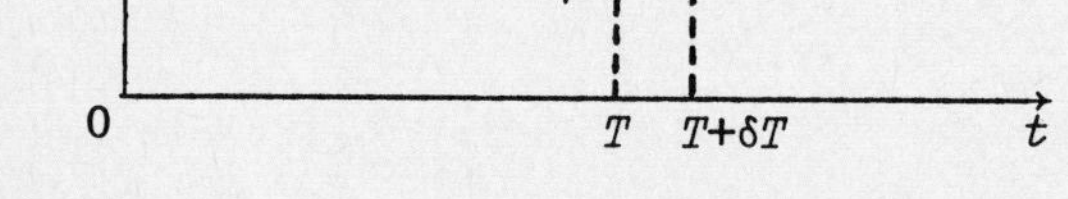

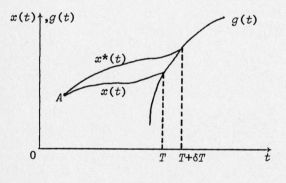

Fig. 3.3

$$\delta x_T = \dot{g}(T)\,\delta T \qquad (15)$$

where $\dot{g}(T) \equiv dg/dt$ at $t = T$.

Substitution into (11) gives, at the terminal time $t = T$, the following condition:

$$\left\{f(.) + [\dot{g}(T) - \dot{x}(T)]f_{\dot{x}}\right\}_{t=T} \delta T = 0 \,. \qquad (16)$$

The same analysis applies, *mutatis mutandis*, to the case in which the initial point $A(x(t_0),t_0)$ is free, i.e. both $x(t_0)$ and t_0 are free. Transversality conditions (11) now become

$$\boxed{[f(.) - \dot{x}f_{\dot{x}}]\delta t\Big|_{t=t_0}^{t=T} + f_{\dot{x}}\,\delta x(t)\Big|_{t=t_0}^{t=T} = 0 \,.} \qquad (17)$$

Similarly, (12), (13) and (14) become

$$f(.)\Big|_{t=t_0}^{t=T} = 0 = f_{\dot{x}}\Big|_{t=t_0}^{t=T} \qquad (18)$$

and (16) becomes

$$[f(.) + (\dot{g} - \dot{x})f_{\dot{x}}]_{t=t_0}^{t=T} \delta T = 0 \,. \qquad (19)$$

These Transversality Conditions are the most basic ones: they cover all other situations as particular cases of their own. Thus

(1) If the end points lie on straight lines $t = t_0$ and $t = T$ then

$\delta t_0 = 0 = \delta T$ and the first term on the LHS of (17) vanishes and (17) is reduced to

$$f_{\dot{x}}\ \delta x \Big|_{t=t_0}^{t=T} = 0$$

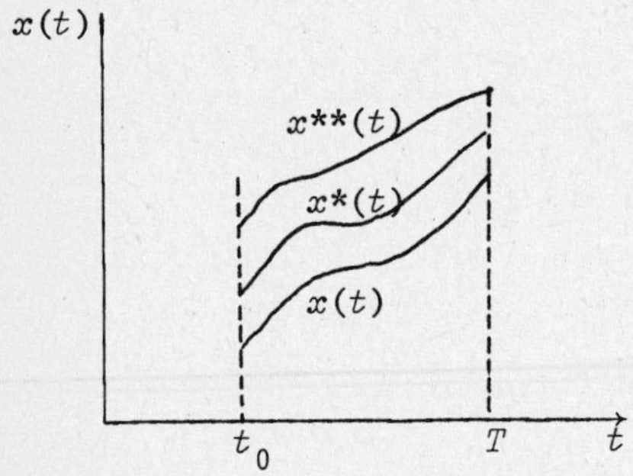

Fig. 3.4a: $x(t_0)$ and $x(T)$ free
t_0 and T specified

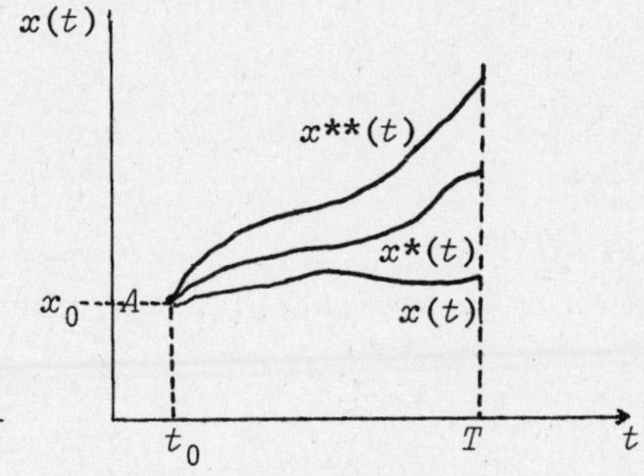

Fig. 3.4b: x_0, t_0, T fixed
x_T free

(2) If both A and B are fixed, i.e. both $x(t_0)$, $x(T)$ and t_0, T are fixed, we have $\delta t_0 = 0 = \delta T$ and $\delta x(0) = 0 = \delta x(T)$: we are back to the two fixed end point case.

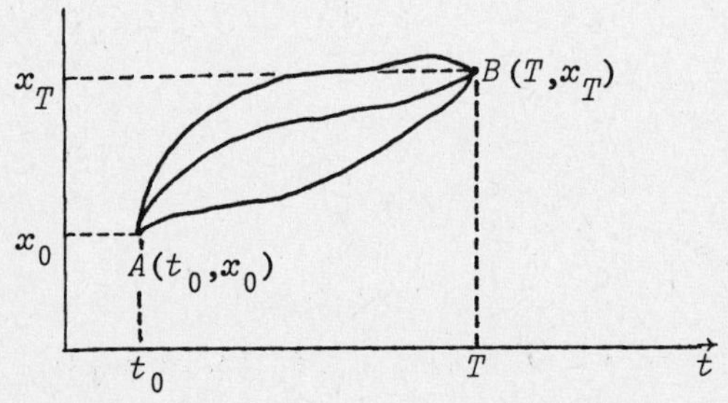

Fig. 3.5 Two fixed end points

(3) If x_0 and x_T are fixed but t_0 and T are free, then $\delta x(t_0) = 0 = \delta x(T)$ but $\delta t_0 \neq 0$ and $\delta T \neq 0$ and (17) gives

$$(f - \dot{x} f_{\dot{x}}) \Big|_{t_0}^{T} = 0$$

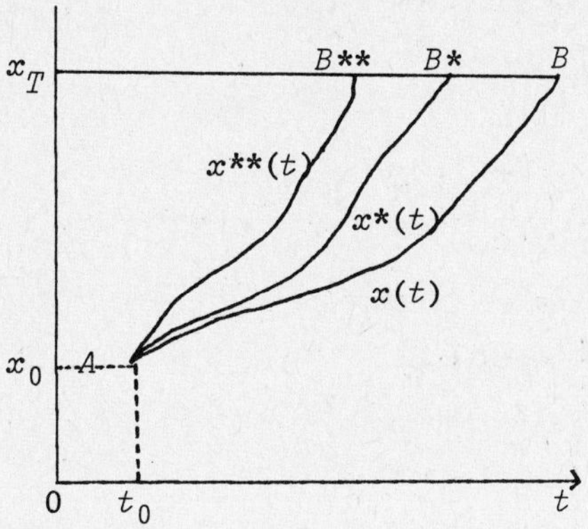

Fig. 3.6a

x_T, x_0, t_0 fixed but T free

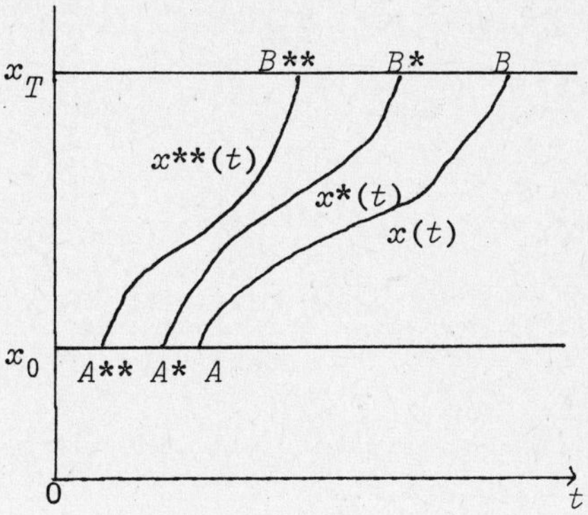

Fig. 3.6b

x_0, x_T fixed but t_0 and T free

(4) Free end points, i.e. t_0, T, x_0 and x_T are all free. Then (17) must be satisfied as it is.

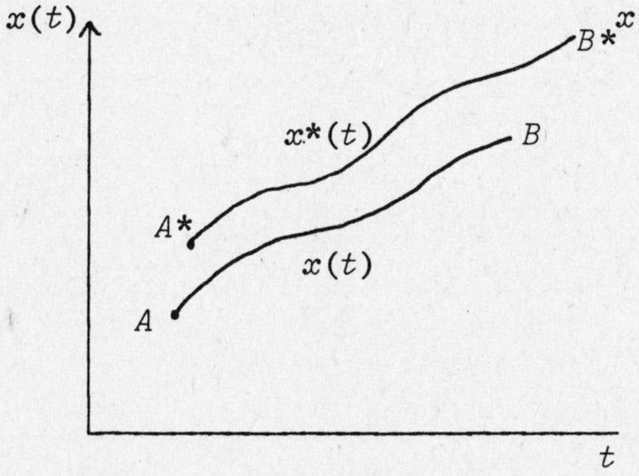

Fig. 3.7a

t_0, T, x_0 and x_T are all free

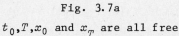

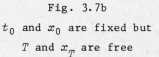

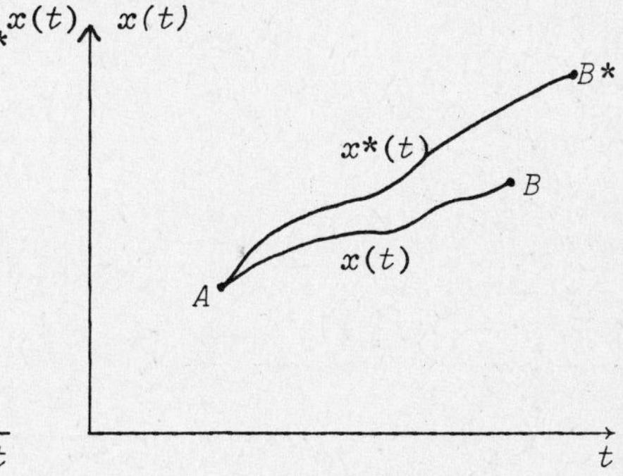

Fig. 3.7b

t_0 and x_0 are fixed but T and x_T are free

In this case,

$$[f(.) - \dot{x} f_{\dot{x}}] \Big|_{t=t_0}^{t=T} = 0$$

and

$$f_{\dot{x}}\Big|_{t=t_0}^{t=T} = 0$$

determine the 4 constants of integration.

(5) Both $x(t_0)$, t_0 and $x(T)$, T are free but $x(t_0)$ and $x(T)$ must move along certain curves $x(t_0) = g_1(t_0)$ and $x(T) = g_2(T)$ (see fig. 3.8). The constants of integration are determined by

$$\left\{f + (\dot{g}_1 - \dot{x})f_{\dot{x}}\right\}\Big|_{t_0} = 0$$

$$\left\{f + (\dot{g}_2 - \dot{x})f_{\dot{x}}\right\}\Big|_{T} = 0$$

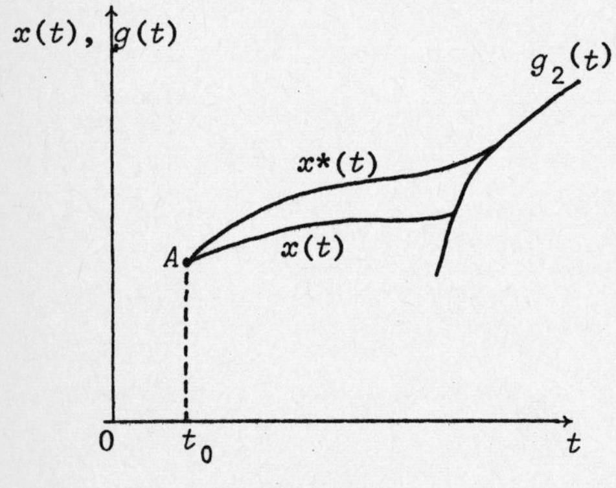

Fig. 3.8a

t_0 and x_0 are fixed
$x(T)$ and T free but
$x(T)$ must move along curve
$x(t) = g_2(t)$

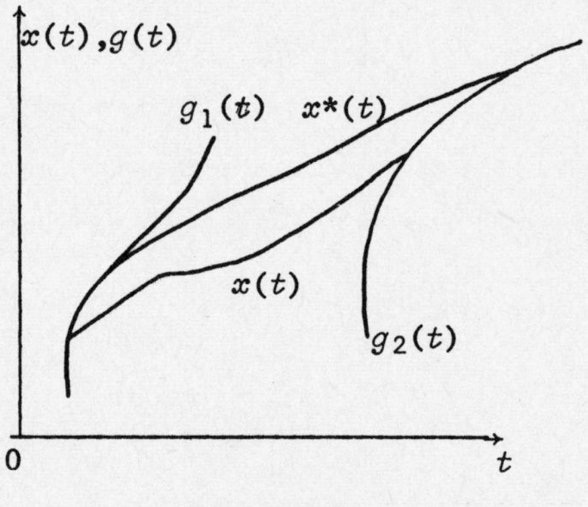

Fig. 3.8b

$t_0, x_0, T, x(T)$ all free but
$x(t_0)$ must move along curve
$x(t) = g_1(t)$ and $x(T)$ must
move along curve $x(t) = g_2(t)$

(6) Any combinations of the above.

Defining the adjoint variable $p(t)$ and the Hamiltonian H as

$$p(t) \equiv - \partial f/\partial \dot{x} \tag{20}$$

$$H \equiv f(x,\dot{x},t) - \dot{x} f_{\dot{x}} \equiv f + p\dot{x} \tag{21}$$

we can write the transversality conditions (17) more compactly as

$$(H\delta t + p\delta x)\Big|_{t=t_0}^{t=T} = 0 \ . \tag{22}$$

Clearly in the case of free initial and final times $t(0)$ and T, $H(t_0) = 0 = H(t)$ and in the case of unspecified $x(0)$ and $x(T)$, $p(t_0) = 0 = p(T)$. These determine the constants of integration.

In the case x and $\dot{x}$ in $f(x,\dot{x},t)$ are n-vectors, (20) is $p_i = -\partial f/\partial \dot{x}_i$ and (21) is $H \equiv f(x,\dot{x},t) + \sum_1^n p_i \dot{x}_i$, all i, and assuming the Jacobian $\frac{\partial(p_1,\ldots,p_n)}{\partial(\dot{x}_1,\ldots,\dot{x}_n)} \equiv \det\left[\frac{\partial^2 f}{\partial \dot{x}_i \partial \dot{x}_j}\right]$ does not vanish, the transversality conditions (22) are

$$(H\delta t + \sum_1^n p_i \ \delta x_i)\Big|_{t=t_0}^{t=T} = 0 \tag{23}$$

i.e. $p\dot{x}$ in (21) and $p\delta x$ in (22) are now inner products. (22) incorporates all cases which will now be summarised in the following table, assuming without loss of generality that the initial point $A(t_0,x_0)$ is fixed i.e. t_0 and x_0 are both specified.

Table 3.1

Boundary Conditions and determination of Constants of integration (cf. eq. 11)

Case	Substitution	Boundary Conditions	Determination of constants
1. $x(T)$ and T both specified	$\delta x_T = 0$ $\delta T = 0$	$x^*(0) = x_0$ $x^*(T) = x_T$	$2n$ equations to determine $2n$ constants
2. $x(T)$ free T specified	$\delta x(T) \neq 0$ $\delta T = 0$	$x^*(0) = x_0$ $f_{\dot{x}_i} = 0$ at $t = T$ $(1 \leq i \leq n)$	$2n$ equations to determine $2n$ constants
3. $x(T) = x_T$ fixed T free	$\delta x_T = 0$ $\delta T \neq 0$	$x^*(0) = x_0$ $x^*(T) = x_T$ $f - \dot{x}f_{\dot{x}} \equiv H = 0$ at $t = T$	$(2n+1)$ equations to determine $2n$ constants and T
4. $x(T)$ and T both free and independent	$\delta x(T) \neq 0$ $\delta T \neq 0$	$x^*(0) = x_0$ $f_{\dot{x}} = 0$ at $t = T$ $H \equiv f - \dot{x}f_{\dot{x}} = 0$ at $t = T$	$(2n+1)$ equations to determine $(2n+1)$ constants x_0, $x(T)$ and T
5. $x(T)$ and T both free but related by $x(T) = g(T)$	$\dot{x}(T) = \dot{g}(T)\delta T$ i.e. $\delta x_i(T) = \dot{g}_i(T)\delta T$ $(1 \leq i \leq n)$	$x^*(0) = x_0$ $x^*(T) = g(T)$ $f + [\dot{g}(T) - \dot{x}(T)]f_{\dot{x}} = 0$ at $t = T$ (see eq. (19))	$(2n+1)$ equations to determine $(2n+1)$ constants x_0, $x(T)$ and T

Source: Kirk (1970), p. 151

Example 3.2.1

Let us examine the problem of finding the extremal for the functional

$$J = \int_0^T (x + \dot{x}^2)\, dt$$

for the following cases:

(a) $x(0) = 1$; $T = 2$; $x(2) = 10$: two fixed end points

(b) $x(0) = 1$; $T = 2$; $x(2)$ free : variable end point

(c) $x(0) = 1$; $x(T) = 4$; T free but $T > 2$: variable terminal time

Solution: Euler equation is

$$1 - 2\ddot{x} = 0$$

giving

$$x^*(t) = \frac{1}{4}\, t^2 + c_1 t + c_2$$

where c_1 and c_2 are arbitrary constants to be determined in each case as follows:

(a) $x(0) = 1 = c_2$

$x(2) = 1 + 2c_1 + 1 = 10$, i.e. $c_1 = 4$.

The solution is

$$x^*(t) = t^2/4 + 4t + 1 .$$

(b) $x(0) = 1 = c_2$

c_1 is determined by the condition (4), i.e. $f_{\dot{x}} = 2\dot{x} = 0$ at $t = T$, i.e. $\dot{x}(T) = 0 = \frac{1}{2}\, T + c_1 = 1 + c_1 = 0$, i.e. $c_1 = -1$. Thus

$$x^*(t) = t^2/4 - t + 1$$

(c) $x(0) = 1 = c_2$

c_1 is determined by condition (11) i.e. $(f - \dot{x}f_{\dot{x}})\Big|_{t=T} = 0$ i.e.

$$-\dot{x}^2(T) + x(T) = 0$$

where $x^*(T) = T^2/4 + c_1T + c_2$ in the solution above and $\dot{x}^*(T)$ its derivative, i.e.

$$-(T/2 + c_1)^2 + T^2/4 + c_1T + c_2 = 0$$

which gives

$$c_1^2 = c_2 = 1 \qquad \text{i.e. } c_1 = \pm 1 \quad .$$

For $c_1 = 1$, $x(T) = T^2/4 + T + 1 = 4$ gives $T = -2 \pm 4 = 2$

For $c_1 = -1$, $x(T) = T^2/4 - T + 1 = 4$ gives $T = 2 \pm 4 = 6$

Since $T > 2$, we have

$$x^*(t) = (\tfrac{1}{4})t^2 - t + 1$$

$$T^* = 6 \quad .$$

Example 3.2.2.

Consider the problem of minimizing the cost of building a road from a given point $A(0,1)$ of a city to an existing highway situated at $g(t) = 11 - 2t$. Assuming constant average construction cost AC (put $AC = 1$ for convenience) per kilometre, this amounts to minimizing

$$J(x) = \int_0^T (1 + \dot{x}^2)^{\frac{1}{2}} \, dt$$

where $x(0) = 1$ but T and $x(T)$ are both unspecified except that $x(T)$ must lie on the line $g(t) = 11 - 2t$. This is a variable end point brachistochrone problem. The Euler equation gives $\frac{d}{dt}(1 + \dot{x}^2)^{-\frac{1}{2}}\,\dot{x} = 0$ i.e. $\ddot{x} = 0$, the solution of which is

$$x(t) = c_1t + c_2$$

The constants c_1 and c_2 are determined by

$x(0) = c_2 = 1$

$f + (\dot{g} - \dot{x})f_{\dot{x}} = 0$ at $t = T$, i.e.

$(1 + \dot{x}^2)^{\frac{1}{2}} - (2 + \dot{x})(1 + \dot{x}^2)^{-\frac{1}{2}}\,\dot{x} = 0$ at $t = T$

or $\dot{x}(t) = 1/2$ for $t = T$. But from the solution $x(t) = c_1 t + c_2$ we have $\dot{x}(t) = c_1$ hence $c_1 = 1/2$ and the solution of the Euler equation is

$$x^*(t) = \frac{1}{2}\,t + 1 \;.$$

The unspecified terminal point T is determined by

$$x^*(T) = g(T)$$

$$\frac{1}{2}\,T + 1 = 11 - 2T$$

giving

$$T = \frac{10}{2.5} = 4 \;.$$

Note that in this case, transversality is reduced to orthogonality: the extremal $x^*(t)$ and $g(t)$ are orthogonal to each other, i.e. the product of their slope is $(-2)\frac{1}{2} = -1$. See fig. 3.9.

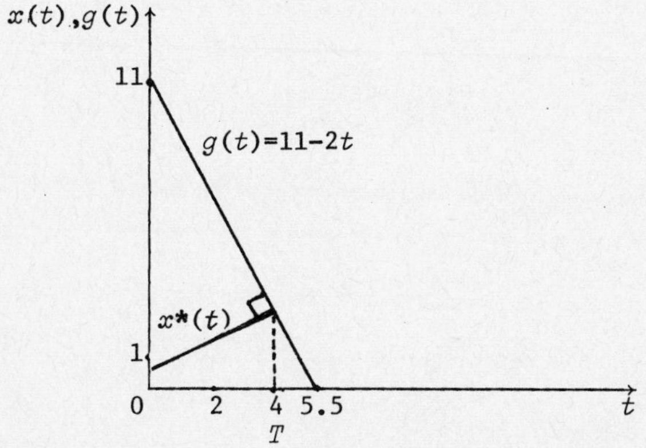

Fig. 3.9

3.3 Broken Extremals and the Erdman-Weierstrass Corner Conditions

The extremals considered so far are assumed to be smooth and continuous. In the real world, many problems have no solutions in the class of smooth functions. For example the shortest road joining A to B with a mountain or a lake between them. If tunnels and ferries are excluded, the shortest road giving minimum cost is a corner extremal which is only piecewise smooth. With the aid of transversality conditions we are now in a position to examine this type of problem.

Consider the problem of extremising the functional

$$\begin{aligned} J(x) &= \int_{t_0}^{T} f(x,\dot{x},t)\,dt \\ &= \int_{t_0}^{t_c^-} f(x,\dot{x},t)\,dt + \int_{t_c^+}^{T} f(x,\dot{x},t)\,dt \\ &\equiv J_1 \qquad\qquad + J_2 \end{aligned} \tag{24}$$

where $t_{\bar{c}}$ and t_{c^+} are variable end points (t_c being unspecified) corresponding to the time before and after the corner at t_c. (See fig. 3.10)

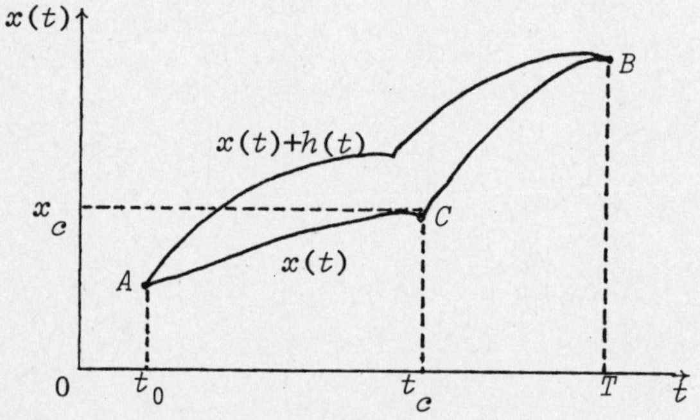

Fig. 3.10

For simplicity, consider the case of two fixed end points A and B with C the "variable end point" between them. Thus we allow for the possibility that the extremal may have a corner at some unspecified point C. The functional J is thus broken into J_1 and J_2 to each of which (8) applies. Since the Euler equation must be satisfied for each functional, the necessary conditions for extremals are reduced to $\delta J \equiv \delta J_1 + \delta J_2 = 0$ where

$$\delta J_1 = (f - \dot{x}f_{\dot{x}})\big|_{t_c^-} \delta t_c + f_{\dot{x}}\big|_{t_c^-} \delta x(t_c)$$

$$\delta J_2 = -(f - \dot{x}f_{\dot{x}})\big|_{t_c^+} \delta t_c - f_{\dot{x}}\big|_{t_c^+} \delta x(t_c) \,. \tag{25}$$

The condition that $x(t)$ is continuous at $t = t_c$ implies that δt_c and $\delta x(t_c)$ in (25) are the same, in δJ_1 and δJ_2 respectively.

$\delta J = \delta J_1 + \delta J_2 = 0$ requires that

$$[(f - \dot{x}f_{\dot{x}})\big|_{t_c^-} - (f - \dot{x}f_{\dot{x}})\big|_{t_c^+}]\, \delta t_c + (f_{\dot{x}}\big|_{t_c^-} - f_{\dot{x}}\big|_{t_c^+})\delta x(t_c) = 0. \tag{26}$$

Since δt_c and $\delta x(t_c)$ are arbitrary, their coefficients vanish and (26) implies the following conditions:

$$(f - \dot{x}f_{\dot{x}})\big|_{t_c^-} = (f - \dot{x}f_{\dot{x}})\big|_{t_c^+}$$

$$f_{\dot{x}}\big|_{t_c^-} = f_{\dot{x}}\big|_{t_c^+} \tag{27}$$

which are called the Erdmann-Weierstrass corner conditions.

In each interval $[t_0, t_c]$ and $[t_c, T]$ Euler equations must be satisfied. Thus we have two second order differential equations with four constants to be determined from $x(t_0) = x_0$, $x(T) = x_T$ and the two Erdmann-Weierstrass corner conditions (27).

In the Hamiltonian notations of (22), these conditions are written more simply as

$$H\Big|_{t_c^-} = H\Big|_{t_c^+}$$

$$p(t_c^-) = p(t_c^+) \ . \tag{28}$$

In the case where x and $\dot{x}$ are n-vectors each, (27) become

$$(f - \sum_1^n \dot{x}_i f_{\dot{x}_i})\Big|_{t_c^-} = (f - \sum_1^n \dot{x}_i f_{\dot{x}_i})\Big|_{t_c^+}$$

$$p_i(t_c^-) = p_i(t_c^+) \qquad (1 \le i \le n) \ . \tag{29}$$

or, in Hamiltonian form

$$H\Big|_{t_c^-} = H\Big|_{t_c^+}$$

$$p_i(t_c^-) = p_i(t_c^+) \ . \tag{30}$$

If in addition, the corner is required to intersect a certain given line, say $g(t)$, then application of (16) gives

$$(\dot{g} - \dot{x}^*) f_{\dot{x}}\Big|_{t_c^-} + f(.)\Big|_{t_c^-} = (\dot{g} - \dot{x}^*) f_{\dot{x}}\Big|_{t_c^+} + f(.)\Big|_{t_c^+} \tag{31}$$

which should be used instead of (27).

Example 3.3.1

Consider Erdman's example, simplified by Bolza (1904), of minimizing the functional

$$\int_0^T (\dot{x} + 1)^2 \dot{x}^2 \, dt .$$

Euler equation gives

$$f_{\dot{x}} = 4\dot{x}^3 + 6\dot{x}^2 + 2\dot{x} = \text{constant}$$

the solution of which is

$$x^*(t) = at + b .$$

Putting $\dot{x}(t_c^-) \equiv p_1$; $\dot{x}(t_c^+) \equiv p_2$, we obtain the Erdmann-Weierstrass conditions

$$\text{(i)} \quad 4(p_1^3 - p_2^3) + 6(p_1^2 - p_2^2) + 2(p_1 - p_2) = 0$$

$$\text{(ii)} \quad -3(p_1^4 - p_2^4) + 4(p_1 - p_2)w + (p_1 + p_2)(p_1 - p_2) = 0$$

where $w \equiv p_1^2 + p_1p_2 + p_2^2$. Putting $u \equiv p_1 + p_2$, remembering that $p_1^3 - p_2^3 \equiv (p_1 - p_2)(p_1^2 + p_1p_2 + p_2^2) \equiv (p_1 - p_2)w$, these become

$$\text{(i)} \quad 2w + 3u + 1 = 0$$

$$\text{(ii)} \quad -3u^3 + 6uw + 4w + u = 0$$

which give one real solution $u = -1, w = 1$ from which we obtain either $p_1 = 0, p_2 = -1$ or $p_1 = -1, p_2 = 0$, i.e. the extremal must be composed of straight horizontal sections (with slope 0) and straight downward sloping sections (with slope -1) or any combinations of these. (For more details, see Bolza (1904).

Example 3.3.2

Consider the problem of building a road from town A to B and touching a straight highway such as to minimize the mileage of the motorist driving from A to B and touching the highway at some point during the trip. The objective is clearly to save petrol in the long run. Let the highway be represented by $y(x) = -x + 3$ and town A be situated at (0,2) and B at (2,0) (see fig. 3.11).

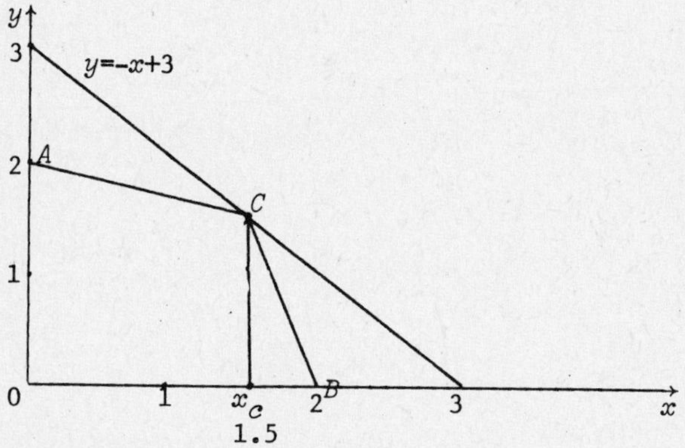

Fig. 3.11

The problem is thus one of finding the shortest piecewise smooth curve joining $A(0,2)$ and $B(2,0)$ and touching $y = -x + 3$. The functional to be minimized is

$$J(y) = \int_0^2 (1 + y'^2)^{\frac{1}{2}} \, dx \; .$$

The solution of the Euler equation is

$$y^* = ax + b \; .$$

Assuming a corner exists and writing $y^*(x)$ on each side of the corner as

$$y_1^*(x) = a_1 x + b_1 \quad \text{for } x \in [0, x_c]$$

$$y_2^*(x) = a_2 x + b_2 \quad \text{for } x \in [x_c, 2]$$

clearly a_1, b_1, a_2, b_2 and x_c are the unknowns which must be determined by the Erdmann-Weierstrass conditions (31). Putting $y'(x_c-) \equiv y_1'$ and $y'(x_c+) \equiv y_2'$ which are the slopes of the extremal on each side of the corner, i.e. $y'(x_c-) = a_1$ and $y'(x_c+) = a_2$ we obtain from (31)

$$\frac{-y_1'(1 + y_1')}{(1 + y_1'^2)^{\frac{1}{2}}} + (1 + y_1'^2)^{\frac{1}{2}} = \frac{-y_2'(1 + y_2')}{(1 + y_2'^2)^{\frac{1}{2}}} + (1 + y_2'^2)^{\frac{1}{2}}$$

or

$$\frac{1 - a_1}{(1 + a_1^2)^{\frac{1}{2}}} = \frac{1 - a_2}{(1 + a_2^2)^{\frac{1}{2}}} .$$

The two fixed end points give

$$y(0) = b_1 = 2$$

$$y(2) = 2a_2 + b_2 = 0 .$$

At corner C,

$$a_1 x_c + b_1 = - x_c + 3$$

$$a_2 x_c + b_2 = - x_c + 3 .$$

The five unknowns are determined by the last five equations as follows:

$$y^*(x) = -x/3 + 2 \quad \text{for} \quad x \in [0, x_c] \quad \text{(line } AC\text{)}$$

$$y^*(x) = -3x + 6 \quad \text{for} \quad x \in [x_c, 2] \quad \text{(line } CB\text{)}$$

$$x_c = 1.5 \ .$$

Thus the extremal is ACB with corner at C with its angle of incidence equal to its angle of reflection. This, of course, is obvious from the start: this problem could be solved by simple geometry as follows. Construct A' symmetric to A about the line $y = -x + 3$, by simply dropping a perpendicular to this line from A and prolong it beyond the line by the same distance (see fig. 3.12).

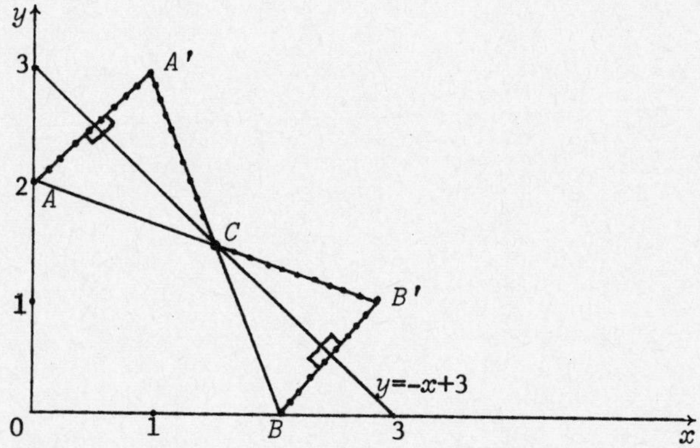

Fig. 3.12

It can then be shown that $\overline{ACB}$ is indeed the shortest line going from A to B, touching the line $-x + 3$ at some point C since $\overline{ACB} = \overline{A'CB}$ and $\overline{A'CB}$ is the straight line joining B and A'. Similarly, a point B' symmetric to B about the line $-x + 3$ could be built and $\overline{ACB} = \overline{ACB'}$ is clearly the shortest line. Furthermore, it is plain that $\overline{ACB} = \overline{A'CB} = \overline{ACB'}$, i.e. the solution is unique. Clearly the purpose of giving an example as obvious as this is to illustrate the corner conditions in less obvious cases.

Example 3.3.3

In Hotelling's Exhaustible Resources' model [1931, p. 160]

$$J = \int_0^T f(x,q,t)\, dt$$

where $f(x,q,t) \equiv pqe^{-rt}$, $x(t)$ is the cumulated amount which has been extracted from a mine, $q = \dot{x}$ = current rate of extraction and the demand function is $p = b - (q-1)^3$. Thus $f = [b - (q-1)^3]\, qe^{-rt}$, $f_q = [b - (4q-1)(q-1)^2]e^{-rt}$ and $f - qf_q = 3q^2(q-1)^2e^{-rt}$ at the corner.

Erdmann-Weierstrass conditions (27) give

$$(4q_1 - 1)(q_1 - 1)^2 = (4q_2 - 1)(q_2 - 1)^2$$

$$q_1^2(q_1 - 1)^2 = q_2^2(q_2 - 1)^2$$

where q_1, the rate of extraction just before the sudden jump and q_2 the rate after it are calculated to be

$$q_1 = (3 + \sqrt{3})/4 = 1.183$$

$$q_2 = (3 - \sqrt{3})/4 = 0.317 .$$

Example 3.3.4

In the problem of finding an extremum of

$$J = \int_0^T (\dot{x}^2 - x^2)\, dt$$

the Erdmann-Weierstrass conditions give

$$f_{\dot{x}}|_{t_c^-} = 2\dot{x} = 2\dot{x} = f_{\dot{x}}|_{t_c^+} ,$$

i.e. the extremal has no corners.

3.4 Canonical Form of the Euler Equation

The solution of the Euler equation $\partial f/\partial x_i - (d/dt)\partial f/\partial \dot{x}_i = 0$, $(i = 1,2,\ldots,n)$, form a set of n second order differential equations. This set could be reduced to a set of $2n$ first order differential equations by defining the Hamiltonian $H(x,p,t) \equiv f(x,\dot{x},t) + \sum_1^n p_i\dot{x}_i$ where $p_i = -\partial f/\partial \dot{x}_i$ $(i = 1,2,\ldots,n)$ as before and differentiating $H(x,p,t)$ where x, p, $\dot{x}$ are n-vectors.

$$dH = \frac{\partial f}{\partial t}\,dt + \sum_1^n \frac{\partial f}{\partial x_i}\,dx_i + \sum_1^n \frac{\partial f}{\partial \dot{x}_i}\,d\dot{x}_i + \sum_1^n p_i\,d\dot{x}_i + \sum_1^n \dot{x}_i\,dp_i \tag{32}$$

But

$$dH(x,p,t) = \frac{\partial H}{\partial t}\,dt + \sum_1^n \frac{\partial H}{\partial x_i}\,dx_i + \sum_1^n \frac{\partial H}{\partial p_i}\,dp_i \tag{33}$$

From (32) and (33) it follows that

$$\frac{\partial H}{\partial t} = \frac{\partial f}{\partial t} \tag{34}$$

$$\frac{\partial H}{\partial p_i} = \dot{x}_i \qquad (i = 1,2,\ldots,n) \tag{35}$$

$$\frac{\partial H}{\partial x_i} = \frac{\partial f}{\partial x_i}\left(= \frac{d}{dt}\frac{\partial f}{\partial \dot{x}_i} = -\dot{p}_i\right)(1 \le i \le n) \tag{36}$$

where $\frac{\partial f}{\partial x_i} = \frac{d}{dt}\frac{\partial f}{\partial \dot{x}_i}$ by Euler equations and $\frac{d}{dt}\frac{\partial f}{\partial \dot{x}_i} = -\dot{p}_i$ by definition.

System (35) and (36) of $2n$ first order differential equations is called the canonical system of the Euler equations and the variables

x_i, p_i $(1 \le i \le n)$ the canonical variables for the functional $J(x)$.

If f is autonomous, i.e. $f(x,\dot{x})$ does not depend on t explicitly, then the canonical system (35), (36) gives

$$\frac{dH}{dt} = \sum_1^n \left(\frac{\partial H}{\partial x_i}\dot{x}_i + \frac{\partial H}{\partial p_i}\dot{p}_i\right)$$
$$= \sum_1^n \left(\frac{\partial H}{\partial x_i}\frac{\partial H}{\partial p_i} - \frac{\partial H}{\partial p_i}\frac{\partial H}{\partial x_i}\right) \equiv 0 \tag{37}$$

i.e. $H(x,p)$ is a constant, or H is a first integral of the Euler equations, i.e. H is a function which has a constant value along each integral curve of the system. These points will be useful in a later chapter when the Maximum Principle is introduced.

Example 3.4.1. Intergenerational Distribution of Exhaustible Resources

Consider the problem of maximizing the present value discounted at rate r, of consumption $q(t)$ of some essential resource whose stock s is fixed (i.e. $s = \bar{s}$) and non-renewable. The utility function $u(q)$ is assumed concave and increasing. The objective functional is thus

$$J = \int_0^T u(q)\, e^{-rt}\, dt$$

subject to

$$\int_0^T q(t)\,dt = \overline{s}$$

Define

$$x(t) \equiv \overline{s} - \int_0^t q(\tau)\,d\tau$$

i.e.

$$\dot{x} = -q(t) \text{ with } x(0) = \overline{s},\ x(T) = 0$$

The Variational approach consists of applying the Euler equation to the function $F(q,\dot{x},t) \equiv u(q)e^{-rt} - p(q + \dot{x})$. This gives, as in chapter 2,

$$F_q - \frac{d}{dt} F_{\dot{q}} = F_q - 0 = e^{-rt}\ u'(q) - p = 0$$

$$F_x - \frac{d}{dt} F_{\dot{x}} = 0 - \frac{d}{dt}(-p) = \dot{p} = 0$$

The Canonical form gives, for $H \equiv u(q)e^{-rt} - pq$,

$$H_q = e^{-rt}\ u'(q) - p$$

$$\dot{p} = -H_x = 0$$

The results are thus identical. They say that $\dot{p} = 0$ i.e. $p(t) = p$, some constant, and $u'(q) = pe^{rt}$ i.e. marginal utility of consumption increases exponentially at rate r over time, implying that future generations must consume less than earlier generations, in view of the concavity assumption, i.e. diminishing marginal utility. If there is no discount, $r = 0$ and $u'(q) = p$ = constant for all $t \in [0,T]$ i.e. each generation consumes the same quantity of the non-renewable resource until exhaustion.

3.5 Some Economic Applications

3.5.1 Dynamic Monopoly

Dynamic Monopoly, to the author's knowledge, constitutes the first economic application of the Calculus of Variations. This pioneering work was done by Evans (1924) who sought to maximize the monopolist's profit over a period of time.

Evans considered a monopolist who produces and sells a single good $x(t)$ for which the demand function is

$$x(t) = ap(t) + b + h\dot{p}(t)$$

and the cost function is

$$C(x) = \alpha x(t)^2 + \beta x(t) + \gamma$$

where a, b, h, α, β, γ are all constants. Only the normal case

$a < 0 < \alpha$ and continuous $p(t)$ having continuous derivatives will be investigated. Note that $\dot{p}(t) \equiv dp(t)/dt$ represents some speculative element in consumers' behaviour.

The monopolist wants to choose an optimal pricing policy $p(t)$ such as to maximize his profit π over a specified planning period of T years, given a historical initial price $p(0) = p_0$ and a target price $p(T) = p_T$ where T is specified. For each such $p(t)$ chosen, the market demand $x(t)$ is determined and given the cost function $C(x)$ above, the profit functional will be obtained. Optimal output policy $x(t)$ is determined by the demand law.

This is a typical two-fixed-end-point problem in the Calculus of Variations and the objective is to maximize the profit functional J defined as

$$J = \int_0^T \pi(p,\dot{p})\, dt$$

where

$$\pi \equiv px - C(x) \text{ which gives, on substitution,}$$

$$\pi(p,\dot{p}) \equiv p(ap + b + h\dot{p}) - \alpha(ap + b + h\dot{p})^2 - \beta(ap + b + h\dot{p}) - \gamma$$

which is not an explicit function of time and the Euler equation is reduced, in this case (see Ch. 2, eq. (28)), to

$$\pi - \dot{p}\pi_{\dot{p}} = \text{constant}$$

i.e.

$$h\, C'(x)\, \dot{p} + ap^2 + bp - C(x) = \text{constant}\ .$$

Differentiating with respect to t, dividing by $\dot{p}$ and substituting

$$C'(x) = 2h\alpha\dot{p} + 2a\alpha p + 2b\alpha + \beta; \qquad C''(x) = 2\alpha$$

yields

$$2h^2\alpha\ddot{p} - 2a(a\alpha - 1)p + b - 2ab\alpha - \alpha\beta = 0$$

or

$$\ddot{p} - b_1 p + c_1 = 0 \quad .$$

The solution of which gives the optimal price policy

$$p(t) = p_e + Ae^{mt} + Be^{-mt}$$

where

$$b_1 \equiv \frac{a(a\alpha - 1)}{h^2\alpha}, \quad c_1 \equiv \frac{b - 2ab\alpha - \alpha\beta}{2h^2\alpha}$$

$$A \equiv \frac{r_1 - r_2 e^{mT}}{1 - e^{2mT}}, \quad B \equiv \frac{r_1 - r_2 e^{-mT}}{1 - e^{-2mT}}$$

$$r_1 \equiv p_0 - p_e \quad , \quad r_2 \equiv p_T - p_e$$

$$p_e = \frac{c_1}{b_1}, \quad m^2 \equiv b_1 > 0 \text{ in view of } a < 0 < \alpha \quad .$$

3.5.2 Optimal Economic Growth

The question "how much should be consumed now and how much should be invested for future consumption?" is the focal point of the theory of optimal economic growth. In this sense, it is also referred to as the theory of optimal saving. Ramsey's Mathematical Theory of Saving (1928) is the pioneering work in this field. It has inspired a voluminous literature. In this section we shall present the optimal growth theory as an application of the Calculus of Variations, using the neo-classical model of Swan (1956) and Solow (1956) as building blocks.

Consider an economy which produces, with the help of capital $K(t)$ and a labour force $L(t)$ growing at a constant exogenous rate n (i.e. $\dot{L}/L = n$), a homogeneous commodity $Y(t)$ called GNP which could either be consumed $C(t)$ or used to build up further capital $\dot{K}(t)$ and replace old ones $\mu K(t)$ (μ constant depreciation rate) i.e.

$$Y(t) = C(t) + \dot{K}(t) + \mu K(t) \quad .$$

The production function $Y(t) = F[K(t), L(t)]$, assumed homogeneous of degree one (i.e. constant returns to scale), could be written as $Y(t) = LF(K/L, 1) \equiv Lf(k)$, or in per capita terms, putting $y \equiv Y/L$, $k \equiv K/L$, as

$$y(t) = f(k)$$

where f is assumed a strictly concave monotonically increasing function of k, with slope decreasing from $+\infty$ at $k = 0$ to 0 at $k = \infty$. With this, net per capita investment $\dot{K}/L$ could be written as per capita output $f(k)$ left over after allowing for per capita consumption $c(t)$ and depreciation $\mu k(t)$ i.e.

$$\dot{K}/L = f[k(t)] - c(t) - \mu k(t) \quad .$$

But from the definition of $k \equiv K/L$, we obtain, by substitution,

$$\begin{aligned} \dot{k} \equiv k(\dot{K}/K - \dot{L}/L) &\equiv \dot{K}/L - nk \\ &= f(k) - (\mu + n)k(t) - c(t) \quad . \end{aligned}$$

We are now ready to apply the Calculus of Variations to this problem by seeking to maximize the functional J

$$J = \int_0^T u(c)e^{-\delta t}\, dt$$

where

$$c(t) = f(k) - \lambda k - \dot{k} \equiv g(k) - \dot{k}$$

with $k(0) = k_0$, $k(T) = k_T$ and T is specified, and where $\lambda \equiv \mu + n$.

The utility function $u(c)$ obeys the law of diminishing marginal utility, i.e. $u''(c) < 0 < u'(c)$ and δ is the constant positive rate of future discount. Substitution gives

$$\varphi(k,k,t) \equiv u(c)e^{-\delta t} = u[g(k) - \dot{k}]e^{-\delta t}.$$

The Euler equation is

$$0 = \varphi_k - \frac{d}{dt}\varphi_{\dot{k}} = e^{-\delta t}u'(c)g'(k) + \frac{d}{dt}u'(c)e^{-\delta t}$$

which gives

$$\dot{c} = -\frac{u'(c)}{u''(c)}[g'(k) - \delta]$$

where $-u'/u'' > 0$. This, together with $\dot{k} = g(k) - c$, determines the behaviour of the system. One solution to this system is given by the (k^*,c^*) at which $\dot{k} = 0 = \dot{c}$. In order to maintain a constant c (i.e. $\dot{c} = 0$), a capital level k^* must exist such that $g'(k^*) = \delta$ (or $f'(k^*) = \lambda + \delta$) and in order to maintain a constant k, (i.e. $\dot{k} = 0$), we must have $c^* = g(k^*)$ ($\equiv f(k^*) - \lambda k^*$). (See Fig. 3.13a and 3.13b.)

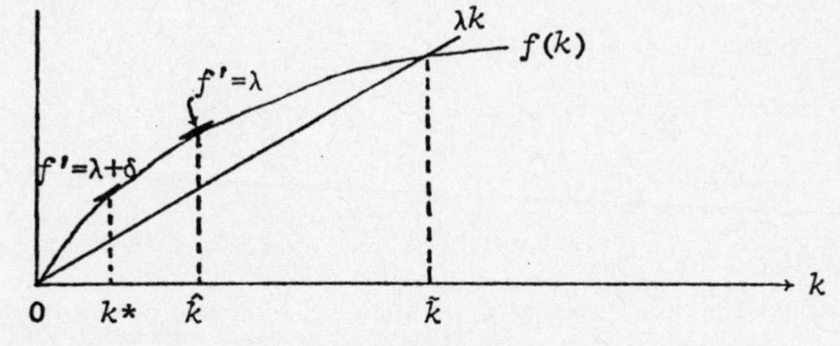

Fig. 3.13a

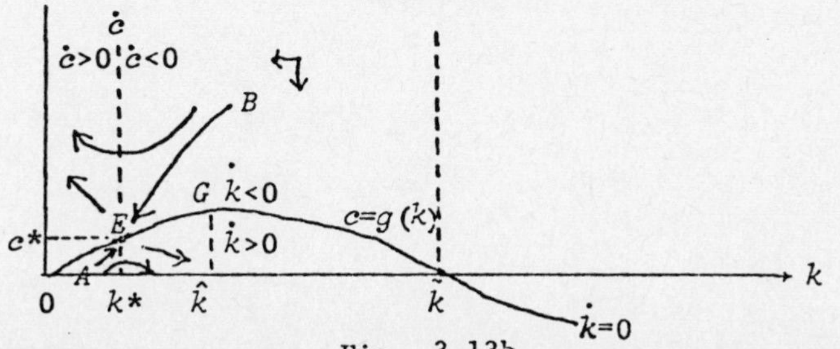

Fig. 3.13b

From Fig. 3.13 , it can be seen that at k^* where the slope of the production function (or the marginal product of capital) is $\lambda + \delta$, consumption per head is constant ($\dot{c} = 0$). This is represented by the vertical line $\dot{c} = 0$: to the right of it, $k > k^*$ and c falls ($\dot{c} < 0$) and to the left of it, $k < k^*$, c rises ($\dot{c} > 0$). The consumption level at which $\dot{k} = 0$ is $c = g(k)$, it is represented by the curve labelled $\dot{k} = 0$ in Fig. 3.13b: above it, $\dot{k} < 0$, i.e. k falls and below it, $\dot{k} > 0$, k rises. Thus equilibrium balanced growth (where $\dot{K}/K = n$ or $\dot{k} = 0$) is obtained at (k^*,c^*) which is the intersection of the curves $\dot{c} = 0$ and $\dot{k} = 0$. The stable branch *AEB* is the locus of all (k,c) points that will eventually reach the equilibrium $E(k^*,c^*)$. To the right of $\dot{c} = 0$ line and below the $\dot{k} = 0$ curve, c falls and k rises and to the left of $\dot{c} = 0$ line and above $\dot{k} = 0$ curve, k falls and c rises, i.e. the system moves away from its equilibrium level $E(k^*,c^*)$. This is the unstable branch. Thus $E(k^*,c^*)$ is a saddle point, a long run equilibrium: from any initial $k(0)$, an appropriate level of consumption can always be chosen such that the system reaches its long run stable equilibrium. It exists and is unique in our model.

The per capita consumption level associated with the capital intensity $\hat{k}$ at which $f'(k) = \lambda$ (or $g'(k) = 0$) and $\dot{k} = 0$ is called the golden rule by Robinson [1956], Swan [1956], Phelps [1966] and others, in the sense that among all balanced growth paths (where capital and labour grow at the same rate) it is the one that maximizes the sustainable level of per capita consumption. It is clear that so long as the rate of future discount δ is positive, the long run equilibrium $E(k^*,c^*)$ is lower than the golden rule level $G(\hat{k},\hat{c})$ at which $g'(k) = 0$.

As $\delta \to 0$, $k^* \to \hat{k}$. For this reason, (k^*,c^*) is often called the modified Golden Rule path, modified to take future discount into consideration.

This constitutes the graphical solution to the Euler equation. An analytical solution would only be possible if the form of $f(k)$ and $u(c)$ are explicitly given. In the Harrod-Domar world of fixed capital output ratio $(1/a)$, $f(k) = ak$, if $u(c) = \frac{c^{1-v}}{1-v}$ $(v > 0,\ v \neq 1$ or $u(c) = \log c$ if $v = 1)$, $-u'/u'' = c/v$ and $f'(k) = a$, the dynamic system is

$$\dot{k} = \alpha k - c$$

$$\dot{c} = \beta c$$

where $\alpha \equiv a - \lambda$, $\beta \equiv (a - \lambda - \delta)/v$, the solution is

$$c(t) = c_0 e^{\beta t}$$

and

$$k(t) = \frac{c_0 e^{\alpha t}}{\beta - \alpha} + \frac{(a - c_0)}{\beta} e^{\beta t}$$

where c_0 = some given level of initial consumption and $\beta \neq \alpha \neq 0$. The cases where $\alpha = \beta$, or $\beta = 0$ or $\alpha \lesseqgtr \beta$ could be further investigated but they are left to the interested reader.

3.5.3 Capital Theory with Exhaustible Resources

As an illustration of the constrained variational problem, let us examine a multisectoral growth model with exhaustible resources (Grandville, 1980).

Consider an economy producing n goods $Y_i(t)$ which could either be consumed $C_i(t)$ or invested $\dot{K}_i(t)$ to increase its stock K_i $(1 \leq i \leq n)$ (Samuelson-Solow 1956, Samuelson 1960) to which are added s non-renewable stocks $\overline{X}_i$ of known and fixed quantities which are extracted at the rate $R_i(t)$ $(1 \leq i \leq s)$ at each period $t \in [0,T]$. More precisely, there are s parametric constraints imposed by non-renewable resources

$$\int_0^T R_i(t)\, dt = \overline{X}_i \qquad (1 \leq i \leq s)$$

which, by a re-definition, could be written as

$$z_i(t) \equiv \int_0^t R_i(\tau)\, d\tau$$

i.e. $\dot{z}_i = R_i(t)$ with $z_i(0) = 0$, $z_i(T) = \overline{X}_i$ and n investment functions

$$\dot{K}_i(t) = Y_i(t) - C_i(t) \qquad (1 \leq i \leq n)$$

where the output $Y_i(t)$ are joint products of a single production process described by the transformation function

$$C_1(t) = f(K_1,\ldots,K_n;R_1,\ldots,R_s;C_2+\dot{K}_2+\ldots+t) - \dot{K}_1 \quad .$$

The objective is to maximize the present value of the consumption utility

$$J = \int_0^T U[C_1(t),\ldots,C_n(t)]e^{-\delta t}\, dt$$

subject to the above $n + s$ differential equations, i.e.

$$\text{Max. } J_a \equiv \int_0^T G(C,K,R,\lambda,z,t)e^{-\delta t}\, dt$$

where

$$G \equiv U[f(K,R;\vec{C} + \vec{\dot{K}}) - \dot{K}_1;\vec{C}]e^{-\delta t} + \lambda(R - \dot{z})$$

and C, K, R, λ, z are vectors and $\vec{C}$, $\vec{\dot{K}}$ are truncated vectors i.e. are C and $\dot{K}$ with the first commodity left out, i.e. $C \equiv (C_1,C_2,\ldots,C_n)$ and $\vec{C} \equiv (C_2,C_3,\ldots,C_n)$ etc.

The Euler Lagrange equations give

$$\frac{\partial G}{\partial C_i} - \frac{d}{dt}\frac{\partial G}{\partial \dot{C}_i} = e^{-\delta t}\left(\frac{\partial U}{\partial C_1}\frac{\partial f}{\partial C_i} + \frac{\partial U}{\partial C_i}\right) = 0$$

implying

$$-\frac{\partial f}{\partial C_i} = \frac{\partial U/\partial C_i}{\partial U/\partial C_1}$$

i.e. the marginal rate of substitution between any pair of goods must be equal to their marginal utility ratios

$$\frac{\partial G}{\partial K_1} - \frac{d}{dt}\frac{\partial G}{\partial \dot{K}_1} = e^{-\delta t}\frac{\partial U}{\partial C_1}\frac{\partial f}{\partial K_1} + \frac{d}{dt}\left(\frac{\partial U}{\partial C_1}e^{-\delta t}\right) = 0$$

$$= e^{-\delta t}\left(\frac{\partial U}{\partial C_1}\frac{\partial f}{\partial K_1} + \frac{d}{dt}\frac{\partial U}{\partial C_1} - \delta\frac{\partial U}{\partial C_1}\right) = 0$$

i.e.

$$\delta = \frac{\partial f}{\partial K_1} + \frac{1}{\partial U/\partial C_1}\frac{d}{dt}\frac{\partial U}{\partial C_1}$$

$$\frac{\partial G}{\partial K_i} - \frac{d}{dt}\frac{\partial G}{\partial \dot{K}_i} = \frac{\partial U}{\partial C_1}\frac{\partial f}{\partial K_i}e^{-\delta t} - \frac{d}{dt}\left(\frac{\partial U}{\partial C_1}\frac{\partial f}{\partial C_i}e^{-\delta t}\right) = 0 \quad .$$

$$\delta = \frac{1}{\partial U/\partial C_i} \frac{\partial U}{\partial C_1} \frac{\partial f}{\partial K_i} + \frac{1}{\partial U/\partial C_1} \frac{d}{dt} \frac{\partial U}{\partial C_1} .$$

Identifying the marginal utility of the first good, $\partial U/\partial C_1$, with $p(t)$ and $\partial f/\partial K_i$ with the nominal yield $q(t)$ of K_i i.e. $\partial f/\partial K_i = q(t)/p(t)$, we have, in an aggregate one-sector model where $C_1 = C_i = C$ $(i = 2,3,\ldots,n)$ Solow-Samuelson's fundamental relation

$$\delta(t) = i = q(t)/p(t) + \dot{p}(t)/p(t)$$

which says that in equilibrium, the discount rate δ = interest rate i = real rental rate q/p plus capital gain $\dot{p}/p$.

Finally, the isoperimetric constraints give

$$\frac{\partial G}{\partial R} - \frac{d}{dt}\frac{\partial G}{\partial \dot{R}} = \frac{\partial U}{\partial C_1}\frac{\partial f}{\partial R} e^{-\delta t} + \lambda = 0$$

$$\frac{\partial G}{\partial z} - \frac{d}{dt}\frac{\partial G}{\partial \dot{z}} = 0 - \dot{\lambda} = 0$$

i.e. the present value of the marginal utility of each depletable resource must be constant, and together these last two equations imply that the current value of its marginal utility must grow at the exponential rate δ over time, implying dwindling non-renewable resources. This is the well-known Hotelling's (1931) result.

Thus, the explicit recognition of non-renewable resources does not change the rules of optimal allocation and the fundamental relation in Capital Theory emerges unscathed, playing the leading role in Optimal Growth Theory.

3.5.4 Optimal Mining with Incomplete Exhaustion

As an illustration of the importance of transversality conditions, let us examine Hotelling's (1931) model with incomplete exhaustion (Levhari and Liviatan 1977). A mining firm having a mineral reserve of a known and fixed quantity s must choose the quantity $q(t)$ to extract at each time period t such as to maximize the present value of his profit $\pi(q,x) = R(q) - C(q,x)$ where $R(q) \equiv$ total revenue (TR), $C(q,x)$ = total cost (TC), $x(t)$ = cumulative output extracted. $R'(q) \equiv dR/dq$ = marginal revenue (> 0), $C_q \equiv \partial C/\partial q$ = marginal cost (MC) of production ($MC > 0$) and $C_x \equiv \partial C/\partial x \geq 0$: rising extraction cost as exhaustion is nearer.

From the definition of $x(t)$, it is clear that

$$x(t) \equiv \int_0^t q(\tau)\, d\tau$$

i.e.

$$\dot{x}(t) = q(t) \qquad \text{with } x(0) = 0,\ x(T) = s \ .$$

The objective functional to be maximized is

$$J = \int_0^T F(x,q,t)\, dt$$

where $F(x,q,t)$ $e^{-rt}\pi(q,x)$ above.

Euler equation gives (writing $F_x \equiv \partial F/\partial x$ etc.)

$$F_x - \frac{d}{dt} F_q = 0$$

whose integration gives

$$R_q(t) = C_q(t) + e^{-r(T-t)} \pi_q(T) - \int_t^T e^{-r(\tau-t)} F_x \, d\tau \equiv MC_t^*$$

i.e. marginal revenue ($MR \equiv R_q(t)$) at each time $t \in (0,T)$ is equal to marginal cost MC_t^* which consists of the MC at t, $C_q(t)$, the present value of marginal profit $\pi_q(t)$ at $t = T$ and the present value of all future additional cost F_x ($\equiv -C_x$) incurred by dwindling stock.

This constitutes Hotelling's optimality conditions in mining theory.

Interesting results are obtained by a close examination of the transversality conditions at the end,

$$[F(x,q,T) - q(T)\, F_q(x,q,T)]\delta T + F_q(x,q,T)\, \delta x(T) = 0 \quad .$$

Two cases of interest are:

(i) Hotelling's complete exhaustion where $x(T) = s$ where s is a known constant, i.e. $\delta x(T) = 0$. The above transversality conditions give, at $t = T$ which is unspecified i.e. $\delta T \neq 0$,

$$(F - qF_q)\Big|_{t=T} = 0$$

i.e. at the final time period T, marginal profit F_q is equal to average profit F/q. In the case of perfect competition in the resource market, total revenue $R(q) = pq(t)$ where price p is constant i.e. is invariant of the output level

$$F_q = F/q \Rightarrow \pi_q = \pi/q \Rightarrow p - C_q = p - C/q$$

i.e. $C_q = C/q$ or $MC = AC$: at the final period, the mine owner produces at minimum AC.

(ii) Levhari-Liviatan's (1977) Incomplete Exhaustion

As costs rise with dwindling reserves, it is not unusual for the firm to close down all operations with some resource left unexploited: the cost of extracting the last drop of oil will not be justified unless the international oil price warrants it. (For example, gold or coal mining in Canada). This means $x(T) < s$ of $\delta x(T) \neq 0$. Then, with unspecified T, $\delta T \neq 0$ and the transversality conditions give at $t = T$

$$(F - qF_q)\Big|_{t=T} = 0 \qquad \text{and} \qquad F_q(T) = 0$$

which jointly imply $F(x,q,T) = 0$ as well. $F_q = 0$ implies $R'[q(T)] = C_q[x(T),q(T)]$ i.e. $MR = MC$ at $t = T$ and $F(x,q,T) = 0$ implies $R[q(T)] = C[q(T),x(T)]$ i.e. $TR = TC$ or $AR = AC$: the firm is back to its breakeven point where AC is tangent to the demand curve, profit vanishes and the firm closes down with an incomplete exhaustion of resources (to be reopened only when world price p rises to a level which is high enough to make the re-opening a profitable business proposition).

The difference between the two cases is thus $F_q = F/q$ in the complete and $F_q = 0 = F$ in the incomplete exhaustion case. Under perfect competition where price p is constant, both cases lead to the same conclusion: at the terminal period, production is at the minimum of average cost. In the case of imperfect competition with downward sloping demand, transversality conditions imply Chamberlin's "tangency solution": $MC = MR < p = AC$.

3.6 Summary

In this chapter, we examined the determination of the constants of integration of the Euler equations. Such determination could only be done when appropriate boundary conditions are given.

From the simplest case of two fixed end points, we proceeded to all the various possibilities of one or both end points are variable. More precisely, with a fixed initial point $A(t_0, x_0)$ where both t_0 and x_0 are fixed, the various possibilities examined are the following cases:

(1) $x(T)$ and T are both fixed in advance;

(2) $x(T)$ is free but T is fixed;

(3) $x(T)$ is fixed but T is free;

(4) $x(T)$ and T are both free and independent;

(5) $x(T)$ and T are both free but related by $x(T) = g(T)$.

The analysis was then applied, *mutatis mutandis*, to the cases where the initial point $A(t_0, x_0)$ is variable, more precisely, where

(6) $x(t_o)$ is free but t_o is specified;

(7) $x(t_o)$ is specified but t_o is free;

(8) $x(t_o)$ and t_o are both free and independent;

(9) $x(t_o)$ and t_o are both free but related by $x(t_o) = h(t_o)$.

Table 3.1 provides a summary of all these cases and shows how the various arbitrary constants are determined.

These conditions are called boundary conditions, or loosely Transversality Conditions. They must be satisfied in every problem. Overlooking them could lead to wrong conclusions. (See, for example, Preston [1972].)

With the help of Transversality Conditions, we discussed the Erdmann-Weierstrass corner conditions which must be satisfied when the extremal has corners.

The chapter is concluded by the introduction of the canonical form of Euler equations. This will prove very useful in Ch. 5 when Pontryagin's Maximum Principle is examined.

CHAPTER 4

SECOND VARIATIONS AND SUFFICIENCY CONDITIONS

4.1 Introduction

In the last two chapters, Necessary conditions for extremals were examined: Euler equations were derived and their solutions as well as the determination of the constants of integration were discussed for each of the various cases of constrained and unconstrained optimisation. This chapter is devoted to the examination of the sufficient conditions of extremals in the Calculus of Variations.

By a relative extremum of a functional is meant the greatest or smallest value a functional takes among those values taken along neighbouring curves. Neighbouring curves are those which are close to the functional under consideration. It will be recalled (see ch. 2) that two curves are close of order zero if the absolute value of their difference is small, i.e. $|x(t) - x^*(t)| < \epsilon$ (for some small positive ϵ) and close of order one if both the absolute value of their difference and of their slopes are small, i.e. $|x(t) - x^*(t)| < \epsilon$ and $|\dot{x}(t) - \dot{x}^*(t)| < \epsilon$. If the curve $x = x^*(t)$ that makes a functional $J[x(t)]$ an extremum (i.e. maximum or minimum) belongs to the class of all curves $x(t)$ which are close of order zero, such extremum is called strong and if it belongs to the class of neighbouring curves which are close of order one, the extremum obtained is called a weak extremum.

4.2 Variations of Functionals

The total variations of a functional, $\Delta J(h)$, is

$$\begin{aligned}\Delta J(h) &\equiv J(x+h) - J(x)\\ &= \int_0^T (f_x h + f_{\dot{x}}\dot{h})dt + \frac{1}{2}\int_0^T (\bar{f}_{xx}h^2 + 2\bar{f}_{x\dot{x}}h\dot{h} + \bar{f}_{\dot{x}\dot{x}}\dot{h}^2)dt + 0(\|h\|)^2\\ &\equiv \delta J(h) \qquad\qquad + \delta^2 J(h) \qquad\qquad + 0(\|h\|)^2 \qquad (1)\end{aligned}$$

= first variation + second variation + higher order terms

where $\quad 0(\|h\|)^2 \to 0 \quad \text{as} \quad h \to 0$

and $\quad \bar{f} \equiv f(t, x+\theta h, \dot{x}+\theta\dot{h}) \quad$ for some $\theta \in (0,1)$.

At an extremum, $\delta J(h) = 0$ and $\Delta J(h)$ must have the same sign as $\delta^2 J(h)$.

Writing $\delta^2 J(h)$ in a more convenient form for later analysis and omitting bars for simplicity, we have

$$\begin{aligned}\delta^2 J(h) &= \frac{1}{2}\int_0^T (f_{xx}h^2 + 2f_{x\dot{x}}h\dot{h} + f_{\dot{x}\dot{x}}\dot{h}^2)\, dt\\ &\equiv \int_0^T (P\dot{h}^2 + Qh^2)\, dt \qquad (2)\end{aligned}$$

where $P \equiv P(t) \equiv \frac{1}{2} f_{\dot{x}\dot{x}}$; $Q \equiv Q(t) = \frac{1}{2}(f_{xx} - \frac{d}{dt} f_{x\dot{x}})$

and $\int_0^T 2f_{x\dot{x}}h\dot{h}\, dt = -\int_0^T (\frac{d}{dt} f_{x\dot{x}})h^2\, dt$ by integration by parts.

For a minimum problem, $\delta^2 J(h) \geq 0$ in (2) and for a maximum problem, $\delta^2 J(h) \leq 0$. For definiteness let us concentrate on the minimum problem and just reverse the signs for the maximum problem. We want to examine the conditions which make $\delta^2 J(h) \geq 0$ in (2).

It will be shown that $\delta^2 J(h) \geq 0$ if and only if $P\dot{h}^2 + Qh^2 \geq 0$ for all $h(t)$ satisfying $h(0) = 0 = h(T)$. Such a function $h(t)$ will be small if $\dot{h}(t)$ $\forall\, t \in (0,T)$ is small but not vice versa. If a function $h(t)$ with the above properties can be found such that $h(t)$ is small but $\dot{h}(t)$ is large for $t \in (0,T)$, then $P\dot{h}^2$ dominates Qh^2 in the determination of the sign of $\delta^2 J(h)$, as can be seen in the following Lemma and Legendre condition.

4.3 The Legendre Condition

Lemma 4.3

A necessary condition for $\delta^2 J(h) = \int_0^T (P\dot{h}^2 + Qh^2)\, dt$ *defined for all differentiable functions* $h(t)$ *in* $t \in (0,T)$ *satisfying* $h(0) = 0 = h(T)$ *to be non-negative is that*

$$P(t) \geq 0 \qquad \forall\, t \in (0,T) \tag{3}$$

This is called the Legendre Condition.

Proof   See Appendix to Chapter 4

This Lemma immediately gives the Legendre theorem:

Theorem 4.3 (Legendre)

A necessary condition for the functional $J(x) = \int_0^T f(x, \dot{x}, t)\, dt$, $x(0) = x_0$, $x(T) = x_T$ *to have a minimum (maximum) for the curve* $x = x(t)$ *is that the Legendre Condition* $P(t) \geq 0)$ *for all* $t \in (0,T)$ *be satisfied.*

Note that, rather unexpectedly, this is still a necessary condition. Legendre's attempt to prove it to be a sufficient condition failed, as pointed out by Lagrange.[1]

Example 4.1

For the problem of minimizing $J(x) = \int_0^T (\dot{x} + x)^2\, dt$ the Legendre condition $P \equiv (1/2)\, f_{\dot{x}\dot{x}} = 1 > 0$ (where $f(x, \dot{x}) \equiv (\dot{x} + x)^2$) is unambiguously safisfied everywhere.

4.4 The Jacobi Condition

Legendre's unsuccessful attempt, however, lead to a linear second order differential equation in v

$$-\frac{d}{dt}(P\dot{v}) + Qv = 0 \qquad (4)$$

If $P > 0$ (< 0) and the solution $v(t)$ of (4) for all differentiable functions $v(t)$ satisfying $v(0) = 0 = v(T)$, then $\delta^2 J > 0$ (< 0) i.e., a minimum (maximum) is obtained. This is called the Jacobi necessary condition.[2]

Example 4.2

Consider the problem of minimizing $J(x) = \int_0^T (\dot{x}^2 - 2\dot{x}t)\, dt$

Here $P \equiv \frac{1}{2} f_{\dot{x}\dot{x}} = 1$, $Q = 0$, the Jacobi equation gives $-\frac{d}{dt}(P\dot{v}) + 0 = -\ddot{v} = 0$ or $\dot{v} = c_1$, $v(t) = c_1 t + c_2 = c_1 t$ since $v(0) = 0$. Thus the Jacobi condition is fulfilled and the solution to the Euler equation, $x(t) = \frac{1}{2}t^2 + at + b$ indeed gives a minimum to the functional $J(x)$.

Note that the Jacobi equation is the Euler equation of the second variation $\delta^2 J(x)$ i.e.

$$F_h - \frac{d}{dt} F_{\dot{h}} = Qh - \frac{d}{dt} P\dot{h} \qquad (5)$$

where $F(h, \dot{h}) \equiv (P\dot{h}^2 + Qh^2)$. In fact, it is the variation of the Euler equation: if $x = x^*(t)$ and an infinitesimally close $x = x^*(t) + h(t)$ are both the solutions of the Euler equation, then

$$f_x(t,\ x + h,\ \dot{x} + \dot{h}) - \frac{d}{dt} f_{\dot{x}}\ (t,\ x + h,\ \dot{x} + \dot{h}) = 0$$

whose Taylor expansion gives the Jacobi equation

$$f_{xx}\ h + f_{x\dot{x}}\ \dot{h} - \frac{d}{dt}\ \left(f_{\dot{x}\dot{x}}\ \dot{h} + f_{\dot{x}x} h\ \right) = 0$$

i.e.

$$Qh - \frac{d}{dt}(P\dot{h}) = 0 \tag{6}$$

Alternatively, the Jacobi equation can be derived from the α-locus of the solution $x(t,\alpha)$ of the Euler equation[3] in a one-fixed-end-point problem say $x(0,\alpha) = x_o$

$$f_x[t,\ x(t,\alpha),\ \dot{x}(t,\alpha)] - \frac{d}{dt} f_{\dot{x}}\ [t,\ x(t,\alpha),\ \dot{x}\ (t,\alpha)]\ = 0 \tag{7}$$

Differentiating (7) with respect to α, putting $v \equiv \partial x/\partial\alpha$, gives

$$f_{xx}\ v + f_{x\dot{x}}\ \dot{v}\ - \frac{d}{dt}\ (f_{x\dot{x}}\ v + f_{\dot{x}\dot{x}}\ \dot{v}) = 0$$

or

$$\left(f_{xx} - \frac{d}{dt}\ f_{x\dot{x}}\ \right)v - \frac{d}{dt}\ \left(f_{\dot{x}\dot{x}}\ \dot{v}\right) = 0 \tag{8}$$

i.e.

$$Qv - \frac{d}{dt}\ (P\dot{v}) = 0 \tag{9}$$

which is indeed the Jacobi equation.

Note that in the case x and $\dot{x}$ in $f(x,\ \dot{x},\ t)$ are n-vectors, (2) becomes

$$\delta^2 J(h) = \int_0^T\ (\dot{h}'\ P\dot{h}\ +\ h'\ Qh)\ dt \tag{10}$$

and (5) is

$$Qh - \frac{d}{dt}\ (P\dot{h})\ =\ 0 \tag{11}$$

where P and Q are symmetric matrices and h, $\dot{h}$ are n-vectors. The conclusions remain unchanged.

Example 4.3

The Jacobi condition of the problem of finding an extremum of $J(x) = \int_0^T f(x, \dot{x})\, dt$ where $f(x, \dot{x}) = \dot{x}^2 + x$, is

$$Qh - \frac{d}{dt} P\dot{h} = 0 - \frac{d}{dt}\dot{h} = 0 - \ddot{h} = 0$$

i.e., $h(t) = at + b = at + 0$ since $h(0) = 0$. Clearly $h(t) \neq 0$ $\forall\ t \neq 0$, and no conjugate points exist.

4.5 The Weierstrass Condition for Strong Extrema

In the case of strong extrema, it is, in general, not possible to use Taylor expansion to examine the sign of ΔJ since the variation $\delta\dot{x}$ may not be small. Another method, developed by Weiestrass, must be applied.

Consider the difference ΔJ between two functionals taken along arc C and C^* respectively (fig. 4.1), namely

$$\Delta J = \int_C f(x, \dot{x}, t)\, dt - \int_{C^*} f(x, \dot{x}, t)\, dt \tag{12}$$

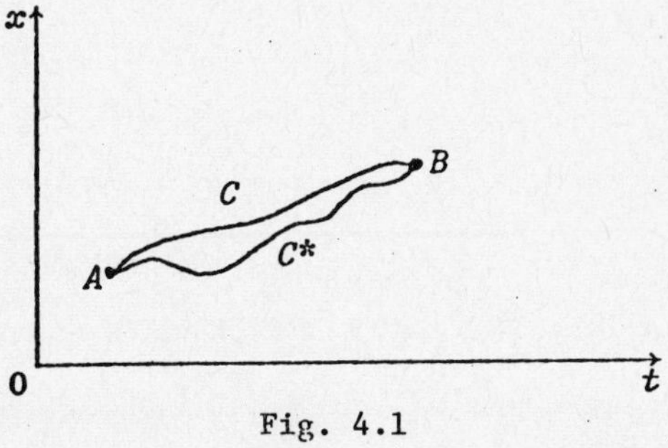

Fig. 4.1

Along an extremal C^* (see ch. 3 eq. 11), the variation is

$$\delta J^* = \left(f - \dot{x}\, f_{\dot{x}}\right) dt + f_{\dot{x}}\, dx \tag{13}$$

whose integral, writing $p(t, x)$ for the slope function of the extremal passing through point (t, x), is

$$J = \int_{C^*} (f - pf_p)\, dt + f_p\, dx \tag{14}$$

which is independent of the path of integration.[4]

Thus ΔJ in (12) becomes

$$\Delta J = \int_C f(x, \dot{x}, t)\, dt - \int_{C^*} [f(x, p, t) + (\dot{x} - p)f_p]\, dt \tag{15}$$

But the second integral being independent of the paths C or C^* of integration, (15) becomes

$$\begin{aligned} \Delta J &= \int_C [f(x, \dot{x}, t) - f(x, p, t) - (\dot{x} - p)\, f_p]\, dt \\ &\equiv \int_C E(x, \dot{x}, p, t)\, dt \end{aligned} \tag{15}$$

where $E(x, \dot{x}, p, t)$, the expression inside the square brackets of (15), is called the Weiestrass E-function (E is the "excess" function).

Provided the Jacobi equation is satisfied,

$$\begin{aligned} & E \leq 0 \text{ for a maximum} \\ \text{and} \quad & E \geq 0 \text{ for a minimum} \end{aligned} \tag{16}$$

Note that for weak extrema, (16) must hold for $\dot{x}$ close to p along C but for strong extrema, (16) must hold for any arbitrary $\dot{x}$ since strong extrema can have widely different slopes. If the Weiestrass condition is satisfied and the Jacobi condition is not, then a local extremum is obtained: sufficiently small segments will admit a strong extremum while the extremal as a whole may not.

Example 4.5.1

Maximize $J(x) = \int_0^T (x^2 - \dot{x}^2)\, dt$ with $x(0) = 0$ and T specified $(0<T<\pi)$.

Euler equation gives

$$2(\ddot{x} + x) = 0$$

the solution of which, account taken of the initial condition $x(0) = 0$, gives

$$x = a \sin t$$

where a is some constant to be determined by $x(T) = x_T$.

A maximum is clearly achieved since $f_{\dot{x}\dot{x}} = -2 < 0$ and $0 < T < \pi$ guarantees the absence of conjugate points.

Weierstrass E-function gives

$$E = (x^2 - \dot{x}^2) - (x^2 - p^2) + 2(\dot{x} - p)\, p$$

$$= -(\dot{x} - p)^2 < 0 \quad \text{everywhere.}$$

Hence a strong maximum is unambiguously obtained.

Example 4.5.2

$$\text{Min} \int_0^1 (x/\dot{x}^2)\, dt$$

$$\text{for } x(0) = 1, \quad x(1) = 1/4.$$

As can readily be seen, the solution of Euler equation is

$$x = (c_1 t + 1)^2$$

where $c_1 = -\frac{1}{2}$.

A weak extremum is achieved since $f_{\dot{x}\dot{x}} = 6x/\dot{x}^4 > 0$ for any value of $\dot{x}$. To investigate strong extrema, the Weierstrass E-function must be examined. The latter gives

$$E = \frac{x}{\dot{x}^2} - \frac{x}{p^2} + \frac{2x}{p^3}(\dot{x} - p)$$

$$= \frac{x(\dot{x} - p)^2(2\dot{x} + p)}{\dot{x}^2 \, p^3} .$$

This may have any sign for arbitrary $\dot{x}$. Hence no strong minima exist.

4.6 The Legendre-Clebsch Condition

The Weierstrass E-function is rather cumbersome to investigate. It would be useful to have a simpler criterion to apply.

If $f(x,\dot{x},t)$ is expanded by Taylor formula, neglecting higher order terms, and substituted into the E-function, a simplified Weierstrass condition is obtained. More precisely, the Taylor expansion of $f(x,\dot{x},t)$ gives

$$f(x,\dot{x},t) = f(t,x,p) + (\dot{x} - p)f_p + \frac{(\dot{x} - p)^2}{2!} f_{\dot{x}\dot{x}}(t,x,q) \qquad (17)$$

where $q = \theta\dot{x} + (1-\theta)p \ (0<\theta<1)$.

Substituting (17) into the Weierstrass E-function yields

$$E \equiv f(t,x,\dot{x}) - f(t,x,p) - (\dot{x} - p)f_p$$

$$= \frac{(\dot{x} - p)^2}{2!} f_{\dot{x}\dot{x}}(t,x,q) \qquad (18)$$

or in the case x is an n-vector, this can be written in matrix form, putting $z' \equiv [\dot{x}_1-p_1, \dot{x}_2-p_2, \ldots, \dot{x}_n-p_n]$ and

$[f_{\dot{x}\dot{x}}] \equiv [\frac{\partial^2 f}{\partial\dot{x}_i \partial x_j}]$ $1 \le i,\ j \le n$

as

$$E = \frac{1}{2} z'[f_{\dot{x}\dot{x}}]z \equiv \frac{1}{2} \sum_{i=1}^{n} \sum_{j=1}^{n} \frac{\partial^2 f}{\partial\dot{x}_i \partial\dot{x}_j} z_i z_j . \qquad (19)$$

This is a simplified Weierstrass condition which is referred to as the Legendre-Clebsch condition, in view of its resemblance to the Legendre condition examined earlier.

In order for $x(t)$ to achieve a strong minimum (maximum), it is sufficient that the simplified Weierstrass or Legendre-Clebsch condition $E \geq 0$ (≤ 0) holds, i.e.

$$f_{\dot{x}\dot{x}} \geq 0 \ (\leq 0) \tag{20}$$

or in general $[f_{\dot{x}\dot{x}}]$ be positive (negative) semi-definite (20) be satisfied as well as the Jacobi condition for all $\dot{x}$, not only at the points of the extremal itself but also in its neighbourhood.

Note that in spite of the apparent similarity between the simplified Weierstrass condition $f_{\dot{x}\dot{x}} \geq 0$ in (20) and the Legendre condition $f_{\dot{x}\dot{x}} \geq 0$ in theorem 4.3, the two are different: In verifying the Legendre condition, $f_{\dot{x}\dot{x}}$ is calculated by taking $\dot{x}$ equal to its value on the extremal, i.e. a weak extremum is investigated, whereas in checking the Weierstrass condition $E \geq 0$ or $f_{\dot{x}\dot{x}} \geq 0$ in (20), the sign must be satisfied not only on the extremal itself but also in its neighbourhood.

Example 4.6.1

Consider the cost minimization problem of Example 3.2.1 where $f(x, \dot{x}) = x + \dot{x}^2$.
Clearly the Legendre and Legendre-Clebsch conditions are satisfied

i.e., $$P \equiv \frac{1}{2} f_{\dot{x}\dot{x}} = 1 > 0.$$

The Jacobi condition is also satisfied

$$-\frac{d}{dt} P\dot{h} + Qh \equiv -\ddot{h} = 0$$

implying $h(t) = at + b = at$ since $h(0) = 0$.

Clearly $h(t) \neq 0 \;\; \forall\, t \neq 0$, i.e. $h(t)$ does not vanish anywhere except at the origin, in other words, no conjugate points exist.

The Weiestrass E-function

$$E \equiv \dot{x}^2 - p^2 - 2\,(\dot{x} - p)\,p$$

$$= (\dot{x} = p)^2 \geq 0$$

is always non-negative, i.e., a strong minimum is obtained.

Example 4.6.2

Examine the extrema of the functional

$$J(x) = \int_0^a \dot{x}^3\, dt \;\; ; \; x(0) = 0 \; ; \; x(a) = b \;\; , \;\; (a,b>0) \; .$$

Euler equation $-\dfrac{d}{dt}\, 3\dot{x}^2 = 0$ gives $\dot{x}$ = some constant c_1, i.e.

$$x = c_1 t + c_2$$

where c_1 and c_2 as determined by $x(0) = 0$ and $x(a) = b$ are 0 and b/a respectively, i.e. extremals could only occur along the line $x = \dfrac{b}{a}\, t$.

Legendre condition is $P \equiv \dfrac{1}{2} f_{\dot{x}\dot{x}} = 3\dot{x}$.

Along the extremal $x = (b/a)t$, $P = 3b/a > 0$.

Jacobi condition is

$$-\frac{d}{dt}\,(P\dot{h}) + Qh = -\frac{d}{dt}\,\left(\frac{3b\dot{h}}{a}\right) + 0$$

since $Q \equiv \dfrac{1}{2}\left(f_{xx} - \dfrac{d}{dt}\, f_{x\dot{x}}\right) = 0$.

Hence

$$\dot{h} = c_3 \text{ (some constant)}$$

$$h = c_3 t + c_4 = c_3 t$$

since $h(0) = 0$. Thus the non-trivial solution $h = c_3 t$ of the Jacobi equation cannot vanish in the interval $(0,a]$, i.e. the interval $(0,a]$ contains no points conjugate to the point $t = 0$ and the Jacobi condition is satisfied.

The Weierstrass E-function gives

$$E \equiv \dot{x}^3 - p^3 - 3p^2 (\dot{x} - p)$$
$$= (\dot{x} - p)^2(\dot{x} + 2p) .$$

Along the extremal $x = (b/a)t$, the slope $p = b/a > 0$. If $\dot{x}$ takes on values close to b/a then $E \geq 0$. However if $\dot{x}$ takes on arbitrary values then $\dot{x} + 2p$ may have arbitrary sign and E may change its sign from $E \geq 0$ to $E \leq 0$, i.e. the condition for a strong minimum is violated.

On the basis of this investigation, we conclude that $J(x)$ has a weak extremum but does not have a strong extremum.

Example 4.6.3

Consider the example 2.5.1 of the profit maximizing producer examined in chapter three. Put $a_2 \equiv 1$ for simplicity so that his costs are $C_1 = a_1 x^2 + b_1 x_1 + c_1$ and $C_2 = \dot{x}^2 + b_2 \dot{x} + c_2 t$ ($a_2 \dot{x}_2 = \dot{x}_2$ as $a_2 = 1$). His profit (π) functional is

$$J(x) = \int_0^T \pi(x,\dot{x},t)\, dt .$$

where $\pi \equiv px - a_1 x^2 - bx - c_1 - \dot{x}^2 - b_2 \dot{x} - c_2 t$

$$\pi_x = p - 2a_1 x - b_1$$

$$\pi_{\dot{x}} = -2\dot{x} - b_2 \,;\; \pi_{\dot{x}\dot{x}} = -2$$

It is clear that Legendre condition $\pi_{\dot{x}\dot{x}} = -2 < 0$ is satisfied.

Weierstrass E condition is also satisfied

$$E \equiv \pi(x,\dot{x},t) - \pi(x,p,t) - (\dot{x} - p)\pi_p$$

$$= - (p - \dot{x})^2 < 0 \text{ everywhere.}$$

The Jacobi condition is also satisfied, i.e.

$$- \frac{d}{dt}(P\dot{h}) + Qh = 0$$

where $P \equiv \frac{1}{2}\pi_{\dot{x}\dot{x}} = -1$,

$$Q \equiv \frac{1}{2}(\pi_{xx} - \frac{d}{dt}\pi_{x\dot{x}}) = - a_1$$

$$- \frac{d}{dt}(P\dot{h}) + Qh = \ddot{h} - a_1 h = 0$$

giving

$$h(t) = Ae^{at} + Be^{-at}$$

where $a \equiv \sqrt{a_1}$

with $h(0) = 0$, $B = - A$ and

$$h(t) = Ae^{at} - Ae^{-at}$$

which cannot vanish in $(0,T]$.

Thus this producer is a "strong" profit maximizer indeed.

4.7 Sufficient Conditions: An Important Special Case

The investigation of sufficient conditions of functionals is very tedious and complicated indeed. However, the case in which $f(x,\dot{x},t)$ is concave or convex in x and $\dot{x}$ constitutes an important special case of remarkable simplicity and usefulness, especially in Economic theory: Euler equation in this case provides both the necessary and sufficient conditions, as can be seen in the following theorem.

Theorem 4.7 (Mangasarian [1966])

Let $f(x,\dot{x},t)$ be twice differentiable with respect to x and $\dot{x}$ and concave (convex) in x and $\dot{x}$. Then a necessary and sufficient condition that x^* maximimize (minimize)

$$\int_0^T f(x,\dot{x},t)\,dt$$

is that it satisfies the Euler equation with $x(0) = x_0$ and $x(T) = x_T$.

Proof. Necessity: Obvious.

Sufficiency: Let x^* and $\dot{x}^*$ satisfy Euler equation and denote

$f^* \equiv f(x^*,\dot{x}^*,t)$ and $f_x^* \equiv \partial f^*/\partial x$,

$$\Delta J \equiv J(x) - J(x^*)$$

$$= \int_0^T (f - f^*)\,dt \leq \int_0^T [(x - x^*)f_x^* + (\dot{x} - \dot{x}^*)f_{\dot{x}}^*]\,dt \text{ by concavity}$$

$$= \int_0^T (x - x^*)(f_x^* - \frac{d}{dt} f_{\dot{x}}^*)\,dt + (x - x^*)f_{\dot{x}}^*\Big|_0^T$$

by integration by parts

$$= (x - x^*)f_{\dot{x}}^*\Big|_0^T \text{ by Euler equation.}$$

$$= 0$$

by fixed end point properties $x(0) = x^*(0) = x_0$ and $x(T) = x^*(T) = x_T$. Thus $\Delta J \leq 0$. If $f(x,\dot{x},t)$ is convex in x and $\dot{x}$, the inequalities are reversed, i.e. $J(x) - J(x^*) \geq 0$, i.e. x^* is the minimal.

N.B. If $f(x,\dot{x},t)$ is strictly concave, the global maximum is unique and if $f(x,\dot{x},t)$ is strictly convex, the global minimum is unique.

Example 4.7.1

The Neo-Classical economic growth model could be simply formulated as (see Ch. 3)

$$\text{Max. } J(k) = \int_0^T u(c)e^{-rt}\, dt$$

where $c = f(k) - \lambda k - \dot{k}$ is per capita consumption,

$k \equiv$ capital per worker, with $k(0) = k_0$; $k(T) = k_T$,

r = constant positive discount rate

λ = constant to take care of constant depreciation rate and population growth

$f(k)$: production function per worker, concave

with $f''(k) < 0 < f'(k)$, $k \geq 0$, $f(0) = 0$; $f'(0) = \infty$; $f'(\infty) = 0$

$u(c)$: utility function, assumed concave to reflect the law of diminishing marginal utility more precisely, $u''(c) < 0 < u'(c)$.

Define $\varphi(k,\dot{k},t) \equiv u(c)e^{-rt} \equiv u[f(k) - \lambda k - \dot{k}]e^{-rt}$ and note that $u(c)$ is a concave function of c and c is a concave function in k and $\dot{k}$ (c is strictly concave in k and linear hence concave in $\dot{k}$), hence φ is concave in k and $\dot{k}$. An application of Theorem 4.7 ensures us that $\Delta J \leq 0$ and hence this is a maximum. In fact a reconstitution of the steps of Theorem 4.7 gives:

$$\Delta J = \int_0^T (\varphi - \varphi^*)\, dt \leq \int_0^T [\varphi_k^*(k - k^*) + \varphi_{\dot{k}}^*(k - \dot{k}^*)]\, dt$$

$$= \int_0^T (k - k^*)(\varphi_k^* - \frac{d}{dt}\varphi_{\dot{k}}^*)\, dt + (k - k^*)\varphi_{\dot{k}}^*\Big|_0^T = 0$$

i.e. $\Delta J \leq 0$ and these results are summarised in the form of a theorem as

"Under the Neo-Classical assumptions listed above concerning the production and utility functions, i.e. if $f(k)$ and $u(c)$ are both concave and well behaved functions, if k^* exists and satisfies the Euler equation with $k(0) = k_0$ and $k(T) = k_T$, then this is sufficient (and of course necessary) for k^* to be optimal, i.e. for an optimal consumption programme c^* to be obtained."

4.8 SUMMARY AND CONCLUSION

Thus, it can be seen that the sufficient conditions for extrema in the Calculus of Variations problem is very complicated indeed. There is not one sufficient condition but a set of necessary conditions which, taken together, provide sufficient conditions. These are summarised as follows.

In order for $x^*(t)$ to be a maximal of the functional

$$J(x) = \int_0^T f(x,\dot{x},t)\,dt$$

with two fixed end points, $A(0,x_0)$ and $B(T,x_T)$, it is sufficient that the following conditions be satisfied:

(i) Euler equation

$$f_x - \frac{d}{dt} f_{\dot{x}} = 0$$

(ii) Legendre conditions

$$f_{\dot{x}\dot{x}} \leq 0$$

(iii) Jacobi conditions of the absence of conjugate points in $(0,T)$, i.e.

$$-\frac{d}{dt}(P\dot{h}) + Qh = 0$$

(iv) Weierstrass E-function

$$E(t,x,\dot{x},p) \leq 0$$

(v) If $f(x,\dot{x},t)$ is concave in x and $\dot{x}$, Euler equation ensures sufficiency, i.e. (i) is both the necessary and sufficient condition for a maximum.

For a minimum, the inequalities in (ii) and (iv) must be reversed, i.e.

$$E(t,x,\dot{x},p) \geq 0 \quad \text{and} \quad f_{\dot{x}\dot{x}} \geq 0$$

and also in (v), the word concave must be replaced by convex, i.e. (v) reads: "If $f(x,\dot{x},t)$ is a convex function in x and $\dot{x}$, the Euler equation provides both the single necessary and sufficient condition for a minimum."

Note, finally, that for a weak extremum, the condition $f_{\dot{x}\dot{x}} \leq 0$ (≥ 0) in (ii) and $E \leq 0$ (≥ 0) in (iv) needs only be satisfied for (t,x) sufficiently close to the extremal yielded by the Euler condition and $\dot{x}$ sufficiently close to p but for a strong extremum, (ii) and (iv) must hold for any arbitrary $\dot{x}$.

Chapter 4: FOOTNOTES

1. See, for example, Gelfand and Fomin (1963), Petrov (1968).

2. Legendre attempted to obtain a complete square of $\delta^2 J(h)$ and hence $P\dot{h}^2 + Qh^2 > 0$, by using some function $w(t)$ satisfying $P(Q + \dot{w}) = w^2$. A solution to this equation exists if and only if the solution $v(t)$ to (4) which vanishes at $t = 0$, T does not vanish anywhere else. In other words, the interval $(0,T)$ contains no points conjugate to 0, a conjugate point being defined as any point in $(0,T)$, at which the solution $v(t)$ to (4) vanishes. The Jacobi condition requires the absence of conjugate points in the open interval $(0,T)$, in addition to $P(t) > 0$ (<0). If the Jacobi condition is violated, the solution $x^*(t)$ cannot impart an extremum to $J(h)$. A pencil of extremals which do not intersect one another except at the centre of the pencil is said to form a central field. This concept is useful to appreciate the Jacobi condition.

3. It will be recalled (see, for example, Gillespie (1954) pp. 66-68) that the curves of the same family which are infinitesimally close to one another intersect on the envelope of the family. For example $f(t, x, \alpha) = 0$ forms a family of curves obtained by giving particular values to the parameter α. A neighbouring curve is $f(t, x, \alpha + \Delta\alpha) = 0$. By Taylor's theorem

$$f(t, x, \alpha + \Delta\alpha) = f(t, x, \alpha) + f_\alpha(t, x, \alpha)\,\Delta\alpha + \frac{1}{2} f_{\alpha\alpha}(t, x, \alpha + \theta\,\Delta\,\alpha)\,.\,\Delta\,\alpha = 0$$

When $\Delta\alpha \to 0$, a point of intersection is obtained where $f(t, x, \alpha) = 0 = f_\alpha(t, x, \alpha)$. The locus of all such intersection points represented by the equation $g(t, x) = 0$ obtained by eliminating α from $f = 0 = f_\alpha$ is called the envelope of the family. Note that the function $f(t, x, \alpha)$ discussed here is the integral $x(t, \alpha)$ of the Euler equation in question where one constant has been determined by the condition of the problem, say that $x(t, \alpha)$ starts from a given point $A(0, x_o)$.

4. To see this, consider any function $f(x, y) = c$ (c = constant). Total differentiation gives

$$f_x\, dx + f_y\, dy \equiv P(x, y)\, dx + Q(x, y)\, dy = 0$$

$$\int \left(P \frac{dx}{dt} + Q \frac{dy}{dt}\right) dt = \int_a^b \frac{df}{dt}\, dt = \int_a^b df = f(b) - f(a)$$

i.e., the integral of an exact differential is independent of the path of integration.

4. cont'd

Note that (14) can also be written as

$$J = \int_{C^*} [f + (\dot{x} - p) f_p] \, dt$$

and along an extremal $x^*(t)$, $\dot{x} = p$ and this integral becomes

$$\int_{C^*} f(x, \dot{x}, t) \, dt.$$

APPENDIX TO CHAPTER 4

<u>Lemma</u>. *A necessary condition for* $\delta^2 J = \int_0^T (P\dot{h}^2 + Qh^2)\,dt \geq 0$ *for all differentiable functions* $h(t)$, $t \in (0,T)$ *satisfying* $h(0) = 0 = h(T)$ *is* $P(t) \geq 0$ *for all* $t \in (0,T)$.

<u>Proof</u>. (Gelfand and Fomin 1963 p. 103)

This proof by contradiction consists in choosing a function $h(t)$ vanishing at $t = 0$, T such that $h(t)$ is small but $\dot{h}(t)$ is large and showing that $P(t) < 0$ will lead to a contradiction. Suppose $P(t') < 0$, say $P(t') = -2\beta < -\beta$ for some $t' \in (0,T)$. Then $P(t') < -\beta$ for some α such that $0 \leq t' - \alpha < t < t' + \alpha \leq T$. Let

$$h(t) \equiv \begin{cases} \sin^2 u & \text{where } u \equiv \pi(t-t')/\alpha \text{ for } t'-\alpha \leq t \leq t'+\alpha \\ 0 & \text{otherwise.} \end{cases}$$

Then $\dot{h} = (\pi/\alpha)\, 2 \sin u \cos u \equiv (\pi/\alpha) \sin 2u$; $\dot{h}^2 = (\pi^2/\alpha^2) \sin^2 (2u)$

$$\left|\frac{\pi^2}{\alpha^2} \int_{t'-\alpha}^{t'+\alpha} P \sin^2(2u)\, dt\right| \leq \frac{\pi^2}{\alpha^2}\int_{t'-\alpha}^{t'+\alpha} P \left|\sin 2u\right| dt \leq \frac{\pi^2}{\alpha^2}\, 2\alpha(1)P(t') < \frac{2\pi^2}{\alpha}(-\beta)$$

$$\left|\int_{t'-\alpha}^{t'+\alpha} Q \sin^4 u\, dt\right| \leq \int_{t'-\alpha}^{t'+\alpha} Q \left|\sin u\right| dt \leq \int_{t'-\alpha}^{t'+\alpha} Q(1)dt \leq 2\alpha M\alpha$$

where $M \equiv \max_{0 \leq t \leq T} |Q(t)|$.

Put together,

$$\int_0^T (P\dot{h}^2 + Qh^2)\, dt = \int_{t'-\alpha}^{t'+\alpha} (P \frac{\pi^2}{\alpha^2} \sin^2 2u + Q \sin^4 u)\, dt < -\frac{2\beta\pi^2}{\alpha} + 2M\alpha$$

For sufficiently small α, $-2\beta\pi^2/\alpha + 2M\alpha < 0$, hence $\delta^2 J < 0$. This contradiction proves the Lemma.

CHAPTER 5

OPTIMAL CONTROL: THE VARIATIONAL APPROACH

5.1 Introduction

Optimal Control was developed in the late 1950's by Pontryagin *et al.* [1962] who called it the Maximum Principle. Although much pioneering work had been carried out by Valentine [1937], McShane [1939, 1940] and Hestenes [1949], Pontryagin and his associates are the first ones to develop and present it in a unified manner. Their work attracted great attention among mathematicians, engineers and economists and spurred wide research activities in the area in the 1960's.

Although Optimal Control theory could be presented in its full generality and the Calculus of Variations, shown to emerge as a special case, as Pontryagin *et al* [1962] have done, we follow most textbook writers (for example Kirk [1970], Hadley & Kemp [1971], Miller [1979] Kamien & Schwartz [1981] and others) in developing it from the classical Calculus of Variations. This approach would put the theory of Optimal Control in its proper historical perspective and allow us to make full use of the results of Calculus of Variations developed in earlier chapters.

5.2 From the Calculus of Variations to Optimal Control

In the Calculus of Variations, we considered the problem of finding an extremal to a certain functional

$$J(x) = \int_{t_0}^{T} f(x,\dot{x},t)\, dt \tag{1}$$

subject, or not, to some constraint written in implicit form as

$$g_i(x,\dot{x},t) = 0 \qquad (1 \leq i \leq r < n) \tag{2}$$

where x and $\dot{x}$ are n-vectors.

In the Optimal Control problem, we seek some control variable $u(t)$ such as to bring the dynamic system

$$\dot{x} = g(x,u,t) \tag{3}$$

from some initial position to some terminal point such as to impart an extremum to some objective functional

$$J(x) = \int_{t_0}^{T} f(x,u,t)\ dt \tag{4}$$

where x and $\dot{x}$ are n-vectors and u is an r-vector. In other words, the optimal control problem consists of finding an extremal of (4) subject to (3). In the special case where (3) takes the special form

$$\dot{x} = u \tag{5}$$

the analogy between the two types of problem becomes obvious.

In fact, the solution of the Calculus of Variations problem is given by the Euler equation

$$f_x - \frac{d}{dt} f_{\dot{x}} = 0 \ .$$

The solution of the Optimal Control problem is obtained by first defining the Hamiltonian H as $H \equiv f + p\dot{x}$ and $F \equiv H - p\dot{x}$ and applying the Euler theorem to it

$$F_x - \frac{d}{dt} F_{\dot{x}} = H_x + \dot{p} = 0 \qquad (6)$$

$$F_u - \frac{d}{dt} F_{\dot{u}} = H_u = 0 \; . \qquad (7)$$

But these are precisely the Euler equations in canonical form (see Ch. 3, section 3.4, eq. (35) and (36)).

Underneath these apparent semblances, however, there lie important differences. The Calculus of Variations has some serious limitations which the Optimal Control theory has successfully overcome. The main ones are in the area of inequality constraints in the state and control variables of the type $|\dot{x}| \le k$, $|x| \le h$, or more generally $\dot{x} \in \Omega(\dot{x})$ and $x \in R(x)$ where Ω and R are closed sets. These inequality constraints arise in a large number of problems. For example, the acceleration α of a car is bounded $0 \le \alpha \le \alpha_{max}$. The saving rate s is in the closed set $0 \le s \le 1$. The Euler equation as well as the Weierstrass E-function then become false, as Pontryagin has pointed out [1962], since they are derived on the assumption of free variations of x and $\dot{x}$. No doubt this problem has been recognized since 1933 by Valentine [1937] and his method has since been improved by McShane [1940] and others but important progress was realized in Optimal Control theory, especially in the linear case. In fact, it can be seen that many problems could not be solved by the Euler equation. For example, for the problem

$$J(x) = \int_0^T (x + \dot{x}) \, dt$$

Euler equation gives the meaningless result $f_x - \frac{d}{dt} f_{\dot{x}} = 1 - 0 = 0$. Optimal Control is well equipped to cope with this type of difficulty.

5.3 Pontryagin's Maximum Principle

In this section, no attempt will be made to prove Pontryagin's Maximum Principle rigorously: such a proof would be beyond the scope of this introductory book. Rather, it will be shown how the Maximum Principle could be derived from the Calculus of Variations for the most general case in which the terminal time and state are unspecified, but the "Scrap function" $S[x(T),T]$ specifying the value of the programme at the terminal time is given and the control and state variables are both in open sets, i.e. they are unbounded. There is no loss of generality to start our exposition with the case in which the initial time (t_0) and state (x_0) are both fixed, i.e. $(t_0,x(0)) = (0,x_0)$. In the case of variable initial point, the transversality conditions for the variable terminal point apply, *mutandis mutatis*.

Consider the problem of choosing a control vector $u(t) = (u_1(t),u_2(t),\ldots,u_r(t))$ from the class of piecewise continuous vector-valued functions such as to bring the dynamic system

$$\dot{x}(t) = f[x(t),u(t),t] \tag{8}$$

from some given initial state (x_0,t_0) to some unspecified final state $(x(T),T)$ (where both $x(T)$ and T are unspecified) such as to impart a maximum or minimum to some objective functional

$$J(u) = S[x(T),T] + \int_0^T f_0(x,u,t)\, dt \tag{9}$$

where $x(t)$ is an n-state vector, $u(t)$ an r-control vector and $S[x(T),T]$ is the "Scrap" function. Both $x(t)$ and $u(t)$ are assumed to be unbounded vectors, $f_0(x,u,t)$ is a scalar function and $f[x(t), u(t)t]$ is an n-vector function, all assumed to be well behaved.

Theorem 5.1 (Pontryagin)

Let $u^*(t)$ be an admissible control vector which transfers (x_0,t_0) to a target $(x(T),T)$ where $x(T)$ and T are not specified in general. Let $x^*(t)$ be the trajectory corresponding to $u^*(t)$. In order for that $u^*(t)$ be optimal, it is necessary that there exist a non-zero, continuous vector function $p^*(t) = (p_1^*(t),p_2^*(t),\dots,p_n^*(t))$ and a constant scalar p_0 such that

(a) $p^*(t)$ and $x^*(t)$ are the solution of the canonical system

$$\dot{x}^*(t) = \frac{\partial H}{\partial p}(x^*,p^*,u^*,t) \tag{10}$$

$$\dot{p}^*(t) = -\frac{\partial H}{\partial x}(x^*,p^*,u^*,t) \tag{11}$$

where

$$H \equiv \sum_0^n p_i f_i(x,p,u,t) \equiv f_0(x,u,t) + \sum_1^n p_i f_i(x,u,t) \tag{12}$$

the usual Hamiltonian, with $p_0 \equiv 1$.

(b) $$H(x^*,u^*,p^*,t) \geqq H(x^*,u,p,t)\,. \tag{13}$$

(c) All boundary conditions be satisfied. (14)

Proof. Pontryagin [1962]

Rather than providing a rigorous proof of this theorem we shall show how this theorem could be developed from the Calculus of Variations, as applied to (8) and (9). The problem is to optimize (say maximize, for definiteness) the objective functional

$$J(u) = S[x(T),T] + \int_0^T f_0(x,u,t)\ dt \tag{9}$$

subject to

$$\dot{x}(t) = f[x(t),u(t),t] \tag{8}$$

where $x(t)$ and $u(t)$ are respectively n- and r-vectors

and $x(0) = x_0$, $t_0 = 0$

$x(T)$ and T are both unspecified.

With the "Scrap" function $S(x(T),T)$ written as

$$S[x(T),T] \equiv S[x_0,0] + \int_0^T \frac{d}{dt} S[x(t),t]\ dt \tag{15}$$

(9) becomes

$$J(u) = S(x_0,0) + \int_0^T [f_0(x,u,t) + \frac{dS}{dt}(x,t)]\ dt \tag{16}$$

$$= S(x_0,0) + \int_0^T [f_0(.) + \frac{\partial S}{\partial x}\dot{x} + \frac{\partial S}{\partial t}]\ dt \tag{17}$$

where $x(t)$, $u(t)$ and $f_0(x,u,t)$, $S(x(t),t)$ are written simply as x, u, $f_0(.)$ and S to alleviate notation whenever no confusion arises. Similar remarks apply to the arguments of other functions.

Omitting $S(x_0,0)$ for simplicity, since $x(0) = x_0$ and $t_0 = 0$ being fixed, are not affected by the optimization process, and writing the augmented functional $J_a(u)$ as

$$J_a(u) \equiv \int_0^T F(x,\dot{x},p,u,t)\ dt \tag{18}$$

where

$$F(x,\dot{x},p,u,t) \equiv f_0(.) + p[f(.) - \dot{x}] + \frac{\partial S}{\partial x}\dot{x} + \frac{\partial S}{\partial t} \qquad (19)$$

$$\equiv H(x,u,p,t) - p\dot{x} + \frac{\partial S}{\partial x}\dot{x} + \frac{\partial S}{\partial t} \qquad (20)$$

where $H(x,u,t) \equiv f_0(x,u,t) + pf(x,u,t)$ is our usual Hamiltonian. Note that inner products are written as scalar products, for example

$$p\dot{x} \equiv p'\dot{x} \equiv \sum_1^n p_i\dot{x}_i$$

$$\frac{\partial S}{\partial x}\dot{x} \equiv \left(\frac{\partial S}{\partial x}\right)'\dot{x} \equiv \sum_1^n \frac{\partial S}{\partial x_i}\dot{x}_i \,, \qquad \text{etc...}$$

From the Variable end point problem in the Calculus of Variations (ch. 3, eqs. 8 and 11) we had, for $J(x) \equiv \int_0^T f(x,\dot{x},t)\,dt$, the necessary conditions

$$\delta J(x) = \int_0^T (f_x - \frac{d}{dt} f_{\dot{x}})h\,dt + \left[f_{\dot{x}}\,\delta x + (f - \dot{x}f_{\dot{x}})\delta t\right]_0^T = 0\,. \qquad (21)$$

Applying this to our augmented functional $J_a(u)$ defined in (18) gives

$$\delta J_a(u) = \int_0^T \left[(F_x - \frac{d}{dt} F_{\dot{x}})\delta x + F_u\,\delta u + F_p\,\delta p\right] dt$$

$$+ \left[F_{\dot{x}}\,\delta x + (F - F_{\dot{x}}\dot{x})\delta t\right]_{t=T} = 0 \;(t_0 = 0 \text{ is fixed}). \qquad (22)$$

From the definition of F in (19) and the fact that the Euler equation must be satisfied, we have

$$F_x - \frac{d}{dt} F_{\dot{x}} = H_x + \frac{\partial}{\partial x} (S_x \dot{x} + S_t) - \frac{d}{dt} (S_x - p)$$

$$= H_x + \cancel{S_{xx}\dot{x}} + \cancel{S_{xt}} - \cancel{S_{xx}\dot{x}} - \cancel{S_{xt}} + \dot{p}$$

$$= H_x + \dot{p} = 0 \; . \tag{23}$$

This gives the Euler equation

$$\boxed{\dot{p} = - H_x} \; . \tag{24}$$

Similarly, δu and δp being arbitrary and independent variations, their coefficients in (22) must vanish, i.e. $F_u = 0$ and $F_p = 0$. But from definition (19)

$$F_u = H_u \quad \text{and} \quad F_p = f(.) - \dot{x} = H_p - \dot{x}$$

hence

$$H_u = 0 \tag{25}$$

$$\dot{x} = f(x,u,t) = H_p \; . \tag{26}$$

Finally, the transversality or boundary conditions given by the remaining terms of (22) are

$$\Big[F_{\dot{x}} \, \delta x + (F - \dot{x} F_{\dot{x}}) \delta t \Big]_{t=T} = 0 \; . \tag{27}$$

But $\quad F_{\dot{x}} = S_x - p$

$$F - \dot{x} F_{\dot{x}} \equiv H - \cancel{p\dot{x}} + \cancel{S_x \dot{x}} + S_t - \cancel{\dot{x} S_x} + \cancel{\dot{x} p}$$

$$= H + S_t \; .$$

Hence (27) becomes

$$\boxed{(S_x - p)\delta x\Big|_{t=T} + [H(t) + S_t]\delta t\Big|_{t=T} = 0} \qquad (28)$$

which will be referred to as boundary or "Transversality" conditions.

If both the initial state vector $x(0)$ and time t are unspecified as well, (28) becomes

$$(S_x - p)\delta x\Big|_{t=t_0}^{t=T} + [H(t) + S_t]\delta t\Big|_{t=t_0}^{t=T} = 0 \qquad (29)$$

which are the results of Pontryagin's theorem.

A number of remarks should be made.

Remark 1. Condition (b) or eq. (13) is called the Maximum Principle by Pontryagin et al. [1962]. It is satisfied by $H_u = 0$ and $H_{uu} < 0$ in our problem where the u_i's ($1 \le i \le r$) are all unbounded. When $u \in U$ and U is a closed set, however, $H_u = 0$ makes no sense unless the maximum of H takes place in the interior of set U. If H is a monotonic increasing function of u everywhere in the closed set U, i.e. $H_{u_i} > 0 \ \forall \ u_i \in U$, optimal control is $u_{i_{\max}}$ for a maximization problem and $u_{i_{\min}}$ for a minimization problem. If H is a monotonic decreasing function of u, i.e. $H_{u_i} < 0 \ \forall \ u_i \in U$, optimal control is $u_{i_{\min}}$ for a maximization and $u_{i_{\max}}$ for a minimization problem. This remark also applies to the special case in which H is linear in u. Optimal control variables u_i are then piecewise continuous and jump from one vertex of the polyhedron to another. This is the typical case of the bang-bang control which will be discussed in greater detail in the next chapter. Note that (13) also covers the sufficient conditions which will be discussed later when the second variations are examined.

Remark 2. The p vector, whose existence is asserted in the theorem, plays the role of the Lagrange multiplier in chapter 3 where the constrained dynamic optimization problems were discussed. It is the shadow price or marginal valuation of x, showing the amount by which a unit increment in x, at time t, contributes to the optimal objective functional J^*. For example, in the optimal economic growth model of chapter 3 where the objective was to maximize $\int_0^\infty ce^{-rt}\,dt$ subject to $\dot{k}(t) = sf(k) - \lambda k(t)$, the Hamiltonian $H \equiv ce^{-rt} + p[sf(k) - \lambda k(t)]$ is the discounted per capita GNP which consists of the discounted value of consumption ce^{-rt} and the discounted valuation $p(t)$ of investment $sf(k) - \lambda k$, in terms of the consumption good c taken as numéraire. Clearly $\dot{p}$ indicates the rate of increase (appreciation, for $\dot{p} > 0$) or decrease (depreciation) in the value of a unit of capital. The Euler equation $-\dot{p} = H_k$ means that capital depreciates $(-\dot{p})$ at the rate at which it imparts its embodied value to the commodities it produces, H_k. In other problems, p has a similar interpretation as a shadow price. (See, for example, Dorfman 1969.)

Note that (p_0,p) is a n+1 non-zero vector, with p_0 a positive constant set equal to one for simplicity. The fact that p_0 is a constant follows from $\dot{p}_0 = -\partial H/\partial x_0 = 0$ but the proof that $p_0 > 0$ as well as Pontryagin's proof that a non zero n+1 vector (p_0,p) exists will not be pursued here. (See Pontryagin et al. 1962.)

Remark 3. $dH/dt = \partial H/\partial t$ as can be seen from chapter 3, equations (32)-(37) or by direct differentiation of the Hamiltonian $H(x,u,p,t)$ where

$H \equiv f_0(x,u,t) + pf(x,u,t)$ and use of $f = \dot{x}$ (26), $\dot{p} = - H_x$ (24) and $H_u = 0$, i.e.

$$\dot{H} \equiv \frac{dH}{dt} = \frac{\partial f_0}{\partial x}\dot{x} + \frac{\partial f_0}{\partial u}\dot{u} + \frac{\partial f_0}{\partial t} + p(f_x\dot{x} + f_u\dot{u} + f_t) + f\dot{p}$$

$$\equiv H_x\dot{x} + H_u\dot{u} + H_t - H_x\dot{x}$$

$$= \frac{\partial H}{\partial t} \equiv H_t \ . \tag{30}$$

Remark 4. Equations (24), (25) and (26) provide the necessary conditions for the problem. They constitute a set of $2n$ first order differential equations (24) and (26) and r algebraic relations (25) which must be satisfied throughout the interval $[0,T]$. Assuming (25) gives u^* explicitly then substitution of u^* in (24) and (25) yields a system of $2n$ equations whose solution contains $2n$ constants of integration. To evaluate these, we have n-equations $x(0) = x_0$, n-equations $S_x - p = 0$ at $t = T$. In addition, when T is unspecified as in this problem, it is determined by the relation $H(t) + S_t = 0$ at $t = T$. Thus, we have enough boundary conditions to specify the arbitrary constants. These are referred to as boundary or Transversality conditions in part (c) of the theorem. Thus (c) is satisfied by (28). Note that in the case of absence of the Scrap function $S[x(T),T]$, (28) becomes simply

$$- p(T)\ \delta x(T) + H(T)\ \delta T = 0$$

and more generally, (29) becomes

$$- p(t)\ \delta x(t)\Big|_{t=t_0}^{t=T} + H(t)\ \delta t\Big|_{t=t_0}^{t=T} = 0 \ . \tag{31}$$

Example 5.3.1

$$\text{Minimize } J = \int_0^1 (x + u^2)\, dt$$

subject to

$$\dot{x}(t) = -u(t); \quad x(0) = 0, \ x(1) \text{ free}$$

The problem is thus to choose an optimal control $u(t)$ to bring the dynamic system $\dot{x}(t) = -u(t)$ from the origin $x(0) = 0$ at time $t = 0$ to a terminal point which is unspecified, at time $t = 1$ such as to minimize the objective functional J.

The Hamiltonian is

$$H = x(t) + u(t)^2 - p(t)\, u(t)$$

$$H_u = 2u(t) - p(t) \Rightarrow u^*(t) = p(t)/2 \ , \ H_{uu} = 2 > 0$$

$\dot{p}(t) = -H_x = -1$, gives

$$p(t) = -t + c_1$$

Since the Scrap function $S(x(1))$ is absent, and $x(1)$ is unspecified, the transversality conditions $p(1)\,\delta x(1) = 0$ imply $p(1) = -1 + c_1 = 0$ i.e. $c_1 = 1$ and

$$p(t) = -t + 1$$

the dynamic system is

$$\dot{x}(t) = -u(t) = -p(t)/2$$

$$= -\frac{1}{2}(-t + 1) = t/2 - \frac{1}{2}$$

whose solution is

$$x(t) = t^2/4 - t/2 + c_2$$

where $c_2 = 0$ in view of $x(0) = 0$

Thus the complete solution is

$$x^*(0) = t^2/4 - t/2$$

$$p^*(t) = -t + 1$$

$$u^*(t) = p^*(t)/2 = -t/2 + 1/2$$

Example 5.3.2

$$\text{Minimize} \quad \frac{1}{2}x(1)^2 + \frac{1}{2}\int_0^1 u^2\,dt$$

subject to

$$\dot{x}(t) = -u(t), \quad x(0) = 1$$

The Hamiltonian is

$$H = \frac{1}{2}u^2 - pu$$

$$H_u = u - p = 0 \Rightarrow u^*(t) = p(t) \;; \quad H_{uu} = 1 > 0$$

$$\dot{p} = -H_x = 0 \quad \Rightarrow p(t) = p\text{- some constant}$$

$$\dot{x} = -u = -p \text{ gives}$$

$$x(t) = -pt + c_1 = -pt + 1 \text{ in view of } x(0) = 1$$

The transversality conditions $p(1) = S_x = x(1)$ give

$$p(1) = x(1) = -p + 1$$

i.e. $2p = 1$, $P^* = \frac{1}{2}$

The solution is thus

$$x^*(t) = -\frac{1}{2}t + 1$$

$$p^* = u^* = \frac{1}{2}$$

5.4 Transversality Conditions

Transversality conditions provide the rules for determining the arbitrary constants contained in the solution of the canonical system (10)-(11).

Several possibilities arise. Without loss of generality we shall assume for the moment that the original point is fixed, i.e. both the initial state variables $x(0) = x_0$ and initial time $t_0 = 0$ are specified and examine the problems with fixed and free final time separately, with reference to (28).

5.4.1 Problems with Fixed Final Time T

With fixed final time T, $\delta T = 0$ and (28) is reduced to

$$(S_x - p)\ \delta x \Big|_{t=T} = 0 \tag{32}$$

and three cases arise.

Case 1: Fixed Final State $x(T) = x_T$

Clearly $\delta x(T) = 0$ and (28) or (32) provides no information. In fact no such information is needed since the $2n$ constants of integration are determined by $x(0) = x_0$ and $x(T) = x_T$.

Case 2: Free Final State $x(T)$

Clearly $\delta x(T) \neq 0$ and application of (32) gives

$$p(T) = S_x \ . \tag{33}$$

If $S[x(T),T]$ is absent, (32) is reduced to $p(T)\ \delta x(T) = 0$, i.e.

$$p(T) = 0 \ . \tag{34}$$

Case 3: Final State lying on Manifold $M[x(t),t] = 0$ (at $t = T$)

If the terminal state lies on some Manifold $M[x(t),t] = 0$ or, put more simply, if the terminal state is constrained by $M[x(t),t] = 0$ at $t = T$, where M is an s-vector, (32) is replaced by

$$(R_x - p)\ \delta x \Big|_{t=T} = 0 \tag{35}$$

where

$$R[x(T),T] \equiv S[x(T),T] + \mu M[x(T),T] \tag{36}$$

μ being an s-vector of Lagrange multipliers, and

$$R_x \equiv \frac{\partial R}{\partial x} = S_x + M'_x \mu$$

or, in full, $$R_{x_i} \equiv \frac{\partial R}{\partial x_i} = \frac{\partial S}{\partial x_i} + \sum_{j=1}^{s} \frac{\partial M_j}{\partial x_i} \mu_j \qquad (i=1,2,\ldots,n) .$$

Clearly

$$p(T) = R_x \equiv S_x + M'_x \mu . \tag{37}$$

and of course, $M[x(T), T] = 0$

5.4.2 Problems with Free Final Time T

With the final time T unspecified, $\delta T \neq 0$ and since $x(T)$ and T are independent, i.e. the time the programme must terminate does not depend on the value the final state $x(T)$ takes at that time, application of (28) gives

$$(R_x - p)\delta x \Big|_{t=T} + [H(t) + R_t]\delta t \Big|_{t=T} = 0 \tag{38}$$

where $R \equiv S + \mu M$ as defined in (36) and $R_x \equiv \frac{\partial R}{\partial x}$; $R_t \equiv \frac{\partial R}{\partial t}$ with $R \equiv S$ when M is absent.

Again the various cases arise as follows.

Case 4: Fixed Final State: $x(T) = x_T$

Clearly $\delta x(T) = 0$ and (28) gives

$$H(T) + \frac{\partial S}{\partial t}\Big|_{t=T} = 0 \tag{39a}$$

where $H(T) \equiv H[x^*(T), u^*(T), p^*(T), T]$ for short and if S is absent,

$$H(T) = 0 . \tag{39b}$$

Thus we have $2n + 1$ relations which are x_0, x_T and one of the equations (39a) or (39b) to determine $2n + 1$ unknowns.

Case 5: Free Final State $x(T)$, i.e. $\delta x(T) \neq 0$

Application of (37) gives

$$p(T) = S_x[x(T),T] \tag{40}$$

$$\text{and} \quad H(T) + S_t = 0 \quad \text{at} \quad t = T \tag{41}$$

or when S is absent, (40) and (41) are simply reduced to

$$p(T) = 0 = H(T) \; . \tag{42}$$

Case 6: Final State free but satisfying the s-vector $M[x(T),T] = 0$.

Application of the transversality conditions (28) gives

$$p(T) = R_x \equiv S_x + M_x'\mu \tag{43}$$

$$H(T) + R_t \equiv H(T) + S_t + \mu M_t = 0 \tag{44}$$

and of course

$$M[x(T),T] = 0 \; . \tag{45}$$

The $n + s + 1$ unknowns $x(T)$, μ and T are determined by the n equations in (43), one equation in (44) and s equations in (45).

In the case where R is absent (28) simply gives

$$p(T) = 0 = H(T) \tag{42}$$

and if S is present but M is not given, (43), (44) and (45) are simply

$$p(T) = S_x \tag{46}$$

$$H(T) + S_t = 0 \quad \text{at} \quad t = T \tag{47}$$

$$S[x(T),T] = 0 \; . \tag{48}$$

This completes the examination of the various cases. These results are now summarized and presented in a table. The standard problem referred to is one of optimizing, for given x_0, t_0, $J(u) \equiv S[x(T),T] + \int_0^T f_0(x,u,t)\, dt$ subject to $\dot{x} = f(x,u,t)$ giving rise to (28) reproduced here for convenience

$$[S_x - p(T)]\ \delta x(T) + [H(T) + S_t]\ \delta t \quad \text{at} \quad t = T\ . \tag{28}$$

Example 5.4.2

Given the following dynamic system

$$\dot{x}_1(t) = x_2(t)\ , \qquad x_1(0) = 0$$

$$\dot{x}_2(t) = u(t)\ , \qquad x_2(0) = 0$$

where both the control and state variables are unbounded, find the optimal control law to minimize

$$J(u) = \int_0^T \frac{1}{2}u^2(t)\ dt\ .$$

The Hamiltonian H is

$$H = \frac{1}{2}u^2(t) + p_1(t)x_2(t) + p_2(t)u(t) \tag{49}$$

$H_u = 0 = u^* + p_2^*$ gives

$$u^*(t) = -\,p_2^*(t) \tag{50}$$

$\dot{p}_1 = -\ \partial H/\partial x_1 = 0$ gives

$$p_1^*(t) = c_1 \tag{51}$$

$\dot{p}_2 = -\ \partial H/\partial x_2 = -\ p_1^* = -\ c_1$ gives

Table 5

Summary of Boundary Conditions in Optimal Control

(Eq. (28) $(S_x - p)\delta x\Big|_{t=T} + [H(t)+S_t]\delta t\Big|_{t=T} = 0$)

	Case	Substitution in eq. (28)	Boundary Condition equations	Determination of Constants
A. Fixed Terminal Time T ($\delta T = 0$)				
1.	Fixed $x(T) = x_T$	$\delta x(T) = 0$ $\delta T = 0$	$x(0) = x_0$; $x(T) = x_T$ (No restrictions on $p(T)$)	$2n$ equations to determine $2n$ constants
2.	Free $x(T)$ i.e. $\delta x(T) \neq 0$	$\delta x(T) \neq 0$ $\delta T = 0$	$x(0) = x_0$ $p(T) = S_x$	$2n$ equations to determine $2n$ constants
3.	Final State $x(T)$ lying on $M[x,t] = 0$	$\delta x(T) \neq 0$ $\delta T = 0$	$x(0) = x_0$ $p(T) = S_x + M'_x\mu$ $M[x(T),T] = 0$	$2n+s$ equations to determine $2n$ constants and s-vector μ
B. Free Terminal Time T ($\delta T \neq 0$)				
4.	Fixed $x(T) = x_T$	$\delta x(T) = 0$ $\delta T \neq 0$	$x(0) = x_0$ $x(T) = x_T$ $H(T) + S_t = 0$ at $t = T$	$2n+1$ equations to determine $2n$ constants and terminal time T
5.	Free $x(T)$	$\delta x(T) \neq 0$ $\delta T \neq 0$	$x(0) = x_0$ $p(T) = S_x[x(T),T]$ $H(T) + S_t = 0$ at $t = T$	$2n+1$ equations to determine $2n$ constants and T
6.	Unspecified $x(T)$ lying on manifold $M[x(T),T] = 0$	$\delta x(T) \neq 0$ $\delta T \neq 0$	$x(0) = x_0$ $p(T) = R_x \equiv S_x + M'_x\mu$ $H(T) + R_t = 0$ $M = 0$	$2n+s+1$ equations to determine $2n$ constants s-Lagrange vector μ and T

Source: Kirk (1970) p. 200

$$p_2^*(t) = - c_1 t + c_2 \tag{52}$$

$\dot{x}_2 = u^*(t) = - p_2^*(t) = c_1 t - c_2$ gives

$$x_2^*(t) = \frac{1}{2} c_1 t^2 - c_2 t + c_3 \, . \tag{53}$$

Finally $\dot{x}_1 = x_2$ gives

$$x_1^*(t) = \frac{c_1}{6} t^3 - \frac{c_2}{2} t^2 + c_3 t + c_4 \tag{54}$$

where c_1, c_2, c_3 and c_4 are arbitrary constants to be determined. Initial conditions $x_1(0) = 0 = x_2(0)$ imply that $c_3 = 0 = c_4$. This reduces the number of constants to be determined to 2. The various cases in Table 5 will now be examined.

<u>Case 1</u>: $x_1(0) = 0 = x_2(0)$; $x_1(1) = 2$; $x_2(1) = 3$; $T = 1$

$x_1(0) = 0 = x_2(0)$ implies $c_3 = 0 = c_4$ in (53) and (54)

$$x_1(1) = c_1/6 - c_2/2 = 2$$
$$x_2(1) = \frac{1}{2} c_1 - c_2 = 3$$

giving
$$\begin{bmatrix} \frac{1}{6} & -\frac{1}{2} \\ \frac{1}{2} & -1 \end{bmatrix} \begin{bmatrix} c_1 \\ c_2 \end{bmatrix} = \begin{bmatrix} 2 \\ 3 \end{bmatrix}$$

i.e., $c_1 = - 6 = c_2$ and (53) and (54) give the optimal trajectory as

$$x_2^*(t) = - 3t^2 + 6t \tag{55}$$

$$x_1^*(t) = - t^3 + 3t^2 \tag{56}$$

and the optimal control $u^* = - p_2^*$, i.e.

$$u^*(t) = - 6t + 6 \, . \tag{57}$$

Case 2: $x_1(0) = 0 = x_2(0)$; $T = 1$; $x_1(1)$ and $x_2(1)$ are unspecified, $S(x(T)) = \frac{1}{2}[x_1(T) - 2]^2$.

Again $c_3 = 0 = c_4$. Case 2 in Table 5.1 gives

$$p_1(T) = \partial S/\partial x_1 = x_1(1) - 2 \tag{58}$$

$$= c_1 \quad \text{by (51)}$$

$$p_2(T) = \partial S/\partial x_2 = 0$$

$$= -c_1 + c_2 \quad \text{by (52)} . \tag{59}$$

By (54) $$x_1(T) = \frac{c_1}{6} - \frac{c_2}{2} = 2 + c_1 \quad \text{by (58)} . \tag{60}$$

Thus c_1 and c_2 are readily found by solving these two equations (59) and (60) as $c_1 = c_2 = -3/2$.

Substituting into (53), (54), (50) and (52) gives

$$x_1^*(t) = -\frac{1}{4}t^3 + \frac{3}{4}t^2$$

$$x_2^*(t) = -\frac{3}{4}t^2 + \frac{3}{2}t$$

$$u^*(t) = -\frac{3}{2}t + \frac{3}{2} .$$

Case 3: $x_1(0) = 0 = x_2(0)$; $T = 1$; $x(T)$ free but lying on $M[x(t)] = x_1(t) + 2x_2(t) - 10 = 0$ at $t = T$, $S = \frac{1}{2}[x_1(T) - 2]^2$.

As before, $c_3 = 0 = c_4$. Case (3) in Table 6.1 gives $p(T) = S_x + M_x\mu$, i.e.

$$p_1(1) = \partial S/\partial x_1 + \mu\partial M/\partial x_1 = x_1(1) - 2 + \mu \tag{61}$$

$$p_2(1) = \partial S/\partial x_2 + \mu\partial M/\partial x_2 = 0 + 2\mu \tag{62}$$

$$M(x(T)) = x_1(1) + 2x_2(1) - 10 = 0 \tag{63}$$

With the substitution of $p_1(1) = c_1$, $p_2(1) = -c_1 + c_2$, $x_1(1) = c_1/6 - c_2/2$ and $x_2(1) = c_1/2 - c_2$ from (51) - (54), the last 3 equations become

$$\begin{bmatrix} -5/6 & -1/2 & 1 \\ -1 & 1 & -2 \\ 7/6 & -5/2 & 0 \end{bmatrix} \begin{bmatrix} c_1 \\ c_2 \\ \mu \end{bmatrix} = \begin{bmatrix} 2 \\ 0 \\ 10 \end{bmatrix} \text{ giving } \begin{bmatrix} c_1 \\ c_2 \\ \mu \end{bmatrix} = - \begin{bmatrix} 1.5 \\ 4.7 \\ 1.6 \end{bmatrix}$$

With these constants thus determined, the solution is

$$u^*(t) = -1.5\ t + 4.7$$

$$x_1^*(t) = -0.25\ t^3 + 2.35\ t^2$$

$$x_2^*(t) = -0.75\ t^2 + 4.7\ t$$

The remaining cases are left to the interested readers.

5.4.3 Transversality Conditions in Infinite Horizon Problems

So far, we have been dealing with the problem of finite horizon, i.e., $T < \infty$. Although most practical problems fall into this category, the arbitrary nature of the decision on the size of the terminal stock $x(T)$ and the horizon T sometimes gives rise to difficulties. The length of the horizon is a crucial factor in the screening of candidate programmes: a shorter horizon favours programmes yielding immediate high returns but growing more slowly, a more distant horizon, on the other hand, will give a higher ranking to programmes growing fastest even if these yield lower returns at first. The time t^* at which the cumulative values of two competing programmes are equal is crucial: which programme will be chosen depends on whether $T \gtrless t^*$. This brings us to the ideas of overtaking and catching up which play an important part in the determination of transversality conditions when $T \to \infty$, especially when the limits $\lim_{t\to\infty} x_i(t)$ do not exist. More precisely, given the difference ΔJ between two admissible programmes

$$\Delta J(t) \equiv \int_0^t f_0\,[x^*(\tau),\ u^*(\tau),\ \tau]dt - \int_0^t f_0[x(\tau),\ u(\tau),\ \tau]dt$$

if there exists a time t^* such that for all $t > t^*$, $\Delta J(t) \geqq 0$, then $x^*(t)$ is said to overtake $x(t)$; if $\lim_{t\to\infty} \Delta J(t) \geqq 0$ then $x^*(t)$ is said to catch up with $x(t)$.

Rose (1977) has shown that in the overtaking case for all $t > t^*$ and in particular for $t = \infty$, the transversality condition is completely analogous to the finite horizon case where $x(T)$ is unspecified, i.e.,

$$p^*(t) \quad \delta x^*(t) \geq 0 \tag{64}$$

This result was obtained by the Taylor expansion of

$$J_a[t,\ x^*(t,\ \varepsilon(t))] - J_a\,[t,\ x^*(t)] \leq 0$$

where $\varepsilon(t)$ is a positive continuous and sufficiently small function and $J_a\,[t,\ x(t)] \equiv \int_0^t [H(u,\ x,\ p,\ \tau) - p\dot{x}]\ dt = \int_0^t [H(u,\ x,\ p,\ \tau) - \dot{p}x]\ dt +$ $p_0x_0 - p(t)x(t)$ account taken of the facts that $H_u = 0$, $\dot{x} = H_p$, $\dot{p} = -H_x$ and hence $\partial J/\partial x = -p(t)$. The result obtained,

$$p^*(t)\ \ \delta x^*(t) \geq 0$$

holds for all t sufficiently large, for both the overtaking and catching up cases.

Example 5.4.3 (Rose 1977)

Consider a family unit, endowed with an initial wealth x_0, which wants to maximize the present value of its consumption utility $u(c) = c^{1-v}/(1-v)$ (where $0 < v < 1$; for the case $v = 1$, $u(c) = \log c$, see Tu 1969), discounted at some positive rate ρ. Ignoring contractual income, the family wealth accumulates at a constant interest rate r minus consumption $c(t)$, i.e.,

$$\dot{x}(t) = rx(t) - c(t),\ x(0) = x_0$$

The objective is to maximize, subject to the above equation, the present value of consumption utility $\int_0^\infty \frac{c^{1-v}}{1-v} e^{-\rho t}\, dt$

The Hamiltonian is

$$H \equiv \frac{c}{1-v} e^{-\rho t} + p(rx - c)$$

$$H_c = c^{-v} e^{-\rho t} - p = 0 \quad \Rightarrow c(t) = c(0)e^{\lambda t} \text{ where } \lambda \equiv (r-\rho)/v$$

$$\dot{p} = -H_x = rp$$

The last two equations jointly imply that

$$p(t) = c(0)^{-v} e^{-\rho t}$$

Substitution into the dynamic system gives

$$\dot{x}(t) = rx(t) - c(0)e^{\lambda t}$$

whose solution, assuming $r > \lambda$, is

$$x(t) = \frac{c(0)}{r-\lambda} e^{\lambda t} + [x_0 - \frac{c(0)}{r-\lambda}]\ e^{rt}$$

from which $\delta x(t) = \frac{\delta c(0)}{r-\lambda} (e^{\lambda t} - e^{rt})$

Clearly

$$p(t)\ \delta x(t) = \frac{\delta c(0)\ c(0)^{-v}}{r - \lambda} [e^{-(r-\lambda)t} - 1]$$

$$\lim_{t\to\infty} p(t)\ \delta x(t) = -\ c(0)^{-v}\ \delta c(0)/(r-\lambda)$$

If $c(0) < (r-\lambda)\ x_0$ then $\delta c(0) > 0$ is admissible and $\lim_{t\to\infty} p(t)\ x(t) < 0$ which is non-optimal. If $c(0) = (r-\lambda)x_0$ then $\delta c(0) < 0$ is admissible and $p(t)\ \delta x(t) > 0$. Thus, if the solution exists, it is $c^*(t) = (r-\lambda)x^*(t)$

and $x^*(t) = x_0 e^{\lambda t}$.

For the particular case where H is concave in x and u, the above result follows from the concavity properties (see Mangasarian theorem (1966) in Ch. 4, Arrow & Kurz (1970) and Seierstad and Sydsaeter (1977)). By a sequence of equalities and inequalities, it can be shown that, if H is concave, then $p(t)\,[x(t) - x^*(t)] \geq 0$ for t sufficiently large, in particular $t = \infty$, is sufficient for $\Delta J_a(x)$ to be non-negative.

Assuming $x(0) = x_0$ and writing $H^* \equiv f_0(x^*, u^*, t) + p.f(x^*, u^*, t)$, $H \equiv f_0\,(x, u, t) + p.f(x, u, t)$; $J_a(t) \equiv \int_0^t (H - p\dot{x})\, dt$,

it can be seen that

$$\Delta J_a(t) \equiv \int_0^t [H^* - H - p\,(\dot{x}^* - \dot{x})]\, dt$$

$$= \int_0^t [H^* - H + \dot{p}\,(x^* - x)]\, d\tau - p(x^* - x)\Big|_0^t$$

$$= \int_0^t [H^* - H - H_x^*(x^* - x)]\, d\tau - p(x^* - x)\Big|_t$$

where the first line is just a definition, the second was obtained by integration by parts, the third line by use of $\dot{p} = - H_x^*$, $H_u = 0$ and $p(x^* - x)\big|_{t=0} = 0$. The integrand in the last line is non-negative because of the concavity of H. Thus, if H is concave, $p(t).[x(t) - x^*(t)] \geq 0$, $\forall t > t^*$ and in particular, for $t = \infty$, $\lim_{t\to\infty} p(t).[x(t) - x^*(t)] \geq 0$, is sufficient for $x^*(t)$ to be optimal, i.e. for $x^*(t)$ to be the solution of the problem of maximizing $\int_0^T f_0\; dt$ subject to $\dot{x} = f(x, u, t)$ for both the cases of finite and infinite T.[1]

It can be shown that this condition is satisfied for both the cases of overtaking and catching up, as well as for the cases where equality and inequality constraints are added. (See Seierstad and Sydsaeter 1977.)

For the case in which the only requirement for $x(T)$ is its non-negativity, i.e. $x_i(T) \geq 0 \; \forall \; i$ and T sufficiently large and in particular, $T = \infty$, Arrow and Kurz (1970) have shown that the Transversality Condition is

$$x_i^* (T), \; p_i^* (T) \geq 0 \quad \text{and} \; p_i^* (T) \; x_i^* (T) = 0 \tag{65}$$

This is obtained by defining a "Scrap" function $S(x)$ as $S(x) \equiv \sum_1^n c_i \; min \; (x_i, 0)$ which gives

$$\frac{\partial S}{\partial x_i} = \begin{cases} c_i & \text{if } x_i < 0 \\ 0 & \text{if } x_i \geq 0 \end{cases}$$

i.e. $0 \leq \partial S/\partial x_i \leq c_i$ $(i = 1, 2, \ldots, n)$. This choice of c_i (a very large positive number) amounts to placing a stiff penalty for negative $x_i(T)$ so that this is eliminated from an optimal programme. Thus, the Transversality condition (33), $p_i(T) = \partial S/\partial x_i(T)$ becomes simply $p_i(T) \geq 0$, $x_i(T) \geq 0$ and $p_i(T) \; x_i(T) = 0$. Although this device is successful in many cases and has been used widely in Economics, it has some difficulties as pointed out by Shell (1969) and Halkin (1974).

Note that the Arrow-Kurz condition $p(\infty)$, $x(\infty) \geq 0$ and $p(\infty) \; x^*(\infty) = 0$ could be considered a particular case of the condition $p\partial x^* \geq 0$ examined above, with equality holding. For sufficiently large t, including $t = \infty$,

$$p(t)\delta x^*(t) \equiv p(t) \; x(t) \; - p(t).x^*(t) \geq 0$$

$$\lim_{t\to\infty} \; p(t).x(t) \; - \; \lim_{t\to\infty} \; p(t).x^*(t) \quad \geq 0$$

A particular case is $\lim_{t\to\infty} p(t)x(t) > 0$ and $\lim_{t\to\infty} p(t)x^*(t) = 0$.

5.5 Second Variations and Sufficient Conditions

Condition (13) in Pontryagin's theorem covers both the necessary and sufficient conditions.

The total variations of the augmented functional $J_a(u)$ are

$$\Delta J_a(u) \equiv J_a(u) - J_a(u^*) = \delta J_a(u) + \delta^2 J_a(u) + 0(u) \tag{66}$$

Neglecting higher order terms $0(u)$ and noting that the first variations $\delta J_a(u)$ vanish for an extremum, it can be seen that the sign of $\Delta J_a(u)$ depends on the sign of the second variations $\delta^2 J_a(u)$ which must be non-positive for a maximum and non-negative for a minimum problem. We shall examine the conditions which bring these about.

Neglecting the Scrap function, i.e. $S(x, t) = 0$, the second variations are

$$\delta^2 J_a(u) = \frac{1}{2}\int_0^T (\delta x \;\; \delta u)\begin{bmatrix} H_{xx} & H_{xu} \\ H_{ux} & H_{uu} \end{bmatrix}\begin{bmatrix} \delta x \\ \delta u \end{bmatrix} dt \tag{67}$$

where $H_{xx} \equiv [\delta^2 H/\delta x_i \, \delta x_j]$, the derivatives of H_x with respect to x, evaluated at (x^*, p^*, u^*, t). Similarly H_{ux} $(= H_{xu})$ and H_{uu}. Finally $\delta x = (\delta x_1, \ldots, \delta x_n)$, the n-variation vector with $\delta x_i(0) = 0 \; \forall \; i$ and $\delta u \equiv (\delta u_1, \ldots, \delta u_r)$, an r-variation vector.

In view of the condition $\delta x(0) = 0$, a small variation δu will be expected to generate a small variation δx but not vice versa: it is possible to find a function $\delta u(t) \equiv h(t)$ such that a large $h(t)$ will cause a small variation in x.[2] This is reminiscent of the remark leading to Lemma 4.3 which immediately gave the Legendre Conditions in Chapter 4. In fact, just as in Legendre's Theorem 4.3 in Chapter 4, the term $\delta u' \, H_{uu} \delta u$ swamps all other terms of $\delta^2 J_a(u)$ in (67): it is the most important in the determination of the sign of $\delta^2 J_a(u)$.[3] Clearly, $\delta^2 J_a(u)$ must be negative for a maximum and positive for a minimum. This is summarised in the following theorem.

<u>Theorem</u>

The control $u^*(t)$ is a local maximum (minimum) of J if it satisfies $\partial H/\partial u = 0 \ \forall \in [0, T]$ and $\delta^2 J_a(u)$ (≤ 0) for all non-zero but otherwise arbitrary vectors $(\delta u, \delta u)$, which implies that H_{uu} evaluated at $u = u^*$ is negative (positive) definite for a maximum (Minimum) problem, and the entire $(n+r) \times (n+r)$ matrix $\begin{bmatrix} H_{xx} & H_{xu} \\ H_{ux} & H_{uu} \end{bmatrix}$ *is negative (positive) semi-definite for a maximum (minimum) problem.*

<u>Proof</u>

See, for example, Athans & Falb (1966), Bryson and Ho (1975).

Example 5.5.1

In the optimal growth problem (see example 4.7.1) where the objective is to maximize the present value of consumption utilities $\int_0^\infty u(c)\, e^{-rt}\, dt$ subject to the the dynamic law of capital accumulation $\dot{k} = f(k) - \lambda k - c$ where $u(c)$ and $f(k)$ are both assumed to be increasing concave functions (i.e. $u''(c) < 0 < u'(c)$, $f''(k) < 0 < f'(k)$), the Hamiltonian is

$$H = e^{-rt} \left\{ u(c) + q\, [f(k) \ - \lambda k(t) - c(t)] \right\}$$

Clearly $H_{cc} = e^{-rt}\, u''(c) < 0$ and the Hessian of the Hamiltonian H at equilibrium values, is

$$\begin{bmatrix} H_{kk} & H_{kc} \\ H_{ck} & H_{uu} \end{bmatrix} = e^{-rt} \begin{bmatrix} qf''(k) & 0 \\ 0 & u''(c) \end{bmatrix}$$

which is clearly negative definite. This is sufficient for the maximization problem.

Note that the condition that H_{uu} be negative (positive) definite is called the Legendre-Clebsch condition (see Chapter 3 eq. 30).

Note also that the Weierstrass E-function is also satisfied. In fact

$$E = f_0(x, \dot{x}, t) - f_0(x, \dot{x}^*, t) - (\dot{x} - \dot{x}^*)\, f^*_{\dot{x}} \tag{68}$$

Identifying $\dot{x}$ with u, $\dot{x}^*$ with u^* and $f^*_{\dot{x}}$ with $-p$, the adjoint variable and writing the Hamiltonian along the optimal path generated by u^* as

$H^* \equiv H(x, u^*, p, t) \equiv f_0(x, u^*, t) + p.f(x, u^*, t)$ and

$H \equiv H(x, u, p, t) \equiv f_0(x, u, t) + p.f(x, u, t)$ along any other admissible path, the E-condition in (68) is precisely

$$E = H(x, u, p, t) - H(x, u^*, p, t) \tag{69}$$

Clearly $E \leq 0$ for a maximum and $E \geq 0$ for a minimum. Thus the condition (b) in Pontryagin's Theorem that $H \leqq H^*$ covers all these cases.

Finally note the important particular case of concavity and convexity of the Hamiltonian examined in Chapter 3.

In the problem of maximizing (minimizing) $\int_0^T f_0(x, u, t)dt$ subject to $\dot{x} = f(x, u, t)$, with $x(0) = x_0$ and $x(T) = x_T$ (two-fixed-end-point problem), if both f_0 and f and hence $H = f_0 + pf$ are concave (convex) in x and u, for given p, then the necessary conditions $\dot{x} = H_p$, $\dot{p} = - H_x$ and $H_u = 0$ are also the sufficient conditions for a maximization (minimization) problem.

This can be shown by writing a sequence of equalities and inequalities as in Chapter 3 (eq. 30) and noting its concavity (convexity) and taking into account the necessary conditions $\dot{x} = H_p$, $\dot{p} = - H_x$ and $H_u = 0$. It is clear from the definition of the Hamiltonian $H(x, u, p, t) \equiv f_0 (x, u, t) + pf(x, u, t)$ and the augmented functional $J_a(u)$, that

$$J_a(u) = \int_0^T [f_0(x, u, t) + p.f(x, u, t) - p\dot{x}]dt$$
$$\equiv \int_0^T (H - p\dot{x})\, dt \tag{70}$$

Denoting $H^* \equiv H(x^*, u^*, p, t)$ and $H \equiv H(x, u, p, t)$ where x^*, u^* are respectively the optimal n-state vector and r-control vector and A^* the resulting Hamiltonian and x, u, $H(x, u, p, t)$ any other admissible ones.

The concavity (convexity) of H in x and u implies

$$\Delta J_a(u) = \int_0^T [(H^* - p\dot{x}^*) - (H - p\dot{x})]\, dt$$
$$= \int_0^T [\, H^* - H - p(\dot{x}^* - \dot{x})]\, dt$$
$$= \int \quad H^* - H + \dot{p}(x^* - x)]\, dt$$

$\geq$ 0 for concavity

$\leq$ 0 for convexity

where the third line was obtained by integrating $\int p(\dot{x} - \dot{x}^*)\, dt$ by parts and using the fixed end points conditions $x(0) = x_0$, $x(T) = x_T$; the fourth line, by using the necessary conditions $\dot{p} = - H_x^*$ and $H_u^* = 0$ in $(u - u^*)\, H_u^*$, not shown. Finally ≥ 0 (and $\leq$) follow from the concavity (convexity) of H. Note also that when H is concave, the maximum is unique, and when it is convex, the minimum is unique.

For further details covering equality and inequality constraints and corners in the case of concave H, see Seierstad and Sydsaeter (1977), for example. The results obtained from the Calculus of Variations thus remain valid.

Example 5.5.2

In the optimal growth model above (Example 3.5.2) where per capita utility $u(c)$ and production $f(k)$ are both concave functions and hence the current Hamiltonian $H \equiv u(c) + q[f(k) - \lambda k - c]$ is concave in c and k, application of the above shows that the first order condition $H_c = 0$ giving $u'(c^*) = q$ is also sufficient for the programme (c^*, k^*) to be optimal.

This can be seen by writing out fully the relation $\int_0^T [H^* - H - H_k^* (k^* - k)]\, dt \geq 0$ out fully term by term, remembering that $H_k^* = q\, [f'(k^*) - \lambda]$ and denoting $u^* \equiv u(c^*)$, $f^* \equiv f(k^*)$ and $H^* \equiv u^* + q\,(f^* - \lambda k^* - c^*)$

$$\Delta J_a \equiv \int_0^T [u^* + q(f^* - \lambda k^* - c^*) - u - q\,(f - \lambda k - c) - q(f' - \lambda)(k^* - k)]dt$$

$$= \int_0^T \{u^* - u - q\,(c^* - c) + q[f^* - f - (k^* - k)f'(k^*)]\}\, dt$$

$$= \int_0^T \{u^* - u - (c^*-c)u'(c^*) + q[f^* - f -(k^*- k)f'(k^*)]\}\, dt$$

$$\geq 0$$

where the second line is just a re-arrangement of the first, the third line is obtained by use of $u'(c^*) = q$ and the last line follows from the concavity of the utility $u(c)$ and production $f(k)$ functions.

5.6 Some Economic Applications

5.6.1 Dynamic Monopoly

Evans' (1924) Dynamic Monopoly examined in Chapter 3 as an illustration of the Calculus of Variations can be formulated as a Maximum Principle problem by seeking the level of production $x(t)$ which maximizes

$$\int_0^T [px - C(x)]\, dt$$

such that

$$\dot{p} = - ap + x - b$$

where $C(x) = \alpha x^2 + \beta x + \gamma$

and $\dot{p} = - ap + x - b$ is Evans' dynamic demand function, with $h = 1$ for simplicity. Note that any realistic monopolist would discount his profit function, i.e. $\pi e^{-rt} = [px - C(x)]e^{-rt}$ where r is some positive discount rate. However, we are not discussing this issue here and assuming the absence of r does not cause any mathematical problem. Taking the production rate $x(t)$ as the control and price $p(t)$ as the state variable respectively, the Hamiltonian is

$$H \equiv px - \alpha x^2 - \beta x - \gamma + \lambda\,(- ap + x - b)$$

$H_x = 0$ gives $p - 2\alpha x - \beta + \lambda = 0$

or $$x = \frac{1}{2\alpha}(p - \beta + \lambda)$$

$$\dot{\lambda} = - H_p = - x + a\lambda\ .$$

Substituting for x, putting $z \equiv \begin{pmatrix} p \\ \lambda \end{pmatrix}$, $c \equiv \begin{pmatrix} c_1 \\ c_2 \end{pmatrix} \equiv - \begin{pmatrix} \beta/2\alpha + b \\ \beta/2\alpha \end{pmatrix}$

$$\dot{z} = Az + c$$

where $$A \equiv \begin{bmatrix} a_{11} & a_{12} \\ a_{21} & a_{22} \end{bmatrix} \equiv \begin{bmatrix} a_{11} & a_{12} \\ -a_{12} & -a_{11} \end{bmatrix} \equiv \begin{bmatrix} (\frac{1}{2\alpha} - a) & \frac{1}{2\alpha} \\ -\frac{1}{2\alpha} & (a - \frac{1}{2\alpha}) \end{bmatrix}$$

whose solution is

$$z(t) = e^{At}\xi - A^{-1}c$$

where ξ is an arbitrary constant vector to be determined by boundary conditions.

Note that the same result could be obtained by differentiating the dynamic system $\dot{p} = -ap + x - b$ with respect to time, which gives, after elimination of x and and grouping appropriate constants as α_{11} and b_1,

$$\ddot{p} - \alpha_{11}p + b_1 = 0$$

whose solution is the same as Evans'. For more details see Simaan and Takayama (1976).

5.6.2 Optimal Growth

The Optimal Growth problem discussed in Chapter 3 (see 3.5.2) could be formulated with all as an optimal control model by choosing a rate of per capita consumption $c(t)$ which satisfies the fundamental neoclassical growth law

$$\dot{k} = f(k) - \lambda k - c \tag{71}$$

and which maximizes

$$\int_0^T u(c)e^{-\delta t}\, dt \tag{72}$$

where, as in Chapter 3, δ is some positive discount rate, and $\lambda \equiv n + \mu$ = constant population growth rate n plus a constant depreciation rate μ. The Hamiltonian is

$$H = e^{-\delta t}\{u(c) + q[f(k) - \lambda k - c]\} \tag{73}$$

where q is the current imputed value of additional capital stock per worker, measured in terms of utility. In terms of our standard notation in this chapter, $qe^{-\delta t} \equiv p(t)$ where $p(t)$ is the co-state variable which in this context means the discounted imputed value of additional investment.

For an interior maximum, $H_c = 0$ gives

$$u'(c) = q \tag{74}$$

$\dot{p} = -H_k$ gives

$$\dot{q} = -[f'(k) - \lambda - \delta]q \; . \tag{75}$$

This, together with (71) gives the behaviour of the model the solution of which could be analysed with the help of a phase diagram in the qk plane. Alternatively $\dot{q}$ in (75) could be changed into $\dot{c}$ and solved with the help of a phase diagram in the ck plane as in Chapter 3. This is done by logarithmic differentiation of (74) and using (75) i.e.

$$\frac{u''(c)}{u'(c)}\dot{c} = \frac{\dot{q}}{q} = -[f'(k) - \lambda - \delta]$$

or, defining elasticity of marginal utility as $\sigma(c) \equiv \frac{c}{u'}\frac{du'}{dc} = -\frac{cu''}{u'}$

$$\dot{c} = \frac{1}{\sigma(c)}[f'(k) - (\lambda + \delta)]c \tag{76}$$

and the results are exactly the same as those obtained in Chapter 3 by use of the Calculus of Variations (see Fig. 5.1 below).

While a more thorough analysis of optimal economic growth is postponed to a later chapter, it can be seen that the solution of system (71) and (76) as given in Fig. 5.1a and 5.1b is identical to the solution obtained by use of the Calculus of Variations (see Ch. 3 Fig. 3.13). The solution of system (71) and (75) is given in Fig. 5.1a and 5.1c. Both methods lead to the same results, namely k^* is the long run equilibrium capital-labour ratio toward which the system moves from the NE or SW of (k^*, c^*) in Fig. 5.1b and from the NW or SE of (k^*, q^*) in Fig. 5.1c. These are the stable branches: from any initial

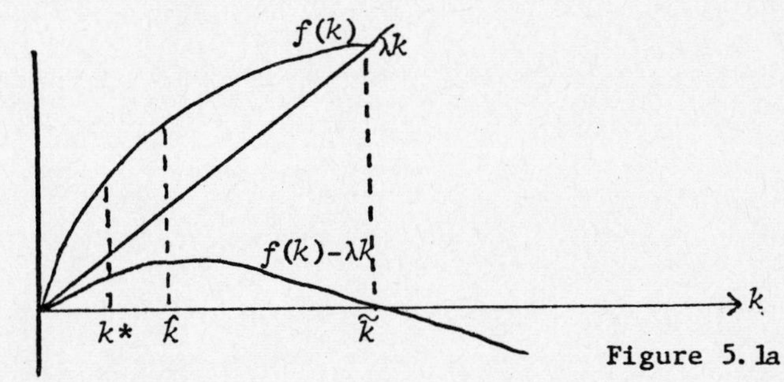

Figure 5.1a

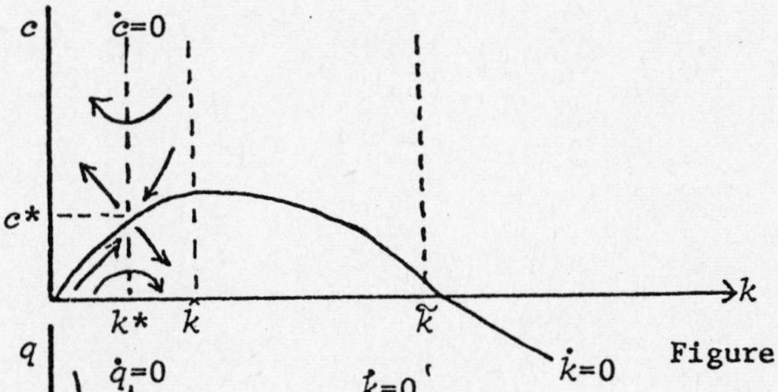

Figure 5.1b

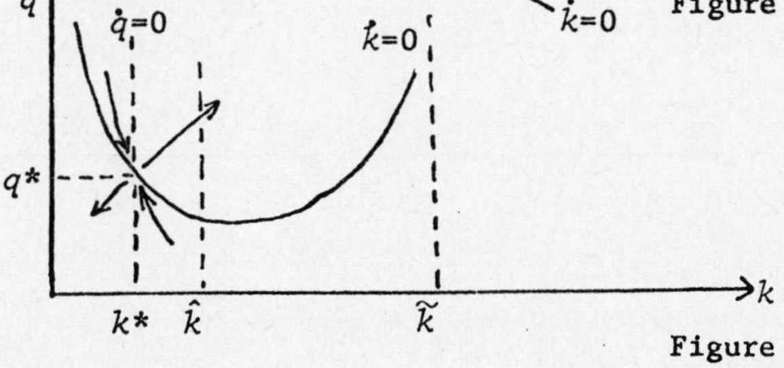

Figure 5.1c

capital per worker $k(0)$ level, an appropriate consumption (in Fig. 5.1b) and imputed price of investment (in Fig. 5.1c) can be chosen such that the system will move to the long run equilibrium k^* with the associated consumption level c^* and investment valuation q^*. Note that unless consumption utility saturates at the horizon ($T = \infty$), the Arrow-Kurz Transversality condition $p(\infty) \geq 0$, $k^*(\infty) \geq 0$ and $\lim_{t\to\infty} p(t)k^*(t) = 0$ $= \lim_{t\to\infty} u'(c)\, e^{-rt}\, k^*(t) = 0$ does not hold, but $\lim_{t\to\infty} p(t)\, \delta k^*(t) \geq 0$ remains valid.

5.6.3 Non-Renewable Resources

As an illustration, let us show how the problem of intergenerational distribution of non-renewable resources mentioned in Chapter 3 (Example 3.4.1) could be formulated as an Optimal Control problem.

Consider a society endowed with a known, finite and non-renewable stock s of some natural resource. Assuming the resource is essential for the production of all goods and services entering the human happiness function, we can consider this utility function as a function of the quantity $q(t)$ of the resource exploited at time t and maximize

$$\int_0^T u(q)e^{-rt}\, dt$$

for $q(t)$ satisfying

$$\int_0^T q(t)dt = s$$

where r is some constant positive rate of future discount and $u(q)$ is assumed concave increasing, i.e. $u''(q) < 0 < u'(q)$ so that the law of diminishing marginal utility is respected.

This is a typical isoperimetric problem of Queen Dido, discussed in Chapter 3.

Define the stock remaining unexploited at time t as $x(t)$ where

$$x(t) \equiv s - \int_0^t q(\tau)\, d\tau$$

with

$$x(0) = s \text{ and } x(T) = 0$$

so that

$$\dot{x}(t) = -\, q(t) \text{ with } x(0) = s;\ x(T) = 0.$$

The Hamiltonian is

$$H \equiv u(q)e^{-rt} - \lambda q(t)\ .$$

The optimal programme is given by

$$H_q = e^{-rt}\, u'(q) - \lambda = 0$$

or

$$\lambda(t) = e^{-rt}\, u'(q)$$

$$\dot{\lambda} = -\, H_x = 0\ , \quad \text{i.e. } \lambda(t) = \text{constant.}$$

This means $e^{-rt}\, u'(q)$ is constant. But e^{-rt} is an exponentially decreasing function of time and in order to preserve the constancy of $e^{-rt}\, u'(q)$, it can be seen that $u'(q)$ must be a compensating exponential function of time. In view of the law of diminishing marginal utility, i.e. $u''(q) < 0 < u'(q)$, this means the optimal quantity $q^*(t)$ exploited should decrease exponentially over time. For example, if

$$u(q) = \frac{q^{1-v}}{1-v} \quad (v > 0,\ v \neq 1)\ .$$

This gives $u'(q)/u''(q) = -\, q/v$, i.e.

$$\dot{\lambda} = 0 = e^{-rt}(\dot{q}u'' - ru') = 0$$

giving

$$\dot{q} = \frac{ru'}{u''} = -\,\frac{r}{v}\, q$$

i.e.

$$q^*(t) = q(0)\, e^{-\frac{r}{v}t}$$

Thus the optimal exploitation of non-renewable resources is exponentially decreasing over time at rate r/v : earlier generations should exploit and consume more non-renewable resources than later generations. It must be pointed out that when $r = 0$, i.e. no future discount is used, q^* is a constant over time: all generations are treated equally.

5.6.4 Optimal Population

The problem of optimal population is an old one but it has not received a rigorous analysis until lately (Meade 1966, 1968, Dasgupta 1969 Sato and Davis 1971, Pitchford 1974, Lane 1975). It could be formulated as an Optimal Control problem. The objective is to maximize the present value of consumption utility

$$\int_0^\infty e^{-\rho t} L\, u(c)\, dt$$

subject to

$$\dot{K} = F(K, L) - C$$

or

$$\dot{K} = L[f(k) - c] \quad , \quad k(0) = k_0$$

where $F(K, L)$ is the neo-classical well-behaved production function, assumed homogeneous of degree one in both capital K and labour L i.e. $F(K, L) = LF(K/L, 1) \equiv Lf(k)$ where $k \equiv K/L$ is capital per worker and $c \equiv C/L$ = per capita consumption. With the usual assumptions of diminishing returns ($f''(k) < 0 < f'(k)$) and diminishing marginal utility of per capita consumption ($u''(c) < 0 < u'(c)$), the dynamic law of capital formation above, and c, L as the control variables and k the state variable, the current value Hamiltonian is

$$H \equiv L\, u(c) + pL[f(k) - c]$$

Pontryagin's theorem gives

$$H_c = u'(c) - p = 0 \qquad (H_{cc} = u''(c) < 0)$$

$$H_L = u(c) + p[f(k) - k\,f'(k) - c] = 0 \quad \left(H_{LL} = \frac{pk^2}{L} f'' < 0\right)$$

$$\dot{p} = [\rho - f'(k)]p$$

$$\lim_{t\to\infty} e^{-\rho t}\, p(t)\, k(t) = 0$$

The results have their usual meaning : $u'(c) = p$ is the Ramsey's rule that marginal utility of consumption should be equal to the imputed value of capital used up ; $\dot{p}/p \left(= \frac{\dot{c}u''(c)}{u'(c)}\right) = p - f'(k)$ says that the rate of decrease in marginal utility $(- \dot{p}/p)$ should be equal to the net marginal product of capital $(f'(k) - \rho)$. Finally the optimal population criterion is given by $u(c) = p\,[c - (f - kf')]$ which is Meade's rule concerning the marginal cost and benefit to society of an additional member. The benefit is in terms of per-capita consumption utility $u(c)$ and the cost is in terms of the reduction in social welfare caused by an additional member of society when he consumes (c) more than his marginal product $(f - kf')$, the loss being measured by the marginal utility $u'(c)$ or shadow price p.

Note that the Hamiltonian H is concave in L and c hence by Mangasarian's theorem, the necessary conditions are also sufficient (see also Dasgupta 1969). Further analysis could be made by examining the implications for the optimal population programme of a subsistence consumption level c_0 which is useless i.e. at which $u(c_0) = 0$ and below which the solution does not exist: mankind is wiped out by starvation (see Lane 1975).

5.6.5 Optimal phasing of deregulation

Price controls imposed in many markets such as utilities, oil and gas, housing rentals and others have long been known to cause undesirable market distortions and deregulation is welcome in many areas. But deregulation, unless properly carried out, may cause inefficient resource allocation and unnecessary wastages. Pindyck (1982) seeks an optimal phasing policy, concentrating on the efficiency aspect.

When prices have been kept at a level below equilibrium, say at $\overline{p}$ in fig. 5.2 (see Pindyck 1982) the quantity supplied $\overline{q} = \alpha\overline{p} + \alpha_0$ is below market clearing level, a full deregulation will cause a sudden price jump to p_0 and stimulate producers to increase supply and price would move to its long-run equilibrium level $\overline{p}^*$. Alternatively, price

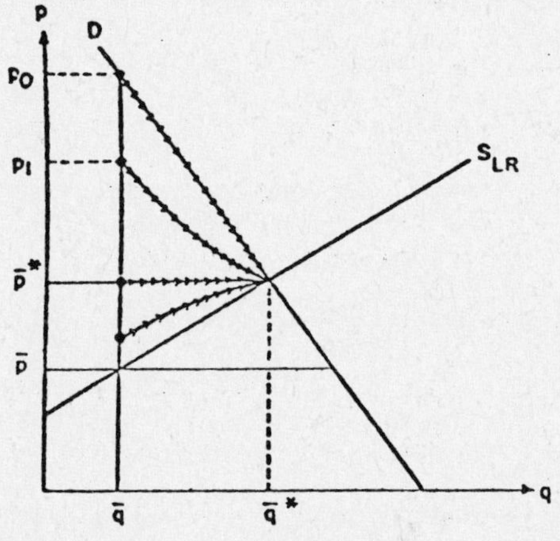

Fig. 5.2 Alternative paths to decontrol

may be allowed to rise more gradually, say to some level below $\overline{p}^*$, or $\overline{p}^*$ itself, or p_1 ($\overline{p}^* < p_1 < p_0$) and eventually be allowed to reach its long-run equilibrium level $\overline{p}^*$.

To facilitate presentation, Pindyck assumes competitive firms with linear production function $q = AK$ where K is capital, quadratic adjustment cost $\frac{1}{2} cI^2$ where I is the rate of sale (or purchase) of Investment at a constant unit price v, and linear demand function $p = a_0 - a_1 q$. The firm's profit function is $\Pi(t) \equiv TR - TC$ is

$$\Pi(t) = pq - C(I) = pAK - vI - \tfrac{1}{2}cI^2$$
$$= a_0 AK - \tfrac{1}{2}a_1 A^2K^2 - vI - \tfrac{1}{2}cI^2$$

and the Investment function is

$$\dot{K} = I - \delta K$$

where δ is the constant depreciation rate. The objective is to maximize

$$\int_0^\infty \Pi(t)\, e^{-rt}\, dt$$

subject to $\dot{K} = I - \delta K$.

The Hamiltonian is

$$H = e^{-rt}\, \Pi(t) + \lambda(I - \delta K)$$

Optimality implies $I = Ap/c(r + \delta) - v/c$

and

$$\dot{q} = A\dot{K} = AI - \delta q = A^2(a_0 - a_1 q)\ /c(r + \delta)\ Av/c$$

or

$$\dot{q} + \alpha q = \beta \quad ; \quad q(0) = \overline{q} \quad ; \quad q^*(t) \rightarrow \overline{q}^* = \beta/\alpha$$

where $\alpha \equiv \delta + a_1 A^2/c(r + \delta)$

$\beta \equiv - [Av/c + a_0 A^2/c(r + \delta)]$

$\overline{q} = A^2\overline{p}/\delta c(r + \delta) - Av/\delta c$, the initial condition

$\overline{q}^* = \beta/\alpha$, the long run equilibrium

With these, the profit function becomes

$$\Pi(p_1 q) \equiv a_0 q - \tfrac{1}{2}a_1 q^2 - A^2p^2/2c(r + \delta)^2 + v^2/\ 2c$$

and the Hamiltonian,

$$H \equiv \Pi(q,\ p)e^{-rt} + \lambda[-\ \delta q + A^2p/c(r + \delta) - Av/c]$$

where output q is the state variable and price p is the control variable.

$$H_p = 0 \Rightarrow \lambda(t) = pe^{-rt}/(r + \delta)$$

Differentiating this and combining with $\dot{\lambda} = -\ H_q$ gives

$$\dot{p} = (r + \delta)(p + a_1 q - a_0)$$

This, together with $\dot{q} = -\ \alpha q + \beta$ above, gives the dynamics of the system. The properties of the solution are best analysed with the help of a phase diagram (see fig. 5.3). Note that the $\dot{q} = 0$ and $\dot{p} = 0$ isoclines give the supply and demand functions, as shown in fig. 5.7. Above (below) the demand curve $\dot{p} = 0$, p increases (decreases) and above (below) the supply curve $\dot{q} = 0$, q increases (decreases). The $\dot{p} = 0$ isocline forms the horizontal manifold and the $\dot{q} = 0$ isocline forms the vertical manifold.

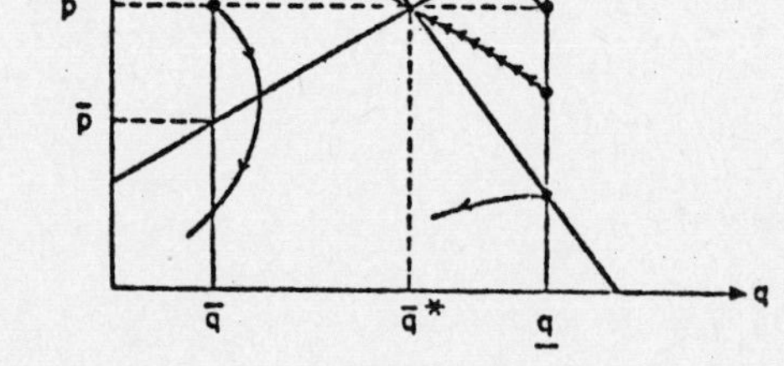

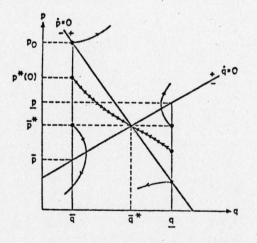

Fig. 5.3 Optimal price policy

Clearly this is a saddle point equilibrium. It is easy to see that deregulation must be such that $p^*(0)$ is on the stable branch if the long run equilibrium $\bar{p}^*$ is to be reached: a full deregulation causing the "wrong" $p(0)$, will cause the system to veer away from equilibrium to the infeasible regions. Thus it is not optimal to deregulate fully all of a sudden.

5.7 Summary and Conclusion

In this chapter, the Optimal Control theory or the Maximum Principle as it is called by Pontryagin and his associates (1962) has been introduced via the Classical Calculus of Variations. The main results have been derived as well as the Transversality Conditions for the various cases, both in the finite and infinite horizons problems. These are summarized in Table 5. Sufficient conditions have also been discussed. Finally, some selected economic applications have been presented mainly to illustrate the various ways the theory has been used in Economics.

The problems of equality and inequality constraints will be discussed in the next chapter.

Footnotes to Ch. 5

1. The sufficient condition $p(t).[x(t) - x^*(t)] \geq 0$ for sufficiently large t and concave $H \equiv f_0(x, u, t) + p(t).f(x, u, t)$ could be shown more directly by

$$H^* + H^*_x (x - x^*) \geq H \text{ by the concavity of } H, \text{ or using } \dot{p} = - H^*_x,$$

$$H^* - \dot{p}(x - x^*) \geq H \text{ which by the definition of } H, \text{ is}$$

$$f^*_0 - f_0 \geq p(\dot{x} - \dot{x}^*) + \dot{p}(x - x^*) \equiv \frac{d}{dt}(p)(x - x^*)$$

Integrating both sides gives

$$\int_0^T f_0 (x^*, u^*, t)dt - \int_0^T f_0 (x, u, t)dt \geqq p(t)[x(t) - x^*(t)]\Big|_0^T$$

i.e.

$$\int_0^T \left(f^*_0 - f_0\right) dt \geq p(T) [x(T) - x^*(T)]$$

in view of $x(0) = x_0$. This holds for all T sufficiently large, and in particular for $T = \infty$. (See Arrow & Kurz 1970.)

2. See, for example, Athans and Falb 1966 p. 262.

3. For further details, see Athans and Falb 1966 pp. 262-270.

CHAPTER 6

CONSTRAINED OPTIMAL CONTROL PROBLEMS

6.1 Introduction

In Chapter 5, we have shown how Optimal Control could be developed from the Calculus of Variations, presented Pontryagin's Maximum Principle and discussed the Transversality conditions for the various cases, both in finite and infinite horizon problems. The second variations and sufficiency conditions have also been examined. In this chapter, we shall continue the exposition of the Optimal Control theory, concentrating on the point, differential equation and isoperimetric equality constraints as well as the control and state variables inequality constraints. Some economic applications will be presented. Finally, Dynamic Programming, Hamilton Jacobi equation and Euler equations will be briefly related to one another. We shall continue to use the results obtained from the Calculus of Variations to save lengthy discussions.

6.2 Optimal Control with Equality Constraints

The presence of equality constraints in Optimal Control problems is handled in much the same way as in the Calculus of Variations. This section will consequently be brief.

Let us consider the problem of maximizing

$$J(x) = \int_0^T f_0 \, (x, u, t) \, dt \tag{1}$$

subject to

$$\dot{x}_i = f(x, u, t) \qquad (1 \leq i \leq n) \tag{2}$$

$$\psi_i \, (x, u, t) = 0 \qquad (1 \leq i \leq q < r) \tag{3}$$

and $$I_i(x) \equiv \int_0^T \varphi_i\ x,\ u,\ t\ \ dt - \ell_i = 0 \qquad (1 \leq i \leq m) \qquad (4)$$

where ℓ_i = constant for all i ($i = 1, 2, \ldots, m$)

f_0, f_i and φ_i are all assumed well behaved

and x is an n-vector of state variables

u is an r-vector of control variables.

This is an isoperimetric and point constraint problem (see Ch.2, section 2.5). As before, let us define

$$y_i(t) \equiv \int_0^t \varphi_i(x, u, t)\ dt \qquad (1 \leq i \leq m) \qquad (5)$$

i.s. $$y_i(0) = 0\ ;\ y_i(T) = \ell_i \qquad (6)$$

$$\dot{y}_i(t) = \varphi_i(x, u, t) \qquad (1 \leq i \leq m) \qquad (7)$$

The m isoperimetric constraint equations (4) have thus been transformed into the m differential equation constraints in (7). In addition, the optimal trajectory must satisfy the q point constraints

$$\psi_i(x, u, t) = 0 \qquad (1 \leq i \leq q < r) \qquad (3)$$

whose Jacobian $[\partial\psi_i/\partial u_j]$ is assumed to have maximum rank q, i.e. the q constraint equations in (3) are assumed to be all independent of one another. The double constrained Hamiltonian $\overline{\overline{H}}$ is

$$\overline{\overline{H}} \equiv f_0\ (x, u, t) + pf(x, u, t) + \lambda\psi(x, u, t) + \mu\psi(x, u, t) \qquad (8)$$

the same analysis leading to theorem 5.1 now gives

Theorem 6.1

If $x^*(t)$ *maximizes (minimizes)* $J(x)$ *in* (87) *subject to* (88), (89) *and* (99) *there exist multipliers* p_0, λ, $p(t)$ *and* $\mu(t)$ *not vanishing simultaneously on* $0 \leq t \leq T$ *and a constrained Hamiltonian* $\overline{\overline{H}}$

$$\bar{\bar{H}}(x, u, p, \mu, t) \equiv f_0 + pf + \lambda\varphi + \mu\psi. \tag{9}$$

such that

(a) p_0, λ_i *are constant* $1 \leq i \leq m$

(b) The n-multiplier vector $p(t)$ *is continuous on each interval of continuity of* $u^*(t)$ *satisfying*

$$\dot{x} = f(x, u, t)$$

$$\dot{p} = -\bar{\bar{H}}_x$$

$$\bar{\bar{H}}_u = 0$$

$$\psi = 0$$

(c) $\bar{\bar{H}}(x, u, p, \mu, t) \leq \bar{\bar{H}}(x, u^*, p, \mu, t)$ *for a maximum*

($\bar{\bar{H}}(x, u, p, \mu, t) \leq \bar{\bar{H}}\, x, u^*, p, \mu, t)$ *for a minimum*)

for all $u^*(t)$ *satisfying* $\psi(x, u^*\, t) = 0$.

(d) The Legendre-Clebsch condition

$$\delta u' \bar{\bar{H}}_{uu} \delta u \begin{cases} \leq 0 & \text{for a maximization problem} \\ \geq 0 & \text{for a minimization problem} \end{cases} \tag{10}$$

holds for all non-zero vector δu *satisfying the inner product*

$$\psi_u \delta u = 0 \tag{11}$$

Remarks

1. If the Legendre-Clebsch condition is satisfied with strict inequality, $\det(H_{uu}) \neq 0$ and the extremal is called non-singular.
2. $\frac{d\bar{\bar{H}}}{dt} = \frac{\partial \bar{\bar{H}}}{\partial t}$ (written as $\dot{\bar{\bar{H}}} = \bar{\bar{H}}_t$) along an extremal (see ch. 3) in all cases and in the particular case in which $\bar{\bar{H}}$ is autonomous, i.e. $\bar{\bar{H}}$ does not depend on time explicitly, $\dot{\bar{\bar{H}}} = 0$ i.e. $\bar{\bar{H}}$ is constant along the extremal. If, furthermore, the terminal time T is unspecified,

$\bar{\bar{H}}(x^*, u^*, p, T)\, \delta T = 0$ implies $\bar{\bar{H}} = 0$ for all $t \in [0, T]$ and not only at T.

3. p_0 is the multiplier associated with f_0. It is a constant since $\dot{\lambda}_0 = - H_{x_0} = 0$, and there is no loss of generality in setting $p_0 = 1$. $\lambda \equiv (\lambda_1, \lambda_2, \ldots, \lambda_m)$ is a constant vector, λ_i $(1 \leq i \leq m)$ being the multipliers associated with the isoperimetric constraints (see Ch. 2).

4. In the case x and u are in open sets, clearly the conditions for a maximum problem are the vanishing of $H_u = 0$ and the negative semi-definiteness of H_{uu}. In the case u and/or x are in closed sets, the more general condition (c) must be used. These will be discussed in the next section.

Example 6.2.1 : Non-Renewable Resources

The Non-renewable resources problem examined in 5.6.3 is an illustration of the differential equation and isoperimetric constraints.

Example 6.2.2 : Capital theory with resources

The problem of capital theory with exhaustible resources in Chapter 3 (see 3.5.3) could easily be formulated as an Optimal Control problem with a number of differential equations constraints as well as isoperimetric constraints. This is left as an exercise for the interested readers.

Example 6.2.3 : Permanent equipment in the extraction industry

As an illustration of the various types of constraints just discussed, consider the problem of increased capital scarcity and hence rental rate as extraction proceeds, investigated by Crémer (1979). The economy has a well behaved production function

$$Q = F(K, R)$$

where Q is product, K is capital, $R = R_1 + R_2$ the two resources, the first has a variable extraction cost $g(z)$ where z is the total amount already exploited, the second, say solar energy, can be supplied in unlimited quantity at a constant cost γ. Let the rate of construction of equipment be G and $h(G)$ the use of resource needed to build it, the problem is to maximize $\int_0^\infty U(C)\, e^{-\delta t} dt$

subject to

$$\dot{K} = F(K, R_1 + R_2) - C - \gamma R_2 - g(z)\, R_1 - h(G)$$

$$\dot{E} = G$$

$$\dot{z} = R_1$$

where F is concave, h is strictly convex and $g'(z) > 0$. (The inequality constraints $E \geq R_1 \geq 0$, $R_2 \geq 0$, $G \geq 0$ are omitted.)

The Hamiltonian is

$$e^{\delta t} H \equiv U(C) + p_1 [F(K, R_1 + R_2) - C - \gamma R_2 - g(z) R_1 - h(G)] + p_2 G + p_3 R_1$$

Pontryagin's theorem gives

$$\dot{p}_1 - \delta p_1 = - e^{\delta t} H_K = - p_1 F_K$$

$$\dot{p}_2 - \delta p_2 = - e^{\delta t} H_E = 0$$

$$\dot{p}_3 - \delta p_3 = - e^{\delta t} H_z = p_1 g'(z) R_1$$

$$H_C = 0 \quad \Rightarrow \quad U'(C) = p_1$$

$$H_{R_1} = 0 \quad \Rightarrow \quad p\,(F_R - g) + p_3 = 0$$

$$H_G = 0 \quad \Rightarrow - p_1 h'(G) + p_2 = 0$$

$$H_{R_2} = 0 \quad \Rightarrow p_1 (F_R - \gamma) = 0$$

These results have their standard equi-marginal economic meaning: marginal utility $U'(c)$ equal to capital's imputed valuation p_1; marginal product, or price, of resource F_R equal to marginal (average) resource cost γ ; marginal utility of G, measured in terms of marginal consumption utility $U'(c)$ (or p_1) equal to its marginal cost i.e. $p_1 h'(=U'(c)h') = p_2$. Finally $H_{R_1} = 0 = H_{R_2}$ together imply $p_1 F_R = p_1 \gamma = p_1 (g - p_3/p_1)$ or $g - p_3/p_1 = \gamma = F_R$ i.e. the net average cost of exploiting resource R_1 equal to γ, the average cost of R_2 which in turn should be equal to the marginal product of resources F_R. For further details, see Crémer (1979) who showed that the above results imply that Hotelling's formula applies only to the price of resources net of the rental of permanent equipment.

6.3 Optimal Control with Inequality Constraints

In this section, we consider the problem of optimal Control with inequality constraints in the control and state vectors. The important implications are that the variations of the state or control variables are no longer free: they must not enter certain forbidden regions of the state space. We shall start with the case of bounded control variables.

6.3.1 Bounded Control Variables

Consider the problem of finding the extremal of

$$\int_0^T f_0 \,(x, u, t)\, dt \tag{12}$$

subject to the usual n-differential equation system

$$\dot{x} = f(x, u, t) \tag{13}$$

where the r-vector u satisfies

$$g_i \,(x, u, t) \geq 0 \qquad (1 \leq i \leq q \leq r < n) \tag{13}$$

where x, $\dot{x}$ are n-vectors.

Note that if q, the number of constraints in (14) is greater than r the number of control variables, there exist no points (x, u, t) which satisfy (14) with more than r constraints active (i.e. holding with strict equalities). In the solution of (14) the rank of matrix g_u is equal to the number of effective constraints. If $q \geq n$, the system depends only on the constraints.

By means of an ingenious device first introduced by Valentine (1937), developed and adopted by McShane (1939), Hestenes (1949, 1965) and Berkovitz (1961) (whom we follow closely in this section, as he has unified the various approaches) this can be transformed into a problem of equality constraints discussed in the last section. This consists in introducing an artifial variable vector $\dot{\xi}$ such that

$$g_i(x, u, t) - \dot{\xi}_i^2 = 0 \qquad (1 \leq i \leq q). \tag{15}$$

Clearly (15) is equivalent to (14) i.e. $\dot{\xi}_i^2$ ensures that g_i is non-negative for each i. With this, the problem is reduced to the classical Bolza variational problem which could be solved by the usual method.

From the Calculus of Variations (multiplier rule, see for example McShane (1939) and Bliss (1946)) we know that the necessary conditions for an extremum is the existence of a constant p_0, an n-vector $p(t)$ and a q-vector $\lambda(t)$ defined on $[0,T]$ such that the multiplier vector (p_0,p,q) never vanishes and a function

$$F \equiv p_0 f_0 + p(f-\dot{x}) + \lambda(g-\dot{\xi}^2) \equiv H + \lambda g - p\dot{x} - \lambda\dot{\xi}^2 \tag{16}$$

such that

(a) The Euler equations hold between corners, i.e.

$$0 = \frac{d}{dt} F_{\dot{x}} - F_x = -\dot{p} - H_x - \lambda g_x$$

$$0 = \frac{d}{dt} F_{\dot{u}} - F_u = 0 - F_u = 0 - H_u$$

$$0 = \frac{d}{dt} F_{\dot{\xi}} - F_{\xi} = -\frac{d}{dt}(2\lambda\dot{\xi}) - 0 \qquad (17)$$

and at corners, the Erdman-Weierstrass conditions hold (see Ch. 3, section 3.3).

(b) The E-Weierstrass conditions are satisfied, i.e. $E \leq 0$ (≥ 0) where, putting $\dot{z} \equiv (u, \dot{x}, \dot{\xi})$ and $\dot{z}^* \equiv (u^*, \dot{x}^*, \dot{\xi}^*)$

$$E \equiv F(x, \dot{z}, t) - F(x, \dot{z}^*, t) - (\dot{z} - \dot{z}^*) F_{\dot{z}^*}$$

Berkovitz (1961) has shown (for details, see Appendix to Ch. 6) that the above conditions are equivalent to the following theorem.

Theorem 6.2

Let $u^*(t)$ *be an admissible control vector which transfers* (x_0, t_0) *to a target* $(x(T), T)$ *while imparting a maximum (minimum) to some functional*

$$J = \int_0^T f_0(x, u, t)\, dt\ .$$

With $\hat{H}$ *defined as* $\hat{H} \equiv p_0 f_0 + pf + \lambda g \equiv H + \lambda g$, *in order for* $u^*(t)$ *to be an optimiser, it is necessary that there exist a constant* p_0, *and n-vector* $p(t)$ *and a q-vector* $\lambda(t)$ *defined and continuous on* $[0, T]$ *except perhaps at corners (where the Erdmann-Weierstrass conditions must hold) such that the vector* (p_0, p) *never vanishes and*

$$\dot{x} = \hat{H}_p \qquad (18)$$

$$\dot{p} = -\hat{H}_x \equiv -(H_x + \lambda g_x) \qquad (19)$$

$$\hat{H}_u = 0 \qquad (20)$$

$\lambda_i \geq 0$ (≤ 0 *for the minimum problem*) $g_i \geq 0$ *and* $\lambda_i g_i = 0$ (21)

Transversality conditions (22) *hold, i.e.* $\hat{H}(T)\delta T + F_z \delta z \big|_T = 0$ (22)

$$\hat{H}(x^*,u,p_0,p) \leq \hat{H}(x^*,u^*,p_0,p) \tag{23}$$

($\geq$ *for the minimum problem*) .

The Clebsch conditions hold, i.e.

$$\delta u\, \hat{H}_{uu}\, \delta u \leq 0 \qquad (\geq 0 \textit{ for the minimum problem}) \tag{24}$$

for all non-zero vectors δu *satisfying the system*

$$g_u\, \delta u = 0 \; . \tag{25}$$

Proof See Appendix to Ch. 6.

Remark 1. (18) through (23) constitute Pontryagin's Maximum Principle.

Remark 2. The general form (23) is applicable to all cases whereas $H_u = 0$ is valid only for the case of unbounded control. In this case $H_u = 0$ and (23) are equivalent to each other.

Remark 3. When the control constraints take the form $m_i \leq u_i \leq M_i$ M_i, m_i are some given numbers (which can be written simply as $|u_i| \leq 1$ by some suitable translation), then for maximization problems

$$\frac{\partial H}{\partial u_i} \begin{cases} \geq 0 & \text{if} \quad u_i^* = M_i \\ = 0 & \text{if} \quad m_i < u_i < M_i \\ \leq 0 & \text{if} \quad u_i^* = m_i \end{cases} \tag{26}$$

and the inequalities are reversed for minimization problems. The constraints g are now

$$\begin{aligned} g_i &\equiv M_i - u_i \quad (i=1,2,\ldots,r) \\ g_j &= u_j - m_j \quad (j=r+1,\ldots,2r) \; . \end{aligned} \tag{27}$$

In the scalar case, for example, where $|u| \leq 1$, these g_i are simply $g_1 \equiv 1 - u$, $g_2 \equiv u + 1$ and $g_i - \xi_i^2 = 0$ are

$$1 - u - \dot{\xi}_1^2 = 0$$

$$u + 1 - \dot{\xi}_2^2 = 0 \,. \tag{27'}$$

Remark 4. Instead of Valentine's device $g_i - \dot{\xi}^2 = 0$, we can use $g_i - \xi_i^2 = 0$, $\forall i$, and achieve the same results.

Remark 5. It is also possible to group $g_i - \xi_i^2$ $(1 \le i \le 2r)$ into pairs as

$$(M_i - u_i)(u_i - m_i) - \alpha_i^2 = 0 \quad (1 \le i \le r)$$

where $\alpha_i^2 \equiv \xi_i^2 \ \xi_{i+r}^2$. For the case of scalar control u, where $|u| \le 1$ for example, this is simply $(1-u)(u+1) = \alpha^2$.

Example 6.3.1

To illustrate the optimization problem with inequality constraints in the control variables and facilitate a comparison with the unconstrained case, let us examine the problem of minimizing

$$J = \frac{1}{2}\,[Tx_2(T) - x_1(T)] + \frac{1}{2}\int_0^T u^2(t)\,dt$$

subject to

$$\dot{x}_1 = x_2 \quad ; \quad x_1(0) = 1\,, \quad x_1(T) \text{ free}$$

$$\dot{x}_2 = -\,u \;; \quad x_2(0) = 2\,, \quad x_2(T) \text{ free}\,.$$

and Case (i) u is unbounded; Case (ii) $|u| \le 1$.

Case (i): Unbounded control $u(t)$

The Hamiltonian of the problem is

$$H = \frac{1}{2}\,u^2 + p_1 x_2 - p_2 u$$

$$H_u = 0 \;\Rightarrow\; u = p_2(t)$$

$\dot{p}_1 = -\,H_{x_1} = 0$, i.e. p_1 is a constant which in view of $p_1(T) = -\frac{1}{2}$ implies

$$p_1 = -\frac{1}{2}$$

$\dot{p}_2 = -H_{x_2} = -p_1$ which, in view of $p_2(T) = \frac{1}{2}T$ implies

$$p_2(t) = \frac{1}{2}t + c_1 = \frac{1}{2}t \ .$$

Hence $u^*(t) = \frac{1}{2}t$

$\dot{x}_2 = -u$ gives

$$x_2^*(t) = -\frac{1}{4}t^2 + 2$$

$\dot{x}_1 = x_2$ gives

$$x_1^*(t) = -\frac{1}{12}t^3 + 2t + 1 \ .$$

<u>Case (ii)</u>: $|u| \leq 1$

Writing the inequality constraint $-1 < u < 1$ as $g_1 \equiv 1 - u \geq 0$ and $g_2 \equiv u + 1 \geq 0$ as in (28), we have as in (16),

$$F \equiv \frac{1}{2}u^2 + p_1x_2 - p_2u - p_1\dot{x}_1 - p_2\dot{x}_2 + \lambda_1(1-u-\dot{\xi}_1^2) + \lambda_2(u+1-\dot{\xi}_2^2)$$

$$\equiv \hat{H} - \sum_1^2 p_i\dot{x}_i - \sum_1^2 \lambda_i\dot{\xi}_i{}^2$$

where $\hat{H} \equiv H + \lambda g \equiv \frac{1}{2}u^2 + p_1x_2 - p_2u + \lambda_1(1-u) + \lambda_2(u+1)$

$$H \equiv \frac{1}{2}u^2 + p_1x_2 - p_2u$$

(11) or (18) through (21) give

$$\frac{d}{dt}F_{\dot{x}_1} - F_{x_1} = -\dot{p}_1 - \hat{H}_{x_1} = -\dot{p}_1 - 0 = 0 \Rightarrow p_1 = \text{const} = -\frac{1}{2} \text{ as in (6.3).}$$

Similarly $\qquad \dot{p}_2 = -p_1$ or

$$p_2(t) = \frac{1}{2}t \quad \text{as in case (i)}$$

$$\frac{d}{dt}F_{\dot{\xi}_i} - F_{\xi_i} = -2\frac{d}{dt}(\lambda_i\dot{\xi}_i) - 0 = 0 \quad (i=1,2)$$

which in view of transversality conditions $H_{\dot{\xi}_i}|_T = 0$, give (see Appendix to Ch. 6)

$$\lambda_i g_i = 0 \quad (i=1,2) \tag{28}$$

$$\frac{d}{dt} F_{\dot{u}} - F_u = 0 = \hat{H}_u = 0 = u - p_2 - \lambda_1 + \lambda_2$$

$$\text{i.e.} \quad u^* = p_2 + \lambda_1 - \lambda_2$$

$F_{uu} = \hat{H}_{uu} = 1$ and the Clebsch conditions are satisfied, i.e. (24) and (25) hold with strict inequality

$$\delta u \, H_{uu} \, \delta u = (\delta u)^2 > 0$$

for all $\delta u \neq 0$ satisfying

$$g_u \, \delta u = (-1)\delta u + (1)\delta u = 0 \, .$$

Weierstrass-E conditions are also satisfied, as can be verified, remembering that $g - \dot{\xi}^2 = 0$ and that Weierstrass conditions are valid only for the unbounded and/or interior optimization case (where $u^* = p_2$) (see, for example, Pontryagin (1962), p. 239).

$$E \equiv F(x,u,\dot{x},t) - F(x,u^*,\dot{x}^*,t) - \sum_1^2 (\dot{x}_i - \dot{x}_i^*) F_{x_i}$$

$$= \frac{1}{2} (u^2 - u^{*2}) - p_2 \, (u - u^*)$$

after cancellation of terms. Substitution of $u^* = p_2$ gives

$$E \equiv \frac{1}{2} (u - u^*)^2 > 0 \, .$$

Thus, a minimum has been obtained.

From $\hat{H}_u = 0$, we have

$$u^*(t) = p_2(t) + \lambda_1(t) - \lambda_2(t) \, .$$

For the case of interior optimization, $g_i > 0$ $(i=1,2)$ and hence $\lambda_1 = 0 = \lambda_2$ by (18), the optimal control is, as in the unconstrained case

$$u^*(t) = p_2(t) \, .$$

For the case of boundary optimization,

(i) If $g_1 = 0$, i.e. $u = 1$ $(= u_{max})$, then clearly $g_2 = u + 1 = 2 > 0$ and by (18), $\lambda_2 = 0$, i.e.

$$u^*(t) = p_2(t) + \lambda_1(t) \leq p_2(t) \text{ since } \lambda_1(t) \leq 0$$

(ii) If $g_2 = 0$, i.e. $u = -1$ $(= u_{min})$, $g_1 = 2 > 0$, hence $\lambda_1 = 0$ and

$$u^*(t) = p_2(t) - \lambda_2(t) \geq p_2(t) \text{ since } \lambda_2(t) \leq 0$$

Thus $u^*(t) = sat\ (-1, 1, p_2) \equiv \begin{cases} -1 \text{ if } p_2 < -1 \\ p_2 \text{ if } -1 < p_2 < 1 \\ 1 \text{ if } p_2 > 1 \end{cases}$

In other words,

$$u^*(t) = \begin{cases} -1 = p_2(t) - \lambda_2(t) & \text{for } t \leq -2 \\ p_2(t) = \tfrac{1}{2}t & \text{for } -2 < t < 2 \\ 1 = p_2(t) + \lambda_1(t) & \text{for } t \geq 2 \end{cases}$$

(See figs. 6.1 and 6.2)

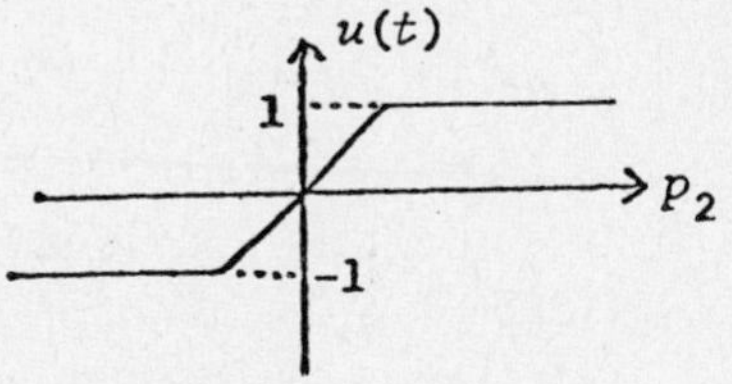

Fig. 6.1 $u = p_2$

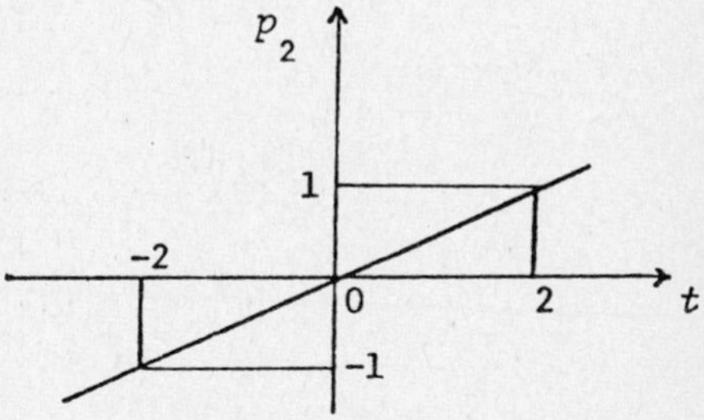

Fig. 6.2 $p_2 = \tfrac{1}{2}t$

It is easy to see from $\hat{H}_u = u - p_2 - \lambda_1 + \lambda_2 = 0$

that $\lambda_1(t) = 0 = \lambda_2(t)$ for $-2 < t < 2$

$\lambda_1(t) = 1 - t/2 \qquad \forall t \geq 2$

$\lambda_2(t) = t/2 + 1 \qquad \forall t \leq -2$

and thus, $\lambda_i(t) \leq 0$ ($i = 1, 2$) for all t.

The state variables are

(1) For $-2 < t < 2$ i.e. $-1 < u < 1$, the interior extremum case, with the initial conditions $x_1(0) = 1$ and $x_2(0) = 2$

$$x_2(t) = -\frac{t^2}{4} + 2$$

$$x_1(t) = -\frac{t^3}{12} + 2t + 1$$

the same as in the unconstrained case above.

(2) For all $t \geq 2$, $u = 1$, $\dot{x}_2 = -u = -1$

$x_2(t) = -t + c_4$, with the "initial condition" from (1) where

$x_2(2) = -t^2/4 + 2 = 1$ i.e. $c_4 = 3$ and

$$x_2(t) = -t + 3$$

$\dot{x}_1 = x_2 = -t + 3$ gives $x_1(t) = -t^2/2 + 3t + c_5$

with the "initial" condition $x_1(2) = 13/3 = -2 + 6 + c_5$

giving $c_5 = 1/3$ and

$$x_1(t) = -t^2/2 + 3t + 1/3$$

(3) For all $t \leq -2$ (extending the analysis to the general case in which $t_0 \neq 0$), $\dot{x}_2 = -u = 1$ giving

$$x_2(t) = t + c_6 = t + 3$$

in view of $x_2(-2) = 1$, and

$x_1(t) = t^2/2 + 3t - 5/3$

in view of the "initial" condition $x_1(-2) = -7/3$

In view of the non-positivity of λ_i $(i = 1, 2)$ (see Appendix to Ch. 6) an alternative way of writing $\hat{H}_u = 0$ is to express it in terms of H_u (cf. $\hat{H} \equiv H + \lambda g$) as follows:

For the minimization problem, provided H is a non-increasing function of u

$$H_u \leq 0 \quad \text{only if } u = u_{max}$$

$$H_u \geq 0 \quad \text{only if } u = u_{min}$$

For the maximization problem, the inequalities are reversed (see Figs. 6.3 and 6.4).

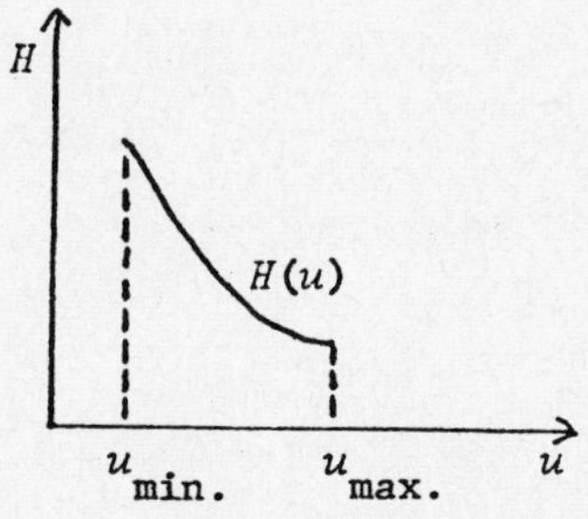

Figure 6.3

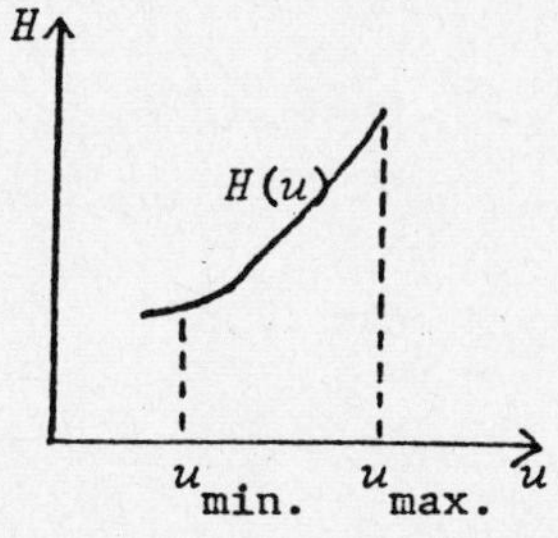

Figure 6.4

It is plain that for given (x^*,p^*), if $H(x^*,p^*,u)$ is a decreasing function of u as in Fig. 6.3, u_{min} should be chosen in a maximization problem (i.e. $H_u \leq 0$) and u_{max} in a minimization problem (i.e. $H_u \leq 0$). When $H(x^*,p^*,u)$ is an increasing function of u, as in Fig. 6.4, u_{min} is the minimizer of H (i.e. $H_u > 0$) and u_{max} is the maximizer (i.e. $H_u > 0$). The equalities elements in $H \geq 0$ or $H \leq 0$ allow for the possibility that the stationary value of function $H(u)$ may happen to take place at the boundary of u, i.e. at u_{max} or u_{min}. This is also Remark 3 of Theorem 6.2 and also holds for the case of linear control, i.e. the case where $H(u)$ is linear in u, to be discussed in the next chapter.

The alternative method of Remark 4 of Theorem 6.2 applied to the same problem gives the same results. In this case

$$F \equiv \frac{1}{2}u^2 + p_1x_2 - p_2u + \lambda_1(1-u-\xi_1^2) + \lambda_2(u+1-\xi_2^2)$$
$$\equiv H + \lambda g - \lambda\xi^2,$$
$$\equiv \hat{H} - \lambda\xi^2$$

$F_u = \hat{H}_u = 0 \Rightarrow u^* = p_2 + \lambda_1 - \lambda_2$ where $\lambda_1, \lambda_2 \leq 0$

$$\dot{x}_i = H_{p_i} \Rightarrow \dot{x}_1 = x_2$$
$$\dot{x}_2 = -u$$

$F_{\xi_i} = 2\lambda_i\xi_i = 0 \Rightarrow \lambda_i g_i = 0 \ (i=1,2)$.

Again, for the case of interior optimization, $g_i > 0$ and $\lambda_1 = 0 = \lambda_2$.

(i) If $g_1 = 0$, i.e. $u = 1$ ($\lambda_1 \neq 0$) then $g_2 > 0 \Rightarrow \lambda_2 = 0$, hence

$$u^*(t) = p_2(t) + \lambda_1(t) \leq p_2 \text{ since } \lambda_1 \leq 0 : H_u \geq 0 .$$

(ii) If $g_2 = 0$, i.e. $u = -1$ ($\lambda_2 \neq 0$), then $g_1 > 0 \Rightarrow \lambda_1 = 0$, hence

$$u^*(t) = p_2(t) - \lambda_2(t) < p_2 \text{ since } \lambda_2 \geq 0 : H_u \leq 0 .$$

Thus, the results are identical.

The second alternative method of Remark 5 of Theorem 6.2 also leads to identical results, as can be shown. In this case, $(1-u)(u+1) - \alpha^2 = 0$ and

$$F \equiv \frac{1}{2}u^2 + p_1x_2 - p_2u + \lambda[(1-u)(u+1) - \alpha^2]$$
$$\equiv \hat{H} - \lambda\alpha^2$$

$$\dot{p}_1 = -\hat{H}_{x_1} = 0$$
$$\dot{p}_2 = -\hat{H}_{x_2} = -p_1$$

$F_u = \hat{H}_u = u - p_2 - 2u\lambda = 0$ which gives

$$u^*(t) = \frac{p_2(t)}{1 - 2\lambda}$$

$F_\alpha = 0 \Rightarrow \lambda\alpha = 0$.

(i) If $\lambda = 0$ then $u^* = p_2(t)$

(ii) If $\lambda \neq 0$, $\alpha = 0 \Rightarrow u \pm 1$ in view of $(1-u)(u+1) - \alpha^2 = 0$ and $\lambda\alpha = 0$: a boundary extremum is obtained.

Economic Application 6.3.1: Permanent Capital in the Resource Industries.

As an application of the various inequalities, let us consider Crémer's model (1979) examined in Example 6.2.3 above with all the inequalities introduced. These are $E - R_1 \geq 0$, $R_1, R_2, G \geq 0$. The Hamiltonian is now

$$e^{\delta t}H = U(C) + p_1[F(K, R_1 + R_2) - C - \gamma R_2 - g(z)R_1 - h(G)]$$

$$+ p_2 G + p_3 R_1 + q_1 (E - R_1) + q_2 R_1 + q_3 R_2 + q_4 G$$

The Maximum Principle gives

$$\dot{p}_1 - \delta p_1 = - e^{\delta t} H_K = - p_1 F_K$$

$$\dot{p}_2 - \delta p_2 = - e^{\delta t} H_E = - q_1$$

$$\dot{p} - \delta p_3 \quad - e^{\delta t} H_z = pg'(z)R_1 = p_1\dot{g}$$

$$H_C = 0 \Rightarrow U'(C) = p_1$$

$$H_{R_1} = 0 \Rightarrow p_1(F_R - g) + p_3 - q_1 + q_2 = 0$$

$$H_G = 0 \Rightarrow -p_1 h' + p_2 + q_4 = 0$$

$$H_{R_2} = 0 \Rightarrow p_1(F_R - \gamma) + q_3 = 0$$

Assuming the first resource, z, is cheaper and alone is first exploited. Then a number of propositions could be proved.

1. If $G > 0$ and $E > R_1$ on $[t_1, t_2]$ then $q_1 = 0 = q_4$ and $\dot{G} > 0$, $H_G = 0$ gives $p_1 h' = p_2$ which, on differentiation and substitution, gives

$$\frac{\dot{p}_1}{p_1} + \frac{\dot{h}'}{h'} = \frac{\dot{p}_2}{p_2}$$

$$\delta - F_K + \frac{\dot{h}''\dot{G}}{h'} = \delta$$

or

$$\dot{G} = (h'/h'')\, F_K > 0$$

since $h(G)$ is strictly increasing convex and $F_K > 0$.

2. At time T when the second resource is first exploited, $q_3 = 0$, $F_R = \gamma$ i.e. the marginal (revenue) product F_R of the second resource is equal to its marginal (average) cost γ.

3. Construction of capacity stops at T_1 before the time T_2 at which the exploitation of resource 1 becomes smaller than capacity. To see this, note that if $T_2 = T_1$ then $q_4(T_2) = 0 = q_1(T_2)$ while $\dot{G} \leq 0$ which contradicts proportion 1.

4. The exploitation of resource 1 stops at the later of the two dates T and T_2 i.e. at T_3 where $T_3 = \max\,(T, T_2)$. To see this, note that after T_3, $q_1 = 0 = q_2 = q_3$, $F_R = \gamma$ and $\partial H/\partial R_1 = 0$ implies

$$p_1(\gamma - g) + p_3 = 0$$

$$\dot{p}_1(\gamma - g) - p_1\dot{g} + \dot{p}_3 = 0$$

which, in view of the relation $\dot{p}_3 = \delta p_3 + p_1\dot{g}$ and $\delta p_3 = -\,\delta p_1\,(\gamma - g)$, $\dot{p}_1 = (\delta - r)p_1$ where $r = F_K$, gives

$$-\,rp_1\,(\gamma - g) = 0$$

Hence $p_3 = 0$ and $p_1\dot{g}$ $(\equiv p_1 g'\dot{z} = p_1 g' R_1) = 0$

which contradicts $g'(z) > 0$.

The various paths can then be examined and it is easy to show that in this model, the Hotelling rule $\dot{p}$ (where $p = F_R$ = the price of the resource in terms of the good) is

$$\dot{p} = r\,(p - g)$$

which applies only in one particular case (the case in which only one resource remains exploited).

This example shows the importance of the inequality constraints and the implications for resource management policy.

6.3.2 Bounded State Variables

Consider the variational problem of finding the extremum of some functional

$$J(x) = \int_{t_0}^{T} f(x, \dot{x}, t)\, dt \tag{29}$$

subject to

$$x(t) \geq \varphi(t) \tag{30}$$

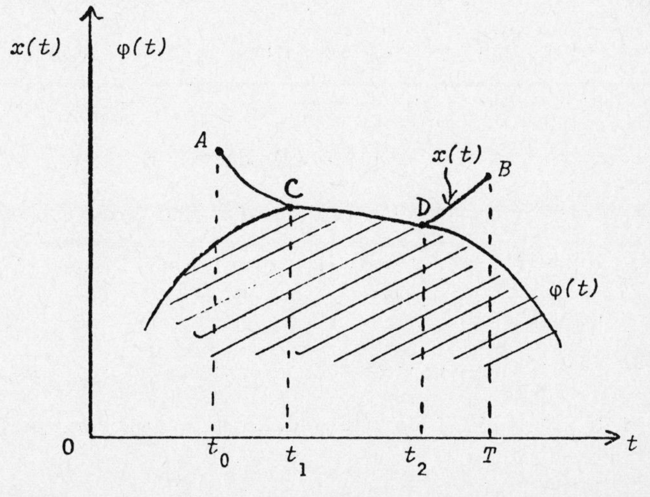

Fig. 6.5 Bounded State variables

As before, the inequality (35) could be turned into an equality by introducing a variable z defined as

$$z^2 \equiv x(t) - \varphi(t) \geq 0 \tag{31}$$

which gives $2z\dot{z} = \dot{x} - \dot{\varphi}$, i.e., $\dot{x} = \dot{\varphi} + 2z\dot{z}$ (or $\dot{x} = 2z\dot{z}$ if $\varphi(t)$ is a constant). Clearly (31) is equivalent to (30). Substituting into (29) gives

$$J = \int_{t_0}^{T} f\left(z^2 + \varphi(t),\ 2z\dot{z} + \dot{\varphi}, t\right) dt \tag{32}$$

The Euler equation gives

$$f_z - \frac{d}{dt} f_{\dot{z}} = 0$$

or

$$2z(f_x - \frac{d}{dt} f_{\dot{x}}) = 0 \tag{33}$$

which means either $z = 0$ in which case $x(t) = \varphi(t)$ and a boundary extremum obtains, or $z \neq 0$ in which case the usual Euler equation holds, i.e.

$$f_x - \frac{d}{dt} f_{\dot{x}} = 0$$

The solution curve thus has some segments on the boundary and some off it. The entry and exit points on this boundary and the corresponding times t_1 and t_2 are determined by the relevant transversality conditions. In Fig. 6.5, assuming A and B are fixed, we can see that C and D are nothing but the variable end points which must move along the curve $\varphi(t)$ and their determination is made by use of transversality conditions (see Chapter 3).

The problem could also be formulated as an optimal control one and solved by the method of the previous section, as follows:

Find the extremum of the functional

$$J(x) = \int_0^T f_0(x,\ u,\ t) \tag{34}$$

subject to

$$\dot{x} = f(x,\ u,\ t) \tag{35}$$

$$g(x) - z^2 = 0 \tag{36}$$

where the scalar function $g(x) \equiv x(t) - \varphi(t)$, i.e., z^2 ensures that $g(x)$ is non-negative.

Thus, (36) is similar to (15) with the exception that $g(x)$ does not involve the control $u(t)$. With this, the problem is much the same as that of section 6.3.1.

More specifically, (16) now becomes

$$\begin{aligned} F &\equiv p_0 f_0 + p(f - x) + \lambda(g - z^2) \\ &\equiv H \quad - p\dot{x} + \lambda(g - z^2) \end{aligned} \tag{37}$$

where $H \equiv f_0 + pf$, the usual Hamiltonian, with $p_0 \equiv 1$.

Euler equations give

$$\frac{d}{dt} F_{\dot{x}} - F_x = \frac{d}{dt}(-p) - H_x - \lambda g_x = 0$$

i.e.,

$$\dot{p} = -(H_x + \lambda g_x) \tag{38}$$

and

$$\frac{d}{dt} F_{\dot{z}} - F_z = 0 + 2\lambda z = 0 \tag{39}$$

i.e., either $z = 0$ in which case $x = \varphi$, i.e., a boundary extremum is reached, or $z \neq 0$, i.e., $g(x) \equiv x(t) - \varphi(t) > 0$ and $\lambda = 0$, i.e., an interior extremum is obtained. Note that with $\lambda = 0$, $\dot{p} = -H_x$ in (38). Thus (38) and (39) spell out the condition (33).

But this is equivalent to setting up, for the problem of finding an extremum of J, subject to $\dot{x} = f(x,\ u,\ t)$ and $g(x,\ t) \geq 0$, the augmented Hamiltonian

$$H = f_0(x, u, t) + p.\ f(x, u, t) + \lambda g(x, t) \tag{40}$$

where λ is the Lagrangean multiplier and $g(x, t)$ is a scalar function, for simplicity. Clearly $\lambda = 0$ for $g > 0$ and $\lambda \neq 0$ for $g \equiv 0$.

The Euler-Lagrange equations are

$$\dot{p} = -H_x \equiv -f_{0x} - p.f_x - \lambda g_x \tag{41}$$

$$H_u = 0 = f_{0u} + p.f_u \tag{42}$$

For the interior extremum or unconstrained portion, i.e., for those periods during which the optimal trajectory lies inside the constraint set (i.e., $\forall\, t \notin [t_1, t_2]$ in fig. 6.5, $g(x, t) > 0$ with strict inequality and $\lambda = 0$, and the Euler-Lagrange equations give

$$\dot{p} = -H_x = -f_{0x} - p.f_x \tag{43}$$

$$H_u = 0$$

i.e., the solution obtained is the same as in the unconstrained case: the constraint being non-binding.

For the boundary extremum portion, i.e., for those periods during which the optimal trajectory lies on the constrained arc, $g(x, t) \equiv 0$ and since $g(x, t)$ vanishes identically on the boundary, $\dot{g} = 0 = \ddot{g} = \dddot{g} = \ldots$ Now $\lambda \neq 0$.

The vector $p(t)$ is, in general, discontinuous at both the entry and exit points (t_1 and t_2 in fig. 7.2.11) of the constraint boundary and the Erdman- Weierstrass conditions must be fulfilled, i.e., for t_i $(i = 1, 2)$ at the entry $(i = 1)$ and exit $(i = 2)$ points,

$$p_c(t_i) = p(t_i) + \lambda g_x|_{t=t_i} \qquad (i = 1, 2)$$

$$H_c(t_i) = H(t_i) - \lambda g_x|_{t=t_i} \tag{44}$$

where p_c, H_c are p and H on the constrained side and

p, H are p and H on the unconstrained side.

For the periods during which the optimal trajectory lies on the boundary, it is clear that $g(t) = 0$ identically, and the purpose of taking repeated time derivatives of $g(t)$, i.e. $d^i g/dt^i = 0$ $(i = 0,1,2,\ldots, k)$ is to bring $u(t)$ out explicitly, i.e.

$$0 = g = \dot{g} = \ddot{g} = \dddot{g} = \ldots. = g^k \qquad (45)$$

where k is the lowest time derivative of g involving the control variable explicitly. For example, first order differentiation gives

$$\begin{aligned} dg/dt \equiv \dot{g} &= g_x \dot{x} + g_t \\ &= g_x f(x, u, t) + g_t \qquad (45') \end{aligned}$$

Since $\dot{g}$ involves $f(x, u, t)$, it may be an explicit function of $u(t)$. If it is not, higher time derivatives may be considered until g^k (kth time derivative of $g(x, t)$) involves $u(t)$ explicitly. This g^k now plays the role of $g(x, t) = 0$ in the determination of $u(t)$ in terms of x and t. Moreover, at entry and exit corners, the following conditions must be met

$$\begin{aligned} g[x(t_i), t_i] &= 0 \\ \dot{g}[x(t_i), t_i] &= 0 \\ &\vdots \\ g^{k-1}[x(t_i), t_i] &= 0 \qquad (i = 1, 2) \qquad (46) \end{aligned}$$

These now play the role of the terminal constraints that at $t_i = t_1$, the unconstrained portion of the system must be on a terminal" manifold $g^i[x(t_1), t_1] \equiv 0$ $(i = 0, 1, \ldots, k-1)$. These are the well known transversality conditions (see Ch.5). And the following Erdman-Weierstrass conditions must also be satisfied

$$p_c(t_i) = p(t_i) + \sum_{i=0}^{k-1} \lambda_i \, \partial g^i/\partial x\Big|_{t=t_i} \qquad (i=1,2)$$

$$H_c(t_i) = H(t_i) - \sum_{i=0}^{k-1} \lambda_i \, \partial g^i/\partial t\Big|_{t=t_i} \qquad (47)$$

Thus, we have $n + 1 + k$ quantities that are determined so as to satisfy $n + 1 + k$ equations.

Economic Application 6.3.2 Optimal investment in physical and human capital

As an economic illustration, consider the problem of optimal allocation of saving between investment in physical (K) and human capital (L) in the neo-classical framework investigated by Dobell and Ho (1967).

The aggregate production function of the economy is

$$Q = F(K, W)$$

with $F(0, W) = 0 = F(K, 0)$

Assuming constant returns to scale, F can be written in terms of output per worker (W) as

$$Q/W = f(k/w)$$

with $f'' < 0 < f'$

where $k \equiv K/L$, $w \equiv W/L$ and L is the total labour force, assumed growing at the constant rate n.

Of the per-capita output $Q/L = wf(k/w)$, a fraction s is invested in physical capital formation, a fraction e, in labour training and the remainder $1 - s - e$ is left for consumption which is to be maximized over the planning horizon. More precisely

$$sQ = \dot{K} + \delta K$$

$$eQ = a(\dot{W} + \mu W)$$

where a = constant average cost of labour training

δ, μ = constant depreciation rate of physical and human capital

Written in per capita terms, the above two equations give the dynamic system

$$\dot{k} = swf(k/w) - (n+\delta)k \text{ , with } k(0) = k_0$$

$$w = (e/a)wf(k/w) - (n+\mu)w \text{ , } w(0) = w_0 \tag{48}$$

The objective is to maximize, subject to the above dynamic system, the functional

$$J = \int_0^T (1-s-e)wf(k/w)\, e^{-\gamma t}\, dt \tag{49}$$

where γ = positive constant rate of discount or rate of interest.

Clearly, the control variables s and e and the state variables w, k are all constrained by the following inequalities

$$\begin{aligned} &0 \le s + e \le 1 \\ &0 \le w \le 1 \\ &0 \le k \end{aligned} \tag{50}$$

This model provides an interesting application of the analysis of inequality constraints discussed above. The various methods discussed in these sections could be used. For example, the Lagrange multipliers λ_i $(1 \le i \le 4)$ and η could be applied to s, e, k, w and $g \equiv 1-w \ge 0$ to ensure each of these remains non-negative. With $f(k/w)$ written as f and $k(t)$, $e(t)$ etc... as k, e to alleviate notations, the Hamiltonian H is

$$H \equiv (1-s-e)wfe^{-\gamma t} + p_1[swf - (n+\delta)k]$$

$$+ (p_2-\eta)\,[(e/a)wf - (n+\mu)w] + \lambda_1 k + \lambda_2 w + \lambda_3 s + \lambda_4 e$$

where $\eta = \begin{cases} 0 & \text{if } w < 1 \\ \geq 0 & \text{if } w = 1 \end{cases}$

Since k and w are always positive in an optimal programme (otherwise $f = 0$ and hence $J = 0$ which is clearly non-optimal), $k \geq 0$ and $w \geq 0$ always hold with inequality and $\lambda_1 = 0 = \lambda_2$, and $\lambda_1 k$, $\lambda_2 w$ could be omitted from H. Similarly $\lambda_3 s$ and $\lambda_4 e$. The authors felt that more insight could be obtained by a direct maximization of H which is now reduced to

$$H = (1-s-e)wfe^{-\gamma t} + p_1[swf-(n+\delta)k] + (p_2-\eta)\,[(e/a)wf-(n+\mu)w] \tag{51}$$

which is $H \equiv f_o + p'f + \eta\dot{g}$

in the general formulation, where $\dot{g}$, the first order inequality constraint (see eq. 45 and 45'), is used since it contains the control e explicitly.

The Euler Lagrange equations are

$$\dot{p}_1 = -H_k = -\,[(1-s-e)e^{-\gamma t} + sp_1 + (e/a)(p_2-\eta)]f' + (n+\delta)p_1$$

$$\dot{p}_2 = -H_w = -\,[(1-s-e)e^{-\gamma t} + sp_1 + (e/a)(p_2-\eta)](f-f'k/w) + (n+\mu)(p_2+\eta) \tag{52}$$

where $\eta = \begin{cases} 0 & \text{if } w<1 \\ \geq 0 & \text{if } w=1 \end{cases}$

Note that the control variables s and e enter the Hamiltonian H linearly. Hence the maximum normally occurs on the boundary of the constraints of the control variables. (See Remark 3 above. For a more detailed analysis of linear control models, see Chapter 7). The maximization rule is given as follows

$$s = \begin{cases} s_{min} = 0 & \text{if } H_s (\equiv \partial H/\partial s) < 0 \\ s_{max} = 1 & \text{if } H_s > 0 \end{cases}$$

$$e = \begin{cases} e_{min} = 0 & \text{if } H_e < 0 \\ e_{max} = 1 & \text{if } H_e > 0 \end{cases}$$

The cases of interior ($w<1$) and boundary ($w=1$) constraint will be examined separately.

1. Interior Extremum Arc: $w<1$, $\eta=0$

$$(H_s, H_e) = (p_1 - e^{-\gamma t}, \; p_2/a - e^{-\gamma t})wf$$

The possible choices are summarised in fig. 6.6.

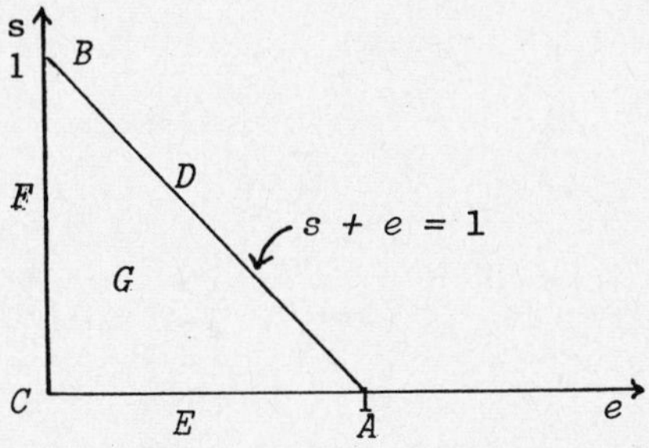

Fig. 6.6 Control space when $w<1$

(i) Case A: $H_e > H_s \Rightarrow p_2/a > p_1$. The optimal choice is $e=1$, $s=0$ which results in the canonical equations

$$\dot{k} = -(n+\delta)k$$

$$\dot{w} = [f/a - (n+\mu)]w$$

$$\dot{p}_1 = -H_k = (f'/a)p_2 + (n+\delta)p_1$$

$$\dot{p}_2 = -H_w = [-\frac{f-f'(k/w)}{a} + (n+\mu)]p_2$$

(ii) Case B: $H_s > H_e \Rightarrow p_1 > p_2/a \Rightarrow s=1$, $e=0$. Substitution into (52) gives the resulting dynamic system.

(iii) Case C: $H_s < 0$, $H_e < 0 \Rightarrow p_1 < e^{-\gamma t}$, $p_2/a < e^{-\gamma t} \Rightarrow s=0=e$. Substitution into (52) gives the corresponding Euler-Lagrange equations.

(iv) Case D: $H_s = H_e > 0 \Rightarrow p_2/a = p_1 \Rightarrow s + e = 1$. This is a singular case (for more detail, see Ch. 7). Differentiating $p_1 - p_2/a \equiv 0$ with respect to time twice gives

$$s = \frac{wf - (\delta-\mu)k}{f\ (w+k/a)}$$

$$e = 1-s = \frac{f(k/a) + (\delta-\mu)k}{f(w + k/a)}$$

from which the corresponding equations of motion are determined. It could be verified that the resulting trajectory is a straight line through the origin of the k-w state space.

(v) Case E: $H_s < 0$, $H_e = 0 \Rightarrow s = 0$, $p_2 = ae^{-\gamma t}$. This is singular in e and differentiating $p_2 - ae^{-\gamma t} \equiv 0$ with respect to time twice gives

$$f - (k/w)f' = (n+\mu+\gamma)a$$

$$e = \frac{a(\mu-\delta)}{f}$$

(vi) Case F: $H_s = 0$, $H_e < 0 \Rightarrow e = 0$, $p_2 \equiv e^{-\gamma t}$. This is singular in s and differentiating $p_1 - e^{-\gamma t} \equiv 0$ twice with respect to time gives

$$f' = (n+\delta+\partial)$$

$$s = \frac{(\delta-\mu)k}{wf}$$

(vii) Case G: $H_s = 0 = H_e \Rightarrow p_1 = e^{-\gamma t} = p_2/a$. This is a double singular case and could not be sustained.

2. Boundary Extremum Arc: $w \equiv 1,\ \eta \geq 0$.

If $w \equiv 1$ for some time, $\dot{w} \equiv 0$. This implies

$$e = (n + \mu)\, a/f \tag{53}$$

Furthermore

$$H_e = [(p_2 - \eta)/a - e^{-\gamma t}]\, wf = 0$$

from which η is determined

$$\eta = p_2 - ae^{-\gamma t} \geq 0$$

$$H_s = (p_1 - e^{-\gamma t})\, wf$$

The values for s is determined according to the sign of H_s. Three more cases are possible (see fig. 6.7).

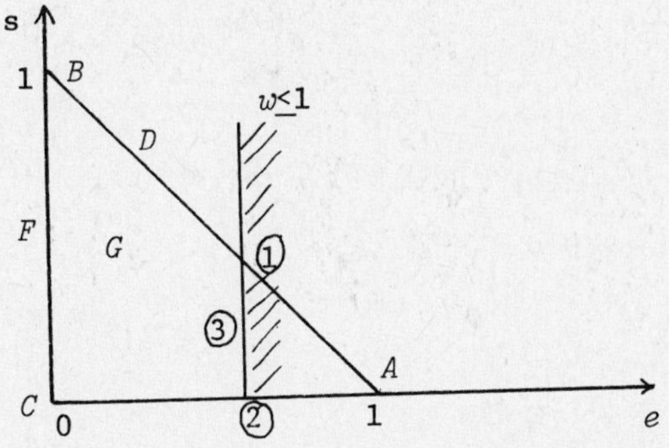

Fig. 6.7 Control space for the case $w \equiv 1$

(viii) Case (1) $H_s > 0 \Rightarrow s = 1 - e = f - (n+\mu)\, a/f$

(ix) Case (2) $H_s < 0 \Rightarrow s = 0$

(x) Case (3) $H_s = 0$. This singular case gives

$$s = (n+\mu)\, k/f$$

This, together with (53), implies $\dot{k} = 0 = \dot{w}$ which is the equilibrium of the system.

Combining fig. 6.6 and 6.7, we obtain a general picture of the solution, in a phase-portrait diagram (see fig. 6.8) which shows not only the directions but also stability of the hajectories.

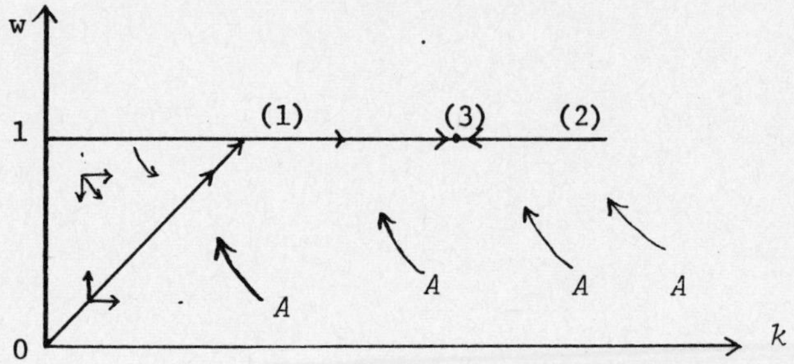

Fig. 6.8 Phase-portrait diagram

It can be seen that (3) is the stable equilibrium of the system: starting from any arbitrary point in the admissible state space, the system moves to (3) as a final resting point.

The sequence of cases that satisfies the given initial conditions, intermediate corner and terminal conditions can easily be established (see eq. 47). They are, at the time (t_1) of entry onto the constraint boundary $w = 1$,

$$w(t_1) = 1$$

$$H(t_1^-) = H(t_1^+)$$

$$p_2(t_1^-) = p_2(t_1^+) + \Pi$$

$$p_1(t_1^-) = p_1(t_1^+)$$

where t_1^- and t_1^+ are the times immediately preceding and following the entry time t_1, and Π is some constant.

This model is thus a natural extension of the optimal economic growth theory to incorporate human capital formation. It is an interesting illustration of the way inequality constraints in both the state and control variables are dealt with. The results are essentially that society should invest so as to attain full employment as quickly as possible. This full employment case is a stable equilibrium towards which all other cases move.

6.4 Dynamic Programming, Hamilton-Jacobi equation and the Euler Equation

Dynamic Programming is a problem of optimizing a sequence of decisions over time. At each stage, a choice of some variables must be made optimally and the state of the system is assumed uniquely determined once such choice has been made. The next stage is determined by a new choice of control variables, taking the final position attained in a period to be the initial conditions for the next stage.

Dynamic Programming was developed by R. Bellman (1957). His optimality Principle states that: "An optimal Policy has the property that whatever the initial state and initial decision are, the remaining decisions must constitute an optimal policy with regard to the state resulting from the final decision".

Example:

A travelling salesman faces the problem of choosing the shortest path from city 1 to say city 6, following the allowable directions given by road regulations and conditions. (See Fig. 6.9). Travelling time and cost being proportional to distances, minimization of mileage amounts to minimization of time and cost.

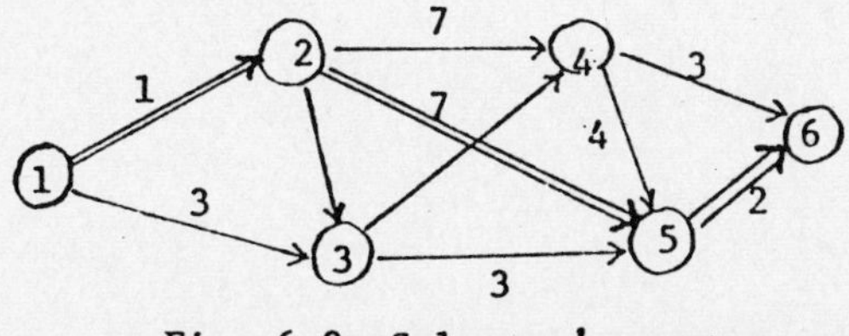

Fig. 6.9 Salesman's routes

Let d_{ij} = the distance between city i and j

V_i = the shortest path to travel from city i to the destination (city 6) along allowable paths

Application of Bellman's Optimality Principle gives

$$V_i = \min (d_{i6})$$

$$= \min (d_{ij} + d_{j6})$$

$$= \min (d_{ij} + V_j) \text{ where } V_6 = 0$$

Starting from destination backward, we have

$$V_6 = 0$$

$$V_5 = \min (d_{56} + V_6) = 2$$

$$V_4 = \min (d_{45} + V_5 , d_{46} + V_6) = \min (4+2, 3+0) = 3$$

$$V_3 = \min (d_{34} + V_4, d_{35} + V_5) = \min (7+3, 3+2) = 5$$

$$V_2 = \min (d_{23} + V_3, d_{24}+V_4, d_{25}+V_5) = \min (2+5, 7+3, 4+2) = 6$$

$$V_1 = \min (d_{12}+ V_2, d_{13} + V_3) = \min (1+6, 3+5) = 7$$

The solution is clearly 1→ 2→ 5→ 6 as indicated by the double arrows in fig. 6.9.

6.4.1 Dynamic Programming and the Hamilton-Jacobi equation

Consider the problem of controlling the dynamic system

$$\dot{x} = f(x,u,t) \tag{54}$$

where $x(t)$ = n - state vector with $x(0)= x_o$ and $x(T)$ free

$u(t)$ = r - control vector, $u(t) \in \Omega\ (u)$

T is fixed but $x(T)$ is unspecified.

The objective is to minimize the cost function

$$J = \int_0^T f_0\ (x,u,t)\ dt \tag{55}$$

Suppose the optimal control and state vectors $u^*(t)$ and $x^*(t)$ have been found, then J is a function of the initial time and state $x(0)$ only. The optimization

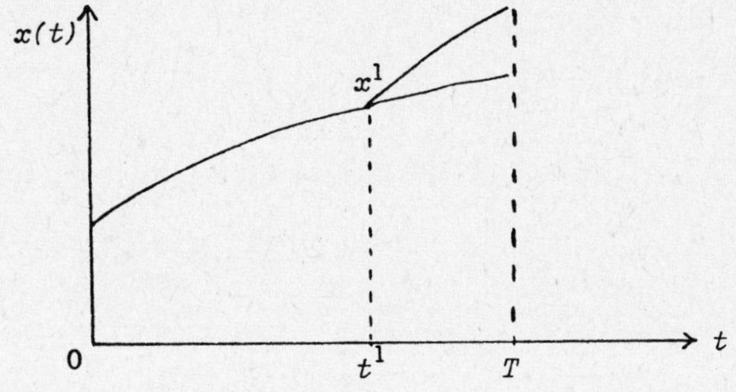

Fig. 6.10 Optimization by stages

is carried out in stages (see fig. 6.10) and the optimality principle implies that no matter how the system arrives at x^1 at the end of stage 1 at $t=t^1$, the subsequent stage must be optimal.

Let us consider only two stages and define

$$V(x,t) \equiv \max_{u \in \Omega(u)} J(x,u)$$

$$= \max_{u \in \Omega(u)} . \left\{ \int_t^{t+\Delta t} f_0(x,u,t)dt + \int_{t+\Delta t}^{T} f_0(x,u,t)dt \right\}$$

$$= \max_{u \in \Omega(u)} . \left\{ f_0(x,u,t)\,\Delta t + V[x(t+\Delta t),\, t+\Delta t] \right\} + 0(\Delta t) \qquad (56)$$

where $\lim_{\Delta t \to 0} \dfrac{0(\Delta t)}{\Delta t} = 0$

Since Δt is small, expansion of the second term on the RHS of (56) gives

$$V(x,t) = \max_{u \in \Omega(u)} \{f_0 (x,u,t) \Delta t + V(x,t) + \frac{\partial V}{\partial t} \Delta t + \frac{\partial V}{\partial x} \cdot \dot{x} \Delta t\} + O(\Delta t) \quad (57)$$

where . indicates dot or inner product. Dividing by Δt and letting $\Delta t \to 0$, we obtain the Hamilton-Jacobi equation

$$\frac{\partial V}{\partial t}(x,t) + \max_{u \in \Omega(u)} \{f_0(x,u,t) + (x,u,t) + \frac{\partial V}{\partial x}(x,t) \cdot f(x,u,t)\} = 0 \quad (58)$$

Defining $p(t) \equiv \partial V(x,t)/\partial x$ which has the economic meaning of the marginal contribution, or shadow price, of $x(t)$, (58) can be written as

$$\frac{\partial V}{\partial t}(x,t) + \max_{u \in \Omega(u)} H(x,p,u,t) = 0 \quad (59)$$

where $H(x,p,u,t) \equiv f_0(x,u,t) + p(t).f(x,u,t)$, the usual Hamiltonian.

Another way to write the Hamiltonian-Jacobi equation is

$$\frac{\partial V}{\partial t}(x,t) + H(x, \frac{\partial V}{\partial x}, t) = 0 \quad (60)$$

If we define

$$V(x_0, 0) \equiv J(x^*,u^*) = \int_0^T f_0 (x^*,u^*,t)dt$$

where x^*,u^* are the optimal values of $x(t)$ and $u(t)$, then clearly

$$V(x(T),T) = 0$$

or, if a "Scrap" function $S[x(T),T]$ is given,

$$V(x(T),T) = S[x(T),T] \quad (61)$$

The Hamilton-Jacobi equation is a partial differential equation which is not easy to solve in general.

6.4.2 Dynamic Programming and Euler Equations

It is possible to derive the Euler equations from the method of Dynamic Programming under the assumptions that $\partial f_0/\partial \dot{x}$ is continuous, x is finite and unbounded.

Let

$$J = \int_0^T f_0 \, (x,u,t)dt$$

the Hamilton-Jacobi equation (58) gives

$$\frac{-\partial V(x,t)}{\partial t} = \max_u . \, \{f_0(x,u,t) + \frac{\partial V}{\partial x} \, . \, f\,(x,u,t)\}$$

$$= \max_u . \, \{f_0(x,u,t) + \frac{\partial V}{\partial x} \, . \, \dot{x}\} \tag{62}$$

But in the Calculus of variations, $u = \dot{x}$, and we have

$$\frac{-\partial V}{\partial x}\,(x,t) = \max_{\dot{x}} . \, [f_0(x,\dot{x},t) + \frac{\partial V}{\partial x} \, . \, \dot{x}] \tag{63}$$

(63) implies two equations

$$-\frac{\partial V(x,t)}{\partial t} = f_0\,(x,\dot{x}^*,t) + \frac{\partial V}{\partial x} \, . \, \dot{x}^* \tag{64}$$

for optimal $\dot{x}^*$, and

$$\frac{\partial f_0}{\partial x_i} + \frac{\partial V(x,t)}{\partial x_i} = 0 \quad (1 \leq i \leq n) \tag{65}$$

Differentiation of (65) with respect to t, using the information $\partial f_0/\partial x_i + \partial V/\partial x_i = 0$, gives

$$\frac{d}{dt}\frac{\partial f_0}{\partial x_i} + \frac{\partial^2 V}{\partial x_i \partial t} + \frac{\partial^2 V}{\partial x_i^2}\dot{x}_i = 0 \tag{66}$$

Differentiation of (64) gives

$$\frac{-\partial^2 V}{\partial x_i \, \partial t} = \frac{\partial f_0}{\partial x_i} + \sum_1^n \frac{\partial^2 V}{\partial x_i^2}\dot{x}_i \tag{67}$$

(66) and (67) together give the Euler equations

$$\frac{d}{dt}\frac{\partial f_0}{\partial x_i} - \frac{\partial f_0}{\partial \dot{x}_i} = 0 \quad (1 \leq i \leq n) \tag{68}$$

From (65), remembering that $\dot{x}$ does not enter $V\,(x,t)$, we obtain, on differentiating with respect to $\dot{x}_i$ and summing,

$$\sum_{i=1}^{n} \sum_{j=1}^{n} \frac{\partial^2 f_0(x,\dot{x},t)}{\partial \dot{x}_i \, \partial \dot{x}_j} \delta x_i \delta x_j \begin{cases} \leq 0 & \text{for a maximum} \\ \geq 0 & \text{for a minimum} \end{cases} \tag{69}$$

which is the Legendre conditions

Also for all $p_i \neq \dot{x}_i$, for $f_0(x,\dot{x},t)$ to take on the absolute maximum, we must have

$$f_0(x,p,t) + \sum_1^n \frac{\partial V}{\partial x_i} p_i \leq f_0(x,\dot{x},t) + \sum_1^n \frac{\partial V}{\partial x_i} \dot{x}_i$$

or

$$\Delta f_0 + \sum_1^n \Delta \dot{x}_i \frac{\partial V}{\partial x_i} \leq 0 \tag{70}$$

where $\Delta f_0 \equiv f_0(x,p,t) - f_0(x,\dot{x},t)$

$\Delta \dot{x}_i \equiv p_i - \dot{x}_i$

But $\dfrac{\partial V}{\partial x_i} = -\dfrac{\partial f_0}{\partial \dot{x}_i}(x,\dot{x},t)$ (85) gives

$$E \equiv \Delta f_0 - \sum_1^n \frac{\partial f_0}{\partial x_i} \Delta \dot{x}_i \begin{cases} \leq 0 & \text{for a maximum} \\ \geq 0 & \text{for a minimum} \end{cases}$$

where $E \equiv E(x,\dot{x},p,t)$ is the Weierstrass E. function (See Ch. 5, eq.21).

6.5 Summary and Conclusion

In this chapter, the problem of equality and inequality constraints in the control and state variables, have been investigated and the various approaches to dynamic optimisation, brought together.

Inequality constraints introduce new difficulties and require new techniques to cope with. No doubt, the inequality problems have been recognized in the 1930's but the classical Calculus of Variations is not suitable to deal with them, especially with the linear case (see ch. 5, section 5.2). The various techniques designed to deal with these inequality problems have been investigated in this chapter and some economic examples involving inequality constraints in both control and state variables has been provided to illustrate the discussion.

Although these techniques are general enough to cover all cases, the particular case of Linear Optimal Control in which the Hamiltonian is linear in the control variables occurs frequently enough in practice to deserve a separate treatment. In general, linear models have neither maxima nor minima unless the (control) variables are in a closed set, in which case, optimization normally, but not always, occurs on the boundaries. This is the subject of bang-bang and singular control techniques which will be discussed in Chapter 7.

APPENDIX TO CHAPTER 6

Proof of Theorem 6.2 (Berkovitz 1961)

Consider the problem of finding the extremal of

$$\int_0^T f_0\,(x,\ u,\ t)\ dt \qquad \text{A.6.1}$$

subject to the n-differential equations

$$\dot{x} = f(x,\ u,\ t) \qquad \text{A.6.2}$$

and the q inequalities

$$g(x,\ u,\ t) \geq 0$$

or

$$g(x,\ u,\ t) - \dot{\xi}^2 = 0 \qquad \text{A.6.3}$$

where x, $\dot{x}$ are n-vectors, u is an r-vector and g is a q-vector $(1 \leq q \leq r \leq n)$.

Define $F \equiv p_0 f_0 + p(f - \dot{x}) - \lambda(g - \dot{\xi}^2)$

$$\equiv H \quad + \lambda g \quad - \quad p\dot{x} \quad - \quad \lambda\dot{\xi}^2 \qquad \text{A.6.4}$$

The Euler equations are

$$0 = \frac{d}{dt} F_{\dot{x}} - F_x = -\dot{p} - H_x - \lambda g_x \qquad \text{A.6.5}$$

$$0 = \frac{d}{dt} F_{\dot{u}} - F_u = 0 - F_u = 0 - H_u \qquad \text{A.6.6}$$

$$0 = \frac{d}{dt} F_{\dot{\xi}} - F_{\xi} = -\frac{d}{dt}(2\lambda\dot{\xi}) - 0 \qquad \text{A.6.7}$$

A.6.7 implies that $F_{\dot{\xi}}$ (or $\lambda\dot{\xi}$) is some constant along an extremal. From the Transversality Condition $F_{\dot{\xi}_i}\ \delta_{\xi_i}\big|_{t=T} = 0$ with unspecified $\xi_i(T)$, it is clear that $F_{\dot{\xi}_i} = 0$ at $t = T$ and hence $F_{\dot{\xi}_i} = 0\ \forall\ t \in (0,\ T)$

i.e. $\lambda_i\dot{\xi}_i = 0\ \forall\ i$. But $\dot{\xi}_i = 0 \Leftrightarrow g_i = 0\ \forall\ i$ and $\dot{\xi}_i \neq 0 \Rightarrow g_i \geq 0$ and $\lambda_i = 0$. Hence, along an extremal, $\lambda_i\dot{\xi}_i = 0$ or

$$\lambda_i\ g_i = 0 \quad \forall\ i \qquad \text{A.6.8}$$

From the vanishing of F_u, $F_{\dot{\xi}}$ and the relation $F_{\dot{x}} = -p$ obtained by differentiating F, we have

$$F - \dot{z}\, F_{\dot{z}} \equiv f - \dot{x}\, F_{\dot{x}} - uF_u - \dot{\xi} F_{\dot{\xi}}$$

$$\equiv H - p\dot{x} \quad - \dot{x}\, F_{\dot{x}} - 0 - 0 \equiv H \qquad \text{A.6.9}$$

where $\dot{z} \equiv (u, \dot{x}, \dot{\xi})$. With this, the Weierstrass E-function is

$$E \equiv F(x, \dot{z}, t) - F(x^*, \dot{z}^*, t) - (\dot{z} - \dot{z}^*)\, F_{\dot{z}^*}$$

$$\equiv H\; x,u,p,t) - H(x,u^*,p,t) \qquad \text{A.6.10}$$

$$\leq 0 \text{ for a maximum}$$

$$\geq 0 \text{ for a minimum}$$

where $\dot{z}^* \equiv (u^*, \dot{x}^*, \dot{\xi}^*)$ and $F_{\dot{z}^*} \equiv \partial F/\partial \dot{z} \equiv \partial F/\partial \dot{z}$ at $\dot{z} = \dot{z}^*$.

Finally, from $F_{\dot{x}} = -p$, we have $F_{\dot{x}\dot{x}} = 0$ and the Legendre-Clebsch condition

$$\delta\dot{x}\; F_{\dot{x}\dot{x}}\; \delta\dot{x} + \delta u\; F_{uu}\; \delta u - 2\delta\dot{\xi}\; \hat{\lambda}\delta\dot{\xi}$$

becomes simply

$$\delta u\; \hat{H}_{uu}\; \delta u - 2\; \delta\dot{\xi}\; \hat{\lambda}\; \delta\dot{\xi} \quad \begin{cases} \leq & \text{for a maximum} \\ \geq & \text{for a minimum} \end{cases} \qquad \text{A.6.11}$$

where $\hat{\lambda} \equiv \text{diag}\,(\lambda_i)$; $F_{uu} = \hat{H}_{uu}$ and the transpose notation ' is omitted, for example $\delta u\; f_{uu}\; \delta u \equiv \delta u'\; F_{uu}\; \delta u$ etc.

Since δu is an arbitrary vector (so long as $\delta z \equiv (\delta u, \delta\dot{x}, \delta\dot{z}) \neq 0$), take $\delta u = 0$, $\delta\dot{x} = 0$ and all but one element of $\delta\dot{\xi}$, say $\delta\dot{\xi}_i$, zero i.e. $\delta\dot{\xi}_i \neq 0$, $\delta\dot{\xi}_j = 0 \;\; \forall\, j \neq i$, then A.6.11 implies

$$2\delta\dot{\xi}\; \hat{\lambda}\; \delta\dot{\xi} = 2\lambda_i\; (\delta\dot{\xi}_i)^2 \geq 0 \quad \text{for a maximum}$$

$$\leq 0 \quad \text{for a minimum}$$

i being arbitrary,

$$\lambda_i \begin{cases} \geq 0 & \forall\, i \text{ for the maximization problem} \\ \leq 0 & \forall\, i \text{ for the minimization problem} \end{cases} \qquad \text{A.6.12}$$

Hence A.6.8 together with A.6.12 imply

$$g_i \geq 0, \ \lambda_i \geq 0 \ (\leq 0) \text{ for a maximum (Minimum)}$$

and

$$\lambda_i \, g_i = 0 \quad (1 \leq i \leq q) \qquad \text{A.6.13}$$

From $g - \dot{\xi}^2 = 0$ we have either $g_i = 0$ and hence $\dot{\xi}_i = 0$ and thus $\lambda_i \, g_i = 0$ or $g_i > 0$ and hence $\dot{\xi}_i \neq 0$ but $\lambda_i \, g_i = 0$ for each i. Hence each term in $2 \, \Sigma_i \, i \, (\delta\dot{\xi}_i)^2$ is zero and the Legendre-Clebsch condition is simply

$$\delta u \, \hat{H}_{uu} \, \delta u \ \leq 0 \ (\geq 0) \text{ for a maximum (minimum)}$$

for all non-zero vectors δu satisfying $g_u \, \delta u = 0$

Finally, the Transversality conditions

$$(F - \dot{z} \, F_{\dot{z}})\big|_T \, \delta T + F_{\dot{z}} \, \delta z\big|_T = 0$$

becomes simply

$$H(T) \, \delta T + F_{\dot{z}} \, \delta z\big|_T = 0 \qquad \text{A.6.14}$$

The above analysis was summarized in Theorem 6.2 which is simple to understand and apply.

CHAPTER 7

LINEAR OPTIMAL CONTROL

7.1 Introduction

In certain problems, control variables enter the Hamiltonian linearly, either via the objective function or the dynamic system or both. This type of problem is called Linear Optimal Control. For example, in the Optimal growth model discussed earlier (see 5.6.2), if the utility function is $u(c) = c$, then the maximization of $\int_0^T ce^{-rt}dt$ subject to $\dot{k} = f(k) - \lambda k - c$ leads to the Hamiltonian $H \equiv (e^{-rt} - p)c + p[f(k) - \lambda k]$ which is linear in the control variable c.

The Hamiltonian in the linear case, could be written as

$$H \equiv \psi(x,p,t) + \sigma(x,p,t)u(t) \tag{1}$$

where $\sigma(x,p,t)$ is the grouping of the coefficients of $u(t)$ called switching function, and $\psi(x,p,t)$ is the collection of all the remaining terms in H not involving $u(t)$. In the optimal growth case above, $\sigma \equiv e^{-rt}-p$ and $\psi \equiv p(f - \lambda k)$. In the case $u = (u_1,\ldots,u_r)$, $\sigma u = \sum_1^r \sigma_i u_i$.

In general, there will be no extremum unless control variables are bounded, in which case they are expected to be at the boundary of the admissible region. This, however, need not always be true as in the degeneracy case.

Let u be bounded, i.e., for all i,

$$m_i \leq u_i \leq M_i \tag{2}$$

where M_i and m_i are respectively the maximum and minimum values u_i can take. If m_i and M_i are constant, by a simple translation, the above can be written as

$$-1 \leq u_i \leq 1 \tag{3}$$

When Pontryagin's Maximum Principle is applied to this type of problem, the optimal control u_i^* is

$$u_i^*(t) = \begin{cases} 1 \text{ (or } M_i) \text{ if } \sigma_i > 0 \\ -1 \text{ (or } m_i) \text{ if } \sigma_i < 0 \end{cases} \tag{4}$$

for the maximization problem and the inequalities are reversed for the minimization problem (see fig. 7.1).

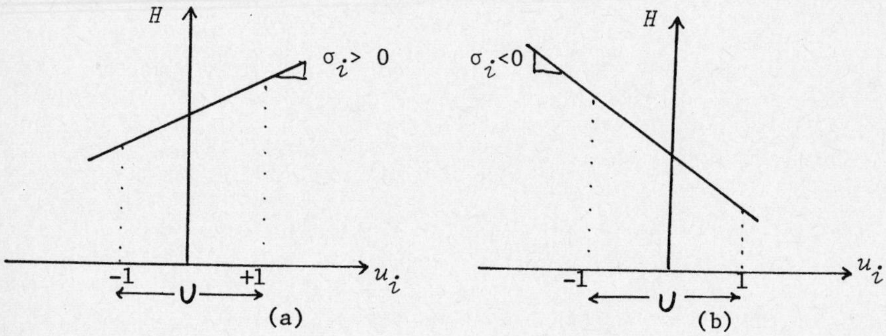

Fig. 7.1 The Switching Function σ_i

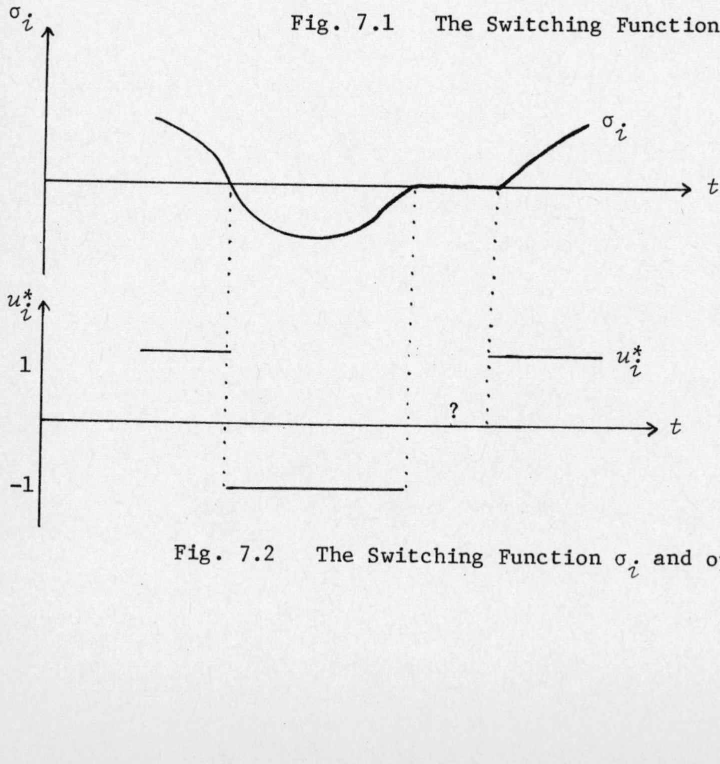

Fig. 7.2 The Switching Function σ_i and optimal u_i^*

Thus, if $u(t)$ appears linearly in H and each component $u_i (1 \leq i \leq r)$ of u is bounded, the optimal control $u_i^*(t)$ is discontinuous: it jumps from a minimum to a maximum and vice versa in response to each change in the sign of σ_i. For this reason, $\sigma(x,p,t)$ is called the switching function and (4) is referred to as bang bang control.

Note that every time the switching function σ_i passes through zero, a switch of some relevant control variable is made either from its minimum to its maximum or vice versa, depending on whether a minimization or maximization problem is involved, the only exception is when $\sigma \equiv 0$ over some non-zero time interval: H is then independent of u, i.e., it cannot be maximized or minimized with respect to u (see fig. 7.2). The Maximum Principle fails to yield a well defined control. This is the case of singular or degenerate control. We shall deal with each case in turn.

7.2 Bang Bang Control and Time Minimum Problem

The time minimum problem consists in devising a control u such as to transfer a system from some given initial state to a specified target (which may be fixed or moving) in minimum time. The Hamiltonian is normally linear in the control variables and the latter are bounded.

The time minimum problem is by no means the only case of Linear Optimal Control, but it is an important one. Although the choice of u^* is dictated by (4) as in every other case, the time minimum problem exhibits additional features which we now discuss.

Consider the following system

$$\dot{x}(t) = Ax(t) + Bu(t) \tag{5}$$

where x is an n-state vector, u an r-control vector

A is nxn and B is nxr constant matrices.

$|u_i| \leq 1$, $(1 \leq i \leq r)$

A control vector u must be chosen such as to transfer the system from its initial state x_0 to the final state $x(T) = 0$ (i.e., target $x(T)$ is set at the origin) in minimum time (i.e., such as to minimize $J = \int_0^T dt$).

The Hamiltonian is (' denotes transposition)

$$H \equiv 1 + p'Ax + p'Bu \tag{6}$$

$$\dot{p} = -H_x = -A'p \tag{7}$$

The switching function $\sigma \equiv p'B \equiv p'[b_1, \ldots, b_n]$ where b_i is column i $(1 \leq i \leq r)$ of matrix B. Since $|u_i| \leq 1 \; \forall i$ and H is linear in u, the control laws are readily found to be

$$u_i^* = \begin{cases} 1 \text{ for } \sigma_i \equiv p'b_i > 0 & : \text{ Bang bang control} \\ -1 \text{ for } \sigma_i \equiv p'b_i < 0 & : \text{ Bang bang control} \\ \text{undetermined for } p'b_i \equiv 0 & : \text{ Singular control} \end{cases} \tag{8}$$

In view of (7), $p(t) = e^{A't}p_0$, it is unlikely that $p = 0$ and hence that $p'b_i \equiv 0$. Thus singular control is ruled out and the remaining possibilities are Bang Bang control.

It is important to note that in time minimum problems, optimal control may or may not exist. If there are no admissable controls $u(t)$ that can bring the system to a target at any time t, then optimal control does not exist. More precisely, if the Reachable set $R(t)$ is defined as the collection of all state values $x(t)$ which could be reached at time t from some given initial state $x(o) = x_0$ when

$x(o)$ is subjected to all admissible controls for the time interval $[o,t]$, then time optimal control does not exist if there are no t for which the reachable set $R(t)$ has at least one point in common with the target set $S(t)$ (i.e., if $R(t) \cap S(t) = \phi \; \forall t$).

In time optimal problems, it is useful to bear in mind the three well known theorems proved by Pontryagin (1962).

Existence Theorem: *If A is a stable matrix (i.e., all the eigen values of A have non positive real parts), then for any point* x_0, *there exists an optimal control which transfers the phase point from* x_0 *to the origin.*

Uniqueness Theorem: *If an optimal control exists, it is unique.*

Switching Theorem: *Suppose the eigen values of the nxn matrix A in* (5) *are all real. Then there exists a unique control vector u. Each* u_i $(1 \leq i \leq r)$, *piecewise constant, takes on only their maximum and minimum values and does not have more than* n-1 *switchings.*

Proof. See Pontryagin (1962) Theorem 11 p. 124 and Theorem 10 p. 120 respectively.

Example 7.2

As an illustration of time optimal control problems, consider the well known double integrator (see, for example, Pontryagin (1962) p. 22) - system $\ddot{x} = u$ and where u is a real control such that $|u| \leq 1$. The system is to be brought from a given initial state x_0 to the origin $(x_1(T), x_2(T) = (0,0))$ as quickly as possible.

The system $\ddot{x} = u$ could be written as a first order differential equation system (see Appendix) by defining $x = x_1$ and

$$\dot{x}_1 = x_2$$

$$\dot{x}_2 = u$$

or in the matrix form of $\dot{x} = Ax + Bu$ as

$$\dot{x} \equiv \begin{bmatrix} \dot{x}_1 \\ \dot{x}_2 \end{bmatrix} = \begin{bmatrix} 0 & 1 \\ 0 & 0 \end{bmatrix} \begin{bmatrix} x_1 \\ x_2 \end{bmatrix} + \begin{bmatrix} 0 \\ 1 \end{bmatrix} u$$

The Hamiltonian is

$$H \equiv 1 + p_1 x_2 + p_2 u$$

Optimal Control as determined by the switching function p_2 is

$$u^* = \begin{cases} -1 & \text{for } p_2 > 0 \\ 1 & \text{for } p_2 < 0 \end{cases}$$

$$\dot{p}_1 = -H_{x_1} = 0$$

$$\dot{p}_2 = -H_{x_2} = -p_1$$

which gives

$$p_1^* = c_1$$

$$p_2^* = -c_1 t + c_2$$

where c_i (i=1,2,...,6) are constant of integration. We know from the three theorems that optimal control exists, is unique and there is at most one switching, the order of A being 2. It can be seen that the switching function $\sigma(t) \equiv p_2(t) = -c_1 t + c_2$ cannot vanish identically for this would imply $c_1 = o = c_2$ i.e., $p_1 = o = p_2$ and H=1 which contradicts the transversality requirement that H=0 at t=T and being autonomous, H=0 for all $t \in [0,T]$. Thus singular control is ruled out, leaving bang-bang control as the only possibility, namely $u^*(t)$ must fall into one of the following 4 cases

$$u^* = \begin{cases} 1 \text{ for } t \in [0,T] \\ -1 \text{ for } t \in [0,T] \\ 1 \text{ for } t \in [0,t_s) \text{ and } -1 \text{ for } t \varepsilon [t_s,T] \\ -1 \text{ for } t \in [0,t_s) \text{ and } +1 \text{ for } t \varepsilon [t_s,T] \end{cases}$$

where T is the terminal period where $(x_1(T), x_2(T) = (0,0))$ and t_s is the switching time. The values of x_1 and x_2 as found by solving the dynamic system with $u = \pm 1$ are

$$x_2(t) = \pm t + c_3$$

$$x_1(t) = \pm \tfrac{1}{2} t^2 + c_3 t + c_4$$

The dynamic system being autonomous, time can be eliminated by squaring the first equation and multiplying the result by ½ and use the second equation to obtain $x_1 = x_1(x_2)$and the corresponding phase diagram (fig. 7.3)

$$x_1 = \tfrac{1}{2} x_2^2 + c_5 \quad \text{for } u = 1$$

$$x_1 = -\tfrac{1}{2} x_2^2 + c_6 \quad \text{for } u = -1$$

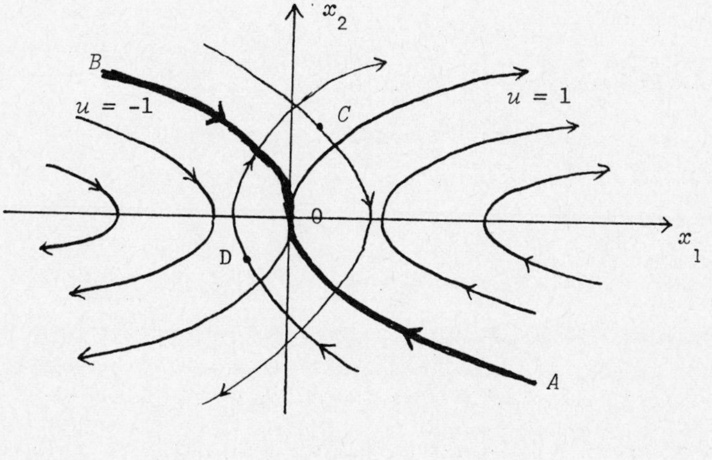

Fig. 7.3 Phase diagram of x_1 and x_2

It can be seen that if the initial state $x(0)$ is on the curve AOB, the control is either $u = 1$ (on OA where $x_1 > 0$) or $u = -1$ (on OB where $x_1 < 0$) and no switching takes place. Note that on the switching curve, $u = \pm 1$ according as $x_1 \gtrless 0$. Everywhere else, there is one switching. For example, if $x(0)$ is at C (see fig. 7.3) the system moves in the direction of the arrow (which is given by the sign of dx_1/dx_2 in each case) with $u = -1$ until a point on OA is reached when u switches to $u = 1$ and brings the system to the origin (0,0) along OA. Similarly at D, $u = 1$ until OB is reached when u switches to $u = -1$ and the system is brought to the origin along OB. The curve AOB is thus called the (bang bang) switching boundary in the sense that it is the locus of all the points at which switching, if any, takes place. More precisely, this switching boundary $s(x)$ is given by the last two equations as

$$s(x) \equiv x_1 + \tfrac{1}{2} x_2 |x_2| = 0 \quad \text{(on } AOB)$$

$$s(x) > 0 \quad \text{above } AOB$$

$$s(x) < 0 \quad \text{below } AOB \quad \text{(see fig 7.4)}$$

The control law in terms of $s(x)$ is given by

$$u^*(t) = \begin{cases} -1 & \text{for } x(t) \text{ such that } s(x) > 0 \\ 1 & \text{" \quad " \quad " \quad " } s(x) < 0 \\ -1 & \text{" \quad " \quad " \quad " } s(x) = 0 \text{ and } x_1 < 0 \\ 1 & \text{" \quad " \quad " \quad " } s(x) = 0 \text{ and } x_1 > 0 \\ 0 & \text{for } x(t) = 0 \end{cases}$$

The last case, $u^* = 0$, is called zero control effort: the initial state being already at the origin, nothing need be done. (See fig. 7.4.)

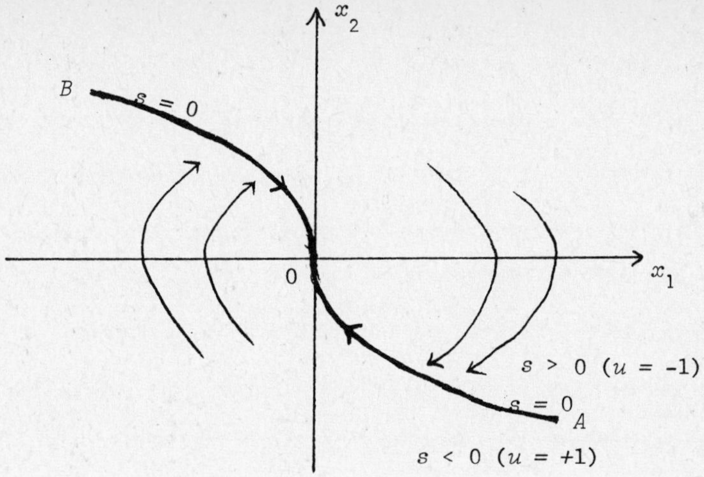

Fig. 7.4 The Switching Curve *AOB*

Finally, note that the terminal time T being unspecified (it is to be minimized), transversality conditions give $H(x,p,u) = 0$ at $t = T$. Together with the boundary conditions $x(T) = 0$, this means

$$1 + p_2(T)u(T) = 0$$

i.e., $p_2(T)$ and $u(T)$ have opposite signs at $t = T$. In view of the bang-bang control law, this implies either $p_2(T) = 1$, $u(T) = -1$ or $p_2(T) = -1$, $u(T) = 1$.

Note also that being autonomous, H is a constant, i.e., $\dot{H} = H_t = 0$ (see ch. 3, eq. 37). This, together with the transversality condition $H|_T = 0$ implies that $H(x,p,u) = 0$ for all $t \in [0,T]$ and not merely at $t = T$.

As a physical example of this, $x(t)$ is the distance at time t, of a spacecraft from a given point of origin, $\dot{x}$ is its speed and $\ddot{x}$, its acceleration which is bounded by the physical capacity of the space vehicle. Thus, acceleration is used as a control variable

($\ddot{x} = u$) in this case, to drive the system home as quickly as possible, i.e., in minimum time.

As an economic example, x may represent foreign indebtedness incurred as a result of an excess of import spending over export receipts, $\dot{x}$ as the debt accumulation or discharge rate and $\ddot{x}$ as the speeding up or slowing down of this rate. The latter is used as a control variable, i.e., $\ddot{x} = u$ to drive the economy home as quickly as possible, to the point where $x_1(T) = 0 = x_2(T)$ i.e., $\dot{x}(T) = x(T) = 0$, the point where the country is absolved from all international indebtedness. Clearly a country's capacity to speed up the borrowing or repayment rate is bounded by its international credibility, economic viability, ability to pay as well as the possible political difficulty caused by an excessive acceleration or deceleration of repayment.

Economic Application: Optimal Monetary Policy

As an economic application, let us consider Peterson & Lerner's Optimal Monetary Policy model (1971). Their dynamic system, based on Friedman's empirical claims and modified, is

$$\frac{a^2}{2}\ddot{x}(t) + a\dot{x}(t) + x(t) = u(t)$$

where x = the proportional rate of growth of money income and

$\dot{x} \equiv dx/dt$, $\ddot{x} \equiv d^2x/dt^2$

$u \equiv m + b\dot{m}$ where $m \equiv \frac{1}{M}\frac{dM}{dt}$ = proportional rate of change of money supply $M(t)$ and b is a constant;

a = a constant representing the length of the business cycle

Defining $x \equiv x_1$, $\dot{x} \equiv x_2$, $\ddot{x} \equiv \dot{x}_2$, we can write the above dynamic system as $\dot{x}(t) = Ax(t) + bu(t)$

or in full as

$$\begin{bmatrix} \dot{x}_1(t) \\ \dot{x}_2(t) \end{bmatrix} = \begin{bmatrix} 0 & 1 \\ -2/a^2 & -2a/a^2 \end{bmatrix} \begin{bmatrix} x_1(t) \\ x_2(t) \end{bmatrix} + \begin{bmatrix} 0 \\ 2/a^2 \end{bmatrix} u(t)$$

The monetary authority would like to achieve a stable rate of growth of national income at some future date, i.e., $x(T) = x_{1T}$; $\dot{x}(T) = 0$ by use of an optimal money supply policy. This objective is equivalent to the minimization of

$$J = \int_0^T dt$$

Since time optimality is not a common practice in Economics, the justification provided is that the failure of x to attain its target represents a social loss: the longer the gap remains, the greater the social loss. It is therefore desirable to reach the desired target in minimum time, i.e., the problem is to minimize $\int_0^T dt$ subject to

$$\dot{x}(t) = Ax(t) + bu(t)$$

with $x(0) \equiv (x_1(0)\ ,\ x_2(0)) = (x_{10},\ x_{20})$ given

and $x(T) \equiv (x_1(T)\ ,\ x_2(T)) = (x_{1T},\ 0)$

$|u| \le .1$ for institutional and economic reasons.

This is a Linear Optimal Control problem where the Hamiltonian H, written in the form (1) is

$$H \equiv 1 - (2p_2/a^2)x_1 + (p_1 - 2p_2/a)x_2 + (2/a^2)p_2 u$$

The switching function $\sigma = (2/a^2)p_2$ and the optimal policy is

$$u^*(t) = \begin{cases} .1 & \text{if } p_2(t) > 0 \\ -.1 & \text{if } p_2(t) < 0 \end{cases}$$

Once $u^*(t)$ is found, m^* and hence M^* are obtained by solving for m.

It is easy to see that the adjoint system is

$$\dot{p} = -H_x = -A'p$$

and the eigen values of $-A'$ are $\mu = 1/a \pm i/a$, giving

$$p_2(t) = e^{t/a} (c_1 \cos t/a + c_2 \sin t/a)$$

This switching function (or the relevant part thereof) being never identically zero for any time interval, the singular control is ruled out, leaving only the bang-bang solution as the optimal policy, i.e., $u^*(t)$ will switch from .1 to -.1 or vice versa every πa years.

The state system is obtained by solving, for $u^* = \pm .1$, $\dot{x} = Ax + bu^* = Ax \pm .1b$. Noting that the eigen values of A being $\mu = -1/a \pm i/a$, the solution is (loc. cit. p. 191, 193)

$$x_1(t) = \pm .1 + e^{-t/a}(c_3 \cos t/a + c_4 \sin t/a) \text{ for } u = \pm .1$$

$$x_2(t) = \frac{e}{a}^{-t/a} \left[(c_4 - c_3) \cos t/a - (c_3 + c_4) \sin t/a \right]$$

Since $a>0$, Re $\mu_i<0$ and μ_1, μ_2 are complex, the phase diagram obtained exhibits a stable focus (see fig. 7.5a.

The optimal policy can be read off fig.7.5b. For example, if initially the economy is at point A where $x_2 < 0 < x_1$, the optimal course is to set $u = -.1 = u_{min.}$ i.e., the tightest monetary policy should be adopted until B is reached when u^* is switched to $u = .1 = u_{max}$, indicating a switch from a contractionary to an expansionary policy. This will carry the economy to the target $(x_T^*, 0)$ where the optimal control u^* is set at x_T^* so that the economy grows at the desired rate x_T^*. Thus the objective of Monetary policy is to move the rate of growth

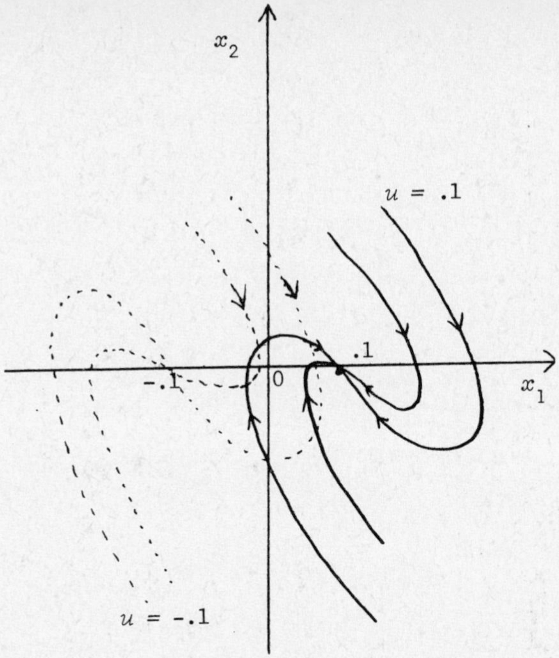

Fig. 7.5a

Locus of (x_1, x_2) when u = .1 (solid lines) and u = -.1 (dotted lines)

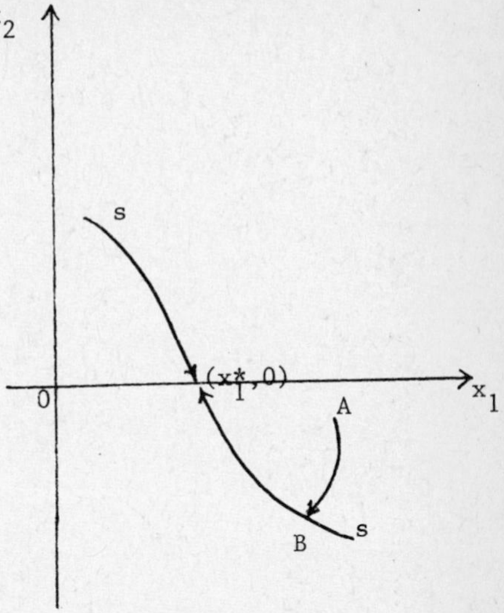

Fig. 7.5b

The switching boundary s

of national income, x_1, from its present position to the desirable one as quickly as possible. Since $u = m + b\dot{m}$ where $m = \dot{M}/M$, the optimal money supply M^* and its optimal rate of change m^* can be determined once u^* is known.

7.3 Singular Control

As has been noted, when the Hamiltonian H is linear in u and u is bounded, if the switching function σ vanishes identically for some time interval (see fig. 7.2), u has no influence on H, the bang-bang control law yields no information on the required optimal control $u^*(t)$: the Maximum Principle fails. This is referred to as degenerate, irregular, ambiguous or singular control.

Singular control may not exist, just as bang-bang control may not exist in a linear problem. If it can be established that σ only vanishes at isolated points of time, then the singular surface does not exist. This is the case discussed in the last section. On the other hand, if it is clear that $\sigma \equiv 0$ for some non zero time interval, which implies $\dot{\sigma} = \ddot{\sigma} = \dddot{\sigma} = \ldots. = 0$, then singular control must be used. Certain problems require a combination of these. This happens when boundary conditions cannot be satisfied by exclusive use of bang-bang or singular control.

In an autonomous system (when H is not an explicit function of time), (1) is written as

$$H \equiv \psi(x,p) + \sigma(x,p)u \tag{9}$$

where H is constant, $\dot{H} = H_t = 0$. In time-optimal problems, T is unspecified (T is to be minimized), $H = 0$ at $t = T$. Together with the fact that H is constant, this implies $H = 0 \ \forall \ t[0,T]$.

If H involves t explicitly, it can be made autonomous by defining $t \equiv x_{n+1}$ with $x_{n+1}(0) = 0$ and $\dot{x}_{n+1} = 1$ and write the Hamiltonian with these augmented vectors x_a and p_a where $x_a \equiv (x_1, x_2, \ldots, x_{n+1})$ and $p_a \equiv (p_1, p_2, \ldots, p_{n+1})$. This modified Hamiltonian is then treated as an autonomous one. If the terminal time T is fixed, a new variable $x_{n+1}(T) \equiv T$ can be introduced to convert the problem with fixed terminal time into one with free terminal time. Thus the autonomous system with free terminal time can be considered the standard case to be examined.

In this standard case, the optimal Hamiltonian (H^*) vanishes at all time $t \in [0,T]$. $H^* \equiv \psi + \sigma u = 0$ implies the following possibilities

(i) $u^* \equiv 0 \quad (\sigma \not\equiv 0)$ and $\psi(t) = 0$ (10)

(ii) $\psi(t) = -\sigma(t)\, u^*(t) \quad (\sigma, \psi \neq 0)$ i.e., $u^* = -\dfrac{\psi(t)}{\sigma(t)}$ (11)

(iii) $\psi(t) \equiv 0 \equiv \sigma(t)$ over a time interval say $[t_1, t_2]$ (12)

Case (i) is the case of zero control effort;

Case (ii) corresponds to the bang-bang control in the last section: $\psi(t)$ and $\sigma(t)$ must vanish simultaneously at each switch;

Case (iii) involves singular control which we shall now examine in more detail with reference to Johnson's (1965) example.

If $\sigma(x,p)$ vanishes identically over a time interval, say $[t_1, t_2]$, then (9) implies $\psi(x,p)$ also vanishes identically, given the fact that $H(x,p) \equiv 0 \;\forall\; t \in [0,T]$. Furthermore, during this singular interval,

$$\frac{d^k}{dt^k}\,\psi(x,p) \equiv 0 \tag{13}$$

$$\frac{d^k}{dt^k}\,\sigma(x,p) \equiv 0\,,\; k = 0, 1, 2, \ldots \tag{14}$$

For $k = 0$, we simply have $\sigma(x,p) \equiv 0$

For $k = 1$, we obtain $\dot{\sigma} = \sigma_x\,\dot{x} + \sigma_p\,\dot{p} \equiv 0$ (15)

Substituting the dynamic and adjoint system $\dot{x} = f(x,u,t)$ and $\dot{p} = -H_x$ into (15), taking into account the fact that $H \equiv 0 \;\forall\; t \in [0,T]$, we may be able to solve (15) explicitly for u^*. If we are, we are finished. If not, we proceed to the cases where $k = 2, 3, 4, \ldots$ i.e., $\ddot{\sigma} \equiv 0 \equiv \dddot{\sigma} \equiv \ddddot{\sigma} \equiv \ldots$, until u^* eventually comes out.

To decide whether a singular interval is involved, a thorough check must be made to ensure that the application of the singularity conditions (13) and (14) does not introduce any inconsistencies into

Example 7.3.1

Consider the classical problem (Johnson and Gibson 1963) of minimizing

$$J = \tfrac{1}{2} \int_0^T x_1^2 \, dt$$

subject to

$$\dot{x}_1 = x_2(t) + u(t) \ ; \quad (x_1(0), x_1(T)) = (a, c)$$

$$\dot{x}_2 = - u(t) \qquad ; \quad (x_2(0), x_2(T)) = (b, d)$$

$|u| \leq 1$ and T is unspecified.

The Hamiltonian is

$$H = \tfrac{1}{2} x_1^2 + p_1 (x_2 + u) - p_2 u$$

$$= \tfrac{1}{2} x_1^2 + p_1 x_2 + (p_1 - p_2)u$$

$$\equiv \psi (x, p) + \sigma(x, p) u$$

where the switching function σ is the coefficient of u and ψ is the remaining terms (not involving u) of H.

$$- H_{x_1} = \dot{p}_1 = - x_1$$

$$- H_{x_2} = \dot{p}_2 = - p_1$$

The solution involves three cases: bang bang B_1 where $\sigma > 0$, bang bang B_2 where $\sigma < 0$ and singular when $\sigma \equiv 0$. We shall examine each one in turn.

1. Bang-Bang arc B_1 when $\sigma \equiv p_1 - p_2 > 0$, $u^* = -1$

   The dynamic system gives

   $\dot{x}_2 = 1$ giving $x_2(t) = t + c_1$ or $t = x_2 - c_1$

   $\dot{x}_1 = x_2 - 1 = t + c_1 - 1$, giving

   $x_1(t) = \tfrac{1}{2} t^2 + (c_1 - 1)t + c_2$, or eliminating t ,

   $x_1(t) = \tfrac{1}{2} x_2^2 - x_2 + c_3$ where $c_3 \equiv - c_1(c_1 - 1) + c_2 + c_1^2/2$
   (see fig. 7.6)

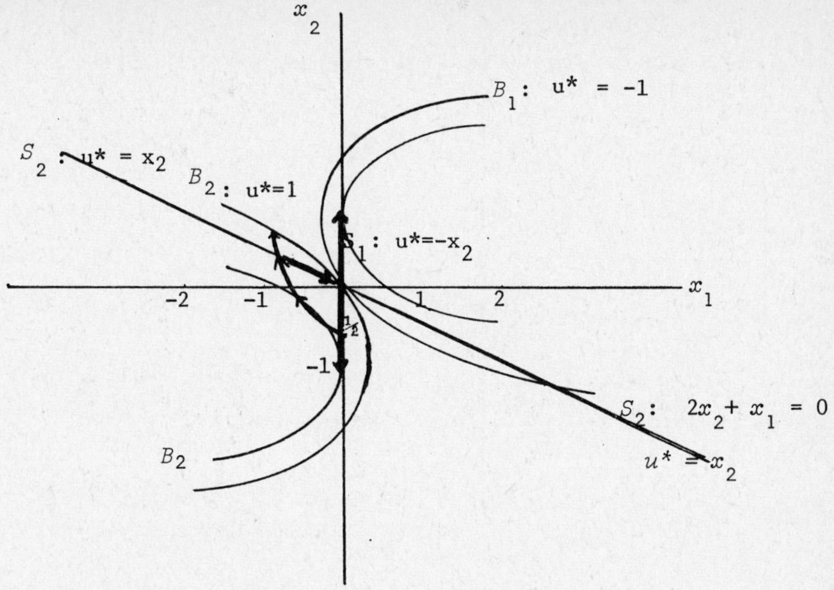

Fig. 7.6 Bang bang B_1, B_2 and Singular S_1, S_2 arcs

2. Bang bang arc B_2 : $\sigma \equiv p_1 - p_2 < 0$, $u^* = +1$

   The dynamic system gives

$$\dot{x}_1 = x_2 + 1$$

$$\dot{x}_2 = -1 \rightarrow x_2(t) = -t + c_1 \text{ or } t = c_1 - x_2$$

   Integrating and eliminating t as before gives

$$x_1(t) = -\tfrac{1}{2}x_2^2 - x_2 + c_4$$

   where $c_4 \equiv c_1(1 + c_1) + c_2 - \tfrac{1}{2}c_1^2$ (See fig. 7.6).

3. Singular arcs S_1, S_2 where $\sigma \equiv p_1 - p_2$

   $\dot{\sigma} = \dot{p}_1 - \dot{p}_2 = -x_1 + p_1 \equiv 0$

   H being autonomous, $\dot{H} = 0$ i.e. H is a constant.

   T being unspecified, $H(T)\delta T = 0 \Rightarrow H(T) = 0$ and hence $H(t) = 0$ for all $t \in [0, T]$. On the singular arc, $\sigma \equiv 0$ and hence $H \equiv \psi + \sigma u = \psi = \frac{x^2}{2} + x_1 x_2 = x_1 (\frac{1}{2} x_1 + x_2) = 0$. This implies

$$x_1 \equiv 0 \qquad \text{on } S_1$$

$$\text{or} \quad 2x_2 + x_1 \equiv 0 \quad \text{on } S_2$$

On S_1, $x_1 \equiv 0 \equiv \dot{x}_1 = x_2 + u \Rightarrow u^* = -x_2$

On S_2, $2x_2 + x_1 \equiv 0 \Rightarrow \dot{x}_2 \equiv -\frac{1}{2}\dot{x}_1 \Rightarrow u^* = x_2$ (see fig. 7.6).

The direction of the arrows in fig. 7.6, given by the sign of $\dot{x}_2/\dot{x}_1 = dx_2/dx_1$, indicates that on S_1 the system is unstable as it moves away from the origin and on S_2 the system is stable : it moves to the origin. the optimal controls $u^* = \pm x_2$ are thus in the feedback form and $|x_2| \leq 1$ in view of $|u| \leq 1$. This gives the truncated fig. 7.6. The state equations $x_1(t)$, $x_2(t)$ for each case are obtained by solving $\dot{x}_1$ and $\dot{x}_2$ with $u^* = \pm x_2$.

If the initial point $x° = (a, b)$ are on the singular arcs S_1 or S_2, the system will stay on them, otherwise it will switch. The interesting question is should a trajectory, starting from an arbitrary initial point not on S_1 or S_2 drifting along a bang-bang path, switch to a singular arc as soon as it intersects it or should it cross the singular arc until it reaches another bang bang surface then switches to it? To answer this question, consider the properties of the switching function σ at the switching time t_s. For the case of autonomous H under discussion, $\dot{H} = 0$ and $H\,\delta T = 0 \Rightarrow H(t) = 0$ and at switching time t_s on the bang-bang surface and at all times on the singular surface, $\sigma = 0$ and $H = \psi + \sigma u = \psi = 0$.

For a minimization problem, $\sigma > 0 \Rightarrow u^* = -1$ and $\sigma < 0 \Rightarrow u^* = 1$ (and vice versa for a maximization problem). In order for the control starting from say $u = -1$ where $\sigma > 0$ to switch to $u = 1$ where $\sigma < 0$, passing through $\sigma(t_s) = 0$, clearly $\dot{\sigma}(t_s) < 0$ at the switching time. Similarly for $u = 1$ to switch to $u = -1$, $\dot{\sigma}(t_s) > 0$.

But $\quad \sigma(t_s) = p_1 - p_2 = 0$

$$\dot{\sigma}(t_s) = \dot{p}_1 - \dot{p}_2 = -x_1 + p_1$$

and $\quad \psi \equiv x_1^2/2 + p_1 x_2 = 0 \Rightarrow p_1 = -x_1^2/2x_2$ i.e.

$$\dot{\sigma}(t_s) = -x_1 - x_1^2/2x_2 = (2x_2 + x_1)\, x_1/2x_2$$

This divides the state space into 6 zones (see fig. 7.7).

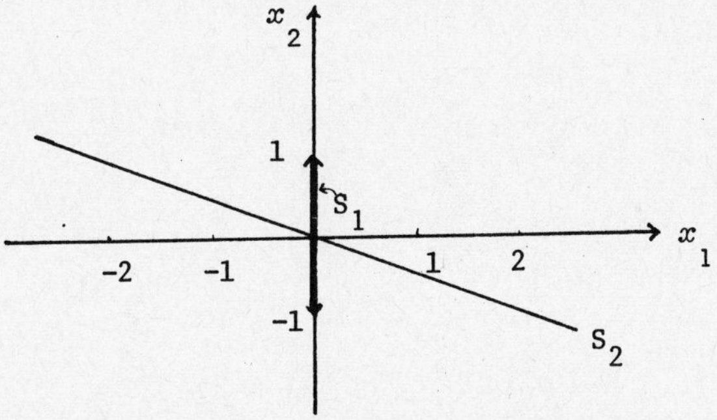

Fig. 7.7 Allowable switches

Table 7.3.1

Zone	Sign of x_1	Sign of x_2	Sign of $\dot{\sigma}(t_s)$	Allowable switches of u
1	+	+	–	from -1 to +1
2	+	–	+	1 to -1
3	+	–	–	-1 to 1
4	–	–	+	1 to -1
5	–	+	–	-1 to 1
6	–	+	+	1 to -1

For example, if the system starts from $x^\circ = (0, -\frac{1}{2})$, clearly the system is on the bang-bang arc B_1 with $u^* = -1$ and $\dot{x}_2 = 1$, and $\dot{x}_2 = 1$, $\dot{x}_1 = x_2 - 1$ give $x_2 = t - \frac{1}{2}$ and $x_1 = \frac{1}{2} t^2 - (3/2)\, t$ or, eliminating t, we have along B_1

$$x_1(t) = \tfrac{1}{2} x_2^2 - x_2 - 5/8$$

At the intersection with the singular arc S_2 where $x_2 = -\frac{1}{2} x_1$, we have

$$x_1^2 - 4\, x_1 - 5 = 0$$

i.e. $x_1 = -1$ and $x_2 = \frac{1}{2}$. At this point, should the system switch to S_2 or keep moving until it intersects B_2 to which it then switches? (See fig. 7.8.)

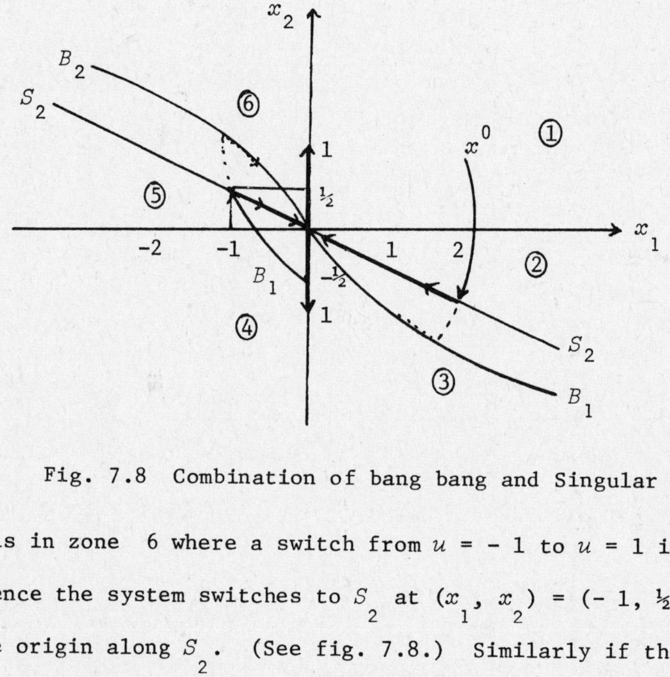

Fig. 7.8 Combination of bang bang and Singular Controls

Clearly B_2 is in zone 6 where a switch from $u = -1$ to $u = 1$ is not allowed. Hence the system switches to S_2 at $(x_1, x_2) = (-1, \frac{1}{2})$ and moves to the origin along S_2. (See fig. 7.8.) Similarly if the system starts from an arbitrary point x° in zone 1 with $u = 1$ the optimal trajectory switches to S_2 as soon as it intersects this singular arc : B_1 being in zone 3 where a switch from $u = 1$ to $u = -1$ is forbidden.

7.4 Singular Control and The Calculus of Variations

The concept of Singular Control would become clearer when compared with the case where the objective functional is linear in $\dot{x}$, discussed in Chapter 2 (section 2.4.4) and Chapter 6 (section 6.2). With u replaced by $\dot{x}$ in the problem of finding an extremum of a functional

$$J = \int_0^T f(x,\dot{x},t)dt$$

where $f(x,\dot{x},t) = \alpha(x(t),\ t) + \beta(x(t),\ t)\ \dot{x}$ (16)

i.e., f is linear in $\dot{x}$ (eq. 2.31).

The Euler equation gives

$$f_x - f_{\dot{x}t} - f_{x\dot{x}}\dot{x} - f_{\dot{x}\dot{x}}\ddot{x} = 0 \tag{17}$$

This is a second order differential equation whose solution contains two arbitrary constants chosen to fit the boundary conditions of a given problem. If $f(x,\dot{x},t)$ is linear in $\dot{x}$, $f_{\dot{x}\dot{x}} = 0$ and (17) simply gives (eq. 2.33 Ch.2)

$$\frac{\partial\alpha}{\partial x} - \frac{\partial\beta}{\partial t} = 0 \tag{18}$$

which is not a differential equation but an algebraic equation. Its solution does not contain any arbitrary constants and in general does not satisfy the boundary conditions (see Ch. 2, examples 2.4.4 and 2.4.3b.

Any function satisfying (18) is a singular solution.

Furthermore (16) can be written as

$$\int_{t_0,x_0}^{t_1,x_1} \alpha(x(t),\ t)dt + \beta(x(t),\ t)dx \tag{19}$$

which, in view of (18), is independent of the path of integration (see Ch. 4, footnote 4).

It is easy to verify that in this case, the Weierstrass excess function E and the Legendre - Clebsch test (≥ 0 for a minimum and ≤ 0 for a maximal optimal control problem, see sections 4.5, 4.6 and 4.7) both give 0, i.e., both hold with equality. They provide no information. This is not unexpected since during the singular interval, the Hamiltonian H is unaffected by u.

Note that the Weierstrass function

$$E = -\Delta H^*$$

where $\Delta H^* \equiv H(x^*, p^*, u^* + \delta u) - H(x^*, p^*, u^*)$. For a more detailed discussion of the relationship between the Weiestrass E function and the Hamiltonian H, see for example Gelfand & Fomin [1963, Appendix].

7.5 Singularity and Controllability

In the discussion of the linear optimal control, it has been noted that when the coefficient of the control variable vanishes identically in a time interval, the Hamiltonian H is not affected by u, in other words, the system is uncontrollable. This seems to suggest that singularity and uncontrollability are closely associated with each other. We are now going to show that this is indeed the case. For simplicity of presentation, only the constant coefficients case will be discussed but the validity of the analysis is not restricted to the time invariant case.

Consider the time optimal problem of section 7.2

$$Min\ J = \int_0^T dt$$

subject to

$$\dot{x}(t) = Ax(t) + bu(t) \tag{20}$$

where A is an nxn constant matrix (' denotes transposition)

b is a constant n. vector and $u(t)$ is a scalar.

The Hamiltonian is

$$H = 1 + p'Ax + p'bu$$

The adjoint system is

$$\dot{p} = -A'p$$

whose solution is

$$p(t) = e^{-A't}c$$

where c is an arbitrary constant vector, more precisely $c = p_0$.

The singularity conditions are

$$\frac{d^k}{dt^k}(p'b) \equiv 0 \;\forall\; t \in [t_1, t_2] \text{ and } k = 0,1,2,3,\ldots \tag{21}$$

For $k = 0$, $\quad p'b \equiv 0$

$k = 1$, $\quad \dot{p}'b = -p'Ab \equiv 0$

$k = 2$, $\quad \ddot{p}'b = -\dot{p}'Ab = -p'A^2b \equiv 0$

$\vdots$

$k = n-1$, $\quad \overset{n-1}{p}{}'b = -p'A^{n-1}b \equiv 0$

or

$$p'[b \vdots Ab \vdots A^2b \vdots \ldots \vdots A^{n-1}b] = 0' \tag{22}$$

or

$$p'M \equiv 0 \tag{23}$$

where $\quad M \equiv [b \vdots Ab \vdots A^2b \vdots \ldots \vdots A^{n-1}b]$

But $p = e^{-At}c \neq 0$, hence M must be singular. Thus we have derived the well known singularity and Controllability Conditions:

The singularity of the matrix M in (23) is a necessary condition both for the existence of singular control and for the uncontrollability of the system.

Conversely, if the system is completely controllable, M is non-singular, i.e., a singular interval cannot exist.

If instead of b in (20), we have an nxn time invariant matrix and instead of the scalar u, we have an r-vector of u, i.e., $\dot{x} = Ax + Bu$, M would become

$$M \equiv [B \vdots AB \vdots A^2B \vdots \cdots \vdots A^{n-1}B]$$

an n x rn matrix, then the controllability and non-singularity (or regularity) conditions require M to be of full rank, i.e., n. Singularity and uncontrollability conditions are that M be of rank less than n. If $A(x)$ and $B(x)$ are both function of x, then $M(x)$ would be more complicated in view of the tedious but straightforward differentations and substitutions. The analysis however remains unaffected.

7.6 Some Economic Applications

7.6.1 Optimal Economic Growth

As an example of Linear Optimal Control, let us consider the neo-classical aggregate economic growth model already examined in section 5.6.2, this time with the specific utility function $u(c) = c$. The problem is to maximize

$$J = \int_0^\infty c(t)\, e^{-\delta t} dt \tag{24}$$

subject to

$$\dot{k}(t) = f(k(t)) - \lambda k(t) - c(t) \tag{25}$$

where, as before, $c(t)$, $k(t)$: per capita consumption and capital respectively, with $k(0) = k_0$; δ, λ are positive constants and $T = \infty$.

The Hamiltonian is

$$H \equiv e^{-\delta t} c + p[f(k) - \delta k - c]$$

or, with the definition $p(t) \equiv q(t)e^{-\delta t}$

$$\begin{aligned} H &\equiv e^{-\delta t}\; [c + q(f(k) - \lambda k - c)] \\ &= e^{-\delta t}(1-q)\, c + e^{-\delta t} q(f - \lambda k) \end{aligned} \tag{26}$$

i.e., H is linear in the control variable $c(t)$. Clearly c is bounded between the subsistence level $\bar{c}$ (which could be set at zero for simplicity), and the maximum level $f(k)$ where total output produced is consumed, i.e., $\bar{c} \leq c^* \leq f(k)$. The switching function $\sigma \equiv e^{-\delta t}(1-q)$ indicates that the optimal policy is to choose

$$c^* = \begin{Bmatrix} \bar{c}\ (=0) \\ 0<c<f(k) \\ f(k) \end{Bmatrix} \quad \text{if } q = \begin{Bmatrix} >1 \\ \equiv 1 \\ <1 \end{Bmatrix} \tag{27}$$

The dynamic state and adjoint systems are

$$\dot{k} = f(k) - \lambda k - c^*(q) \tag{28}$$

$$\dot{q} = [\lambda + \delta - f(k)]\, q \tag{29}$$

which are solved by substituting the optimal c^*. Three cases are to be considered:

(1) Bang-bang policy B_1 should be used when $q < 1$, $c^* = f(k) = c_{max}$. The above dynamic system becomes

$$\dot{k} = -\lambda k$$

$$\dot{q} = (\lambda + \delta - f')\, q$$

i.e., the capital stock decreases at the rate λ, all output being consumed.

(2) Bang-bang policy B_2 should be used when $q > 1$, $c^* = \bar{c} = c_{min} = 0$ and the dynamic system is now

$$\dot{k} = f(k) - \lambda k$$

$$\dot{q} = (\lambda + \delta - f')\, q$$

(3) Singular Control S would be the appropriate policy if $q \equiv 1$ for some time interval during which $(d^k/dt^k)\,\sigma \equiv 0$, $k = 0,1,2,\ldots$ (see eq. 14)

i.e.,

$$\sigma \equiv 0 \quad \Rightarrow \quad q \equiv 1$$

$$\dot{\sigma} = -e^{-\delta t} \; [\delta \; (1-q) + \dot{q}] \equiv 0 \quad \Rightarrow \quad \dot{q} = 0 \; (q \equiv 1)$$

$$\ddot{\sigma} = 0 \quad \Rightarrow \quad \ddot{q} = 0 \quad \Rightarrow \quad \dot{k} \equiv 0$$

This is the equilibrium case where $k \to k^*$ and $q \to q^*$, i.e., $k(t)$ and $q(t)$ have approached and reached their long run equilibrium values (k^*, q^*) where $\dot{q}^* = 0 = \dot{k}^*$.

These three cases may be summarised in the following phase

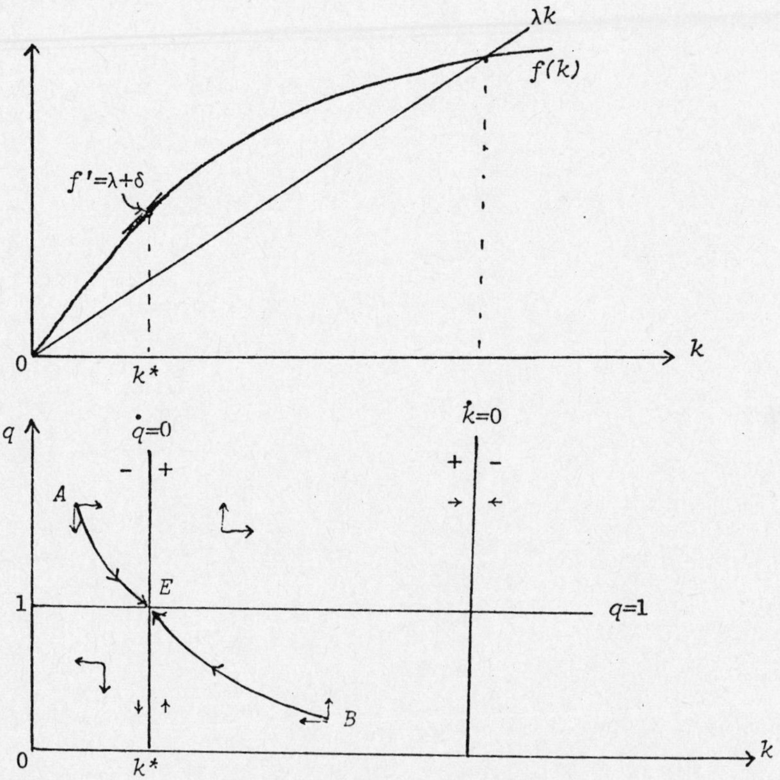

Fig. 7.9 Optimal growth and Linear Optimal Control

The line $q = 1$ divides the diagram into the bang-bang region B_1 which lies below it, where k falls and the bang-bang region B_2 which lies above it. On the line $q = 1$, if q remains equal to one for a non zero time interval, at E, singular control is the optimal policy. This is the long run equilibrium position $k = k^*$, $q = q^* = 1$. Thus the solution is a saddle point: for each k, there is one optimal q, such as A and B in fig. 7.9 which would drive the system to its long run equilibrium (k^*, q^*) at E. A wrong choice of q will cause the system to veer away from its equilibrium causing instability and infeasibility in the long run.

7.6.2 Resource Economics

The Economics of Natural Resources received a rigorous treatment for the first time by Hotelling (1931) and subsequently by Scott (1955), Gordon (1954, 1967), Smith (1968, 1969), Clark (1976) and others.

The central problem in Resource Economics is to find an optimal rate of resource consumption over time, bearing in mind that more today means less tomorrow in the case of non-renewable resources and over-consumption now may lead to an eventual extinction of some species in future, in the case of reproducible resources. Stated this way, the problem is clearly cast in the framework of Capital theory, a well developed area in Economic theory.

We shall treat the problem of renewable resources first and examine the case of fisheries for definiteness, but the analysis applies to the case of any reproducible resource such as forestry. Exhaustible resources are just the special case where reproduction is zero. Our interest being in the application of Linear Optimal Control, we shall not cover all the aspects of Resource Economics. The model presented in this section closely follows Hotelling (1931), Clark (1976), Clark and Munro (1982). See also Mirman & Spulber (1982).

Reproducible Resources

Consider a fish population $x(t)$ with a growth function $f(x)$ in the absence of harvesting $h(t)$, i.e.,

$$\dot{x}(t) = f(x) - h(t) \tag{30}$$

where $0 \leq h \leq h_{max}$. It is reasonable to expect that $f(0) = 0 = f(K)$, $f(x) > 0 \;\forall\; x \in (0, K)$ and $f(x) < 0 \;\forall\; x > K$ and also $f''(x) < 0$ i.e., $f(x)$ is concave and there is some saturation level K above which overcrowding will drive down the rate of growth of the fish population. A typical example is the logistic curve (see fig. 8.10).

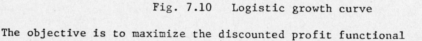

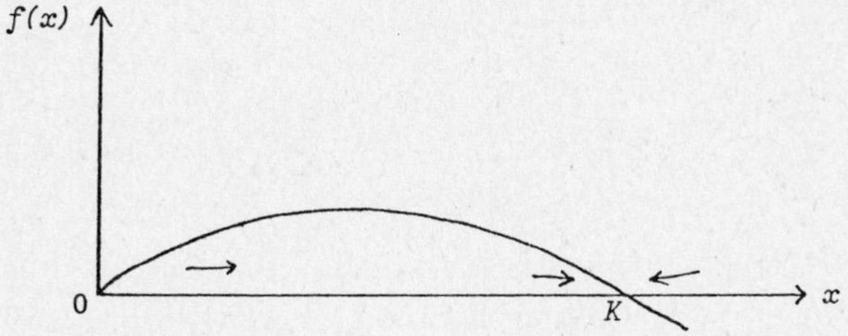

Fig. 7.10 Logistic growth curve

The objective is to maximize the discounted profit functional

$$J = \int_0^\infty e^{-\delta t} \; [p - c(x)]h(t) \; dt \tag{31}$$

subject to

$$\dot{x} = f(x) - h(t)$$

where δ = positive rate of future discount

p = price of fish

$c(x)$ = unit cost of harvesting, $c'(x) < 0$

The Hamiltonian H is linear in the control $h(t)$

$$H \equiv e^{-\delta t}[p - c(x)] \; h(t) + \lambda(t) \; [f(x) - h(t)]$$

$$\equiv \sigma(t) \; h(t) + e^{-\delta t} \; \lambda(t) \; f(x) \tag{32}$$

where $\sigma(t) \equiv e^{-\delta t}\,[p - c(x) - \lambda(t)]$, the switching function.

The state and adjoint dynamic system is

$$\dot{x} = f(x) - h(t) \qquad = H_{\lambda} \tag{33}$$

$$\dot{\lambda} = e^{-\delta t}\, hc'(x) - \lambda f'(x) = -H_{x} \tag{34}$$

If $\sigma(t) > 0$, $h^* = h_{max}$; if $\sigma(t) < 0$, $h^* = 0 = h_{min}$

and if $\sigma(t) \equiv 0$ for some non-zero time interval, singular control takes place. We shall examine the latter case first and show that it is the equilibrium case.

$$\sigma(t) \equiv 0 \quad \Rightarrow \lambda(t) = p - c(x)$$

$$\dot{\sigma}(t) \equiv 0 = -e^{-\delta t}\,[(p-c)\,\delta + c'(x)\,\dot{x}] - \dot{\lambda} = 0 \tag{35}$$

Substituting $\dot{x}$ and $\dot{\lambda}$ in the above and dividing by $e^{-\delta t}\,(p-c)$, we obtain the well known equilibrium relation in Capital theory (see Clark 1976)

$$f'(x) - \frac{c'f(x)}{p - c(x)} \quad = \delta \tag{36}$$

$f'(x)$ is the marginal productivity--or marginal reproduction--of the fish population and $c'(x)f(x)/(p-c(x))$ is the marginal stock effect. The latter constitutes a link between the present and future: it is important when more than one period must be taken into account. Together, the above relation means that in an optimal programme, the marginal productivity of fish (capital in general) net of the stock effect must be equal to the social discount rate δ. If unit cost is stock invariant, i.e., $c'(x) = 0$, we have the well known rule that the marginal product of capital is equal to the rate of future discount which is also equal to the rate of interest under equilibrium conditions.

Assuming the above equilibrium relation has a unique solution x^* (a strong assumption) then $\sigma(t) > 0$ whenever $x(t) > x^*$ and $\sigma(t) < 0$

whenever $x(t) < x^*$, i.e., the optimal harvesting policy is

$$h^*(t) = \begin{cases} h_{max} & \text{whenever } x(t) > x^* \\ 0 & \text{whenever } x(t) < x^* \end{cases}$$

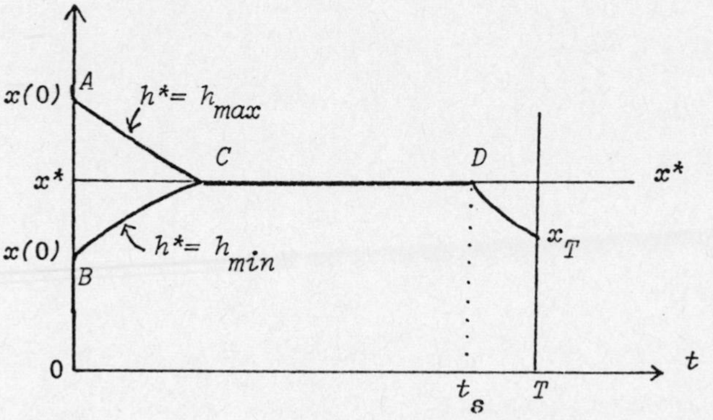

Fig. 7.11 Optimal harvesting policy

Thus, if the initial stock $x(0)$ is larger than the optimal x^*, say at A in fig. 8.11, $h^*(t) = h_{max}$ is applied until $x(t)$ is reduced to its optimal level x^*, say at C. Similarly, the bang-bang control $h^*(y) = \dot{0} = h_{min}$ is applied, should $x(0) < x^*$, say at B, until x^* is reached, say at C. Where will the programme end? If $x(T) = x^*$, it will end on the singular line $x = x^*$ at $t = T$. If $x(T) = x_T \neq x^*$, the system will leave the singular path $x = x^*$ at some time $t_s < T$, to be driven by the bang-bang control to x_T at $t = T$ as prescribed. This is the "turnpike" (Samuelson 1958, 1965) property of the model.

Assuming perfect competition and absence of all stock effect on $p(t)$, i.e., $p(t)$ is not dependent on $q(t)$ and $x(t)$, we have

$$\dot{\lambda}(t) = -E_x = 0 \Rightarrow \lambda(t) = \lambda, \text{ a constant}$$

The optimal $q^*(t)$ at each period is determined by $\sigma(t)$, i.e.,

$$q^*(t) = q_{max} \quad \text{whenever } \sigma(t) > 0 \quad \text{i.e., } p(t) > \lambda e^{\delta t}$$

$$q^*(t) = 0 \quad \text{when } \sigma(t) < 0 \quad \text{i.e., } p(t) < \lambda e^{\delta t}$$

$$\sigma(t) \equiv 0 \quad \Rightarrow p(t) = \lambda e^{\delta t}$$

$$\dot{\sigma}(t) \equiv 0 \quad = e^{-\delta t}(\dot{p} - \delta p) \Rightarrow \dot{p}/p = \delta \tag{37}$$

This is Hotelling's fundamental price rule (1931): in an optimal programme, the price of a non-renewable resource can be expected to increase exponentially at the rate of discount δ which is equal to the rate of interest in equilibrium. If $\dot{p}(t)$ increases faster than interest rate, it would pay the mine owner to postpone extraction, i.e., the optimal policy is $q^* = 0$. If on the other hand, $p(t)$ increases less rapidly, it would be in his interest to sell his stock immediately and deposit the proceeds in his bank to earn interest income, i.e., his optimal policy is $q^*(t) = q_{max}$ (see fig. 7.12, also Solow (1974) and Clark (1976)).

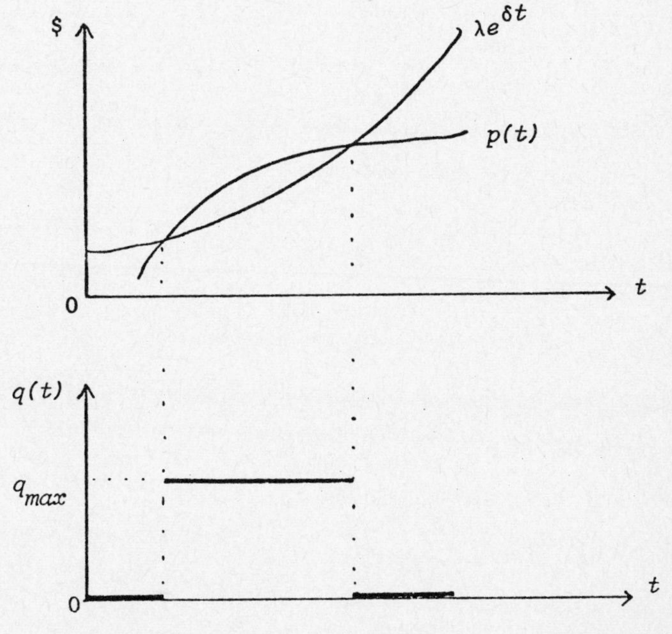

Fig. 7.12 Optimal policy of extracting NRR

Non Renewable Resources

Non renewable resources are by definition the ones which are available in fixed quantities and are not reproducible: more consumption today means less available for consumption tomorrow. The owner of a non renewable resource faces the problem of optimal allocation of this fixed quantity over time such as to maximize his profit functional. The performance index is thus to maximize

$$J = \int_0^T e^{-\delta t} p(t)\, q(t)\, dt$$

subject to

$$\int_0^T q(t)\, dt = \overline{Q}$$

where $q(t)$ = quantity of NRR extracted at time t and $p(t)$ its price

δ = positive rate of future discount

$\overline{Q}$ is a fixed quantity assumed known with certainty

T is unspecified.

Defining $x(t)$ as the remaining NRR at t, we can write the above constraint as

$$x(t) \equiv \overline{Q} - \int_0^t q(t)\, dt$$

$$\dot{x}(t) = -\, q(t)$$

$$x(0) = \overline{Q}$$

$$x(T) = 0$$

Abstracting from the extraction cost, and considering J above as the present value of his resource, we have the Hamiltonian

$$H \equiv \left[e^{-\delta t} p(t) - \lambda(t)\right] q(t) \equiv \sigma(t)\, q(t) \qquad (38)$$

where $\sigma(t) \equiv e^{- t}p - \lambda$ = the usual switching function

$q(t)$ is the control variable, assumed bounded $0 \leq q(t) \leq q_{max}$.

When the assumption of perfect competition is relaxed, $p = p(q)$ with $p'(q) < 0$, i.e., we have the monopolist's downward sloping demand curve and what was said about price (which is also marginal revenue under perfect competition) must now be said about marginal revenue (MR). With this only modification, the analysis remains the same. The Hamiltonian is

$$H = \left[e^{-\delta t} p(q) - \lambda(t) \right] q(t)$$

However, this is no longer a case of linear optimal control. The optimal plan is

$$\partial H / \partial q = e^{-\delta t}(p + qp') - \lambda = 0$$

i.e., $\qquad MR = \lambda e^{\delta t}$

where $\qquad MR \equiv p(q) + qp'(q)$

$$\dot{\lambda}(t) = -H_x = 0 \Rightarrow \lambda(t) = \lambda \text{ (constant)}$$

In the light of our transversality conditions (Ch. 6, eq. 40)

$H(t)\, \delta T = 0 \quad \Rightarrow \quad H(T) = 0$ i.e., either $q(T) = 0$ or

$$\lambda(T) = p[q(T)]\, e^{-\delta T}$$

i.e. $p\,[q(T)] = MR\,[q(T)] \equiv p[q(T)] + q(T)\, p'\,[q(T)]$

But this means $q(T) = 0$ and thus $\lambda = \lambda(T) = e^{-\delta T}p(0)$ where $p(0)$ is the highest price which could be expected for the resource, i.e., the price the public would be willing to pay for the last unit. For example, for a linear downward sloping demand curve $p(q) = a + bq$ (see Hotelling 1931), $p(0) = a$. Thus, at $t = T$, price or $AR = MR$ and $q(T) = 0$ (see fig. 7.13).

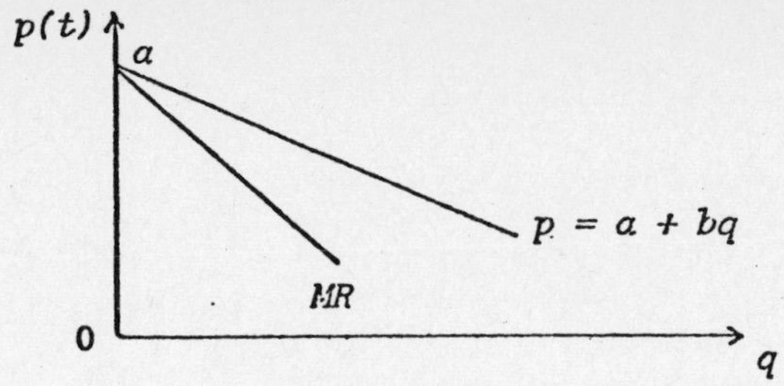

Fig. 7.13

7.6.3 Optimal domestic and foreign investment

As an illustration of the allowable switches discussed in section 7.3 above, in connection with example 7.3.1, let us consider Hamada's model (1969) of optimal capital accumulation of a small country facing a perfect international capital market. Although Hamada's model could be changed into a linear one (by simply assuming a linear utility function $u(c) = c$ with $u'(c) = 1$ and $u''(c) = 0$ rather than a general one with diminishing marginal utility $u''(c) < 0 < u'(c)\ \forall\ c > 0$ with $\lim_{c \to 0} u'(c) = \infty$) and apply the analysis of 7.3 to it, it would be more interesting to examine the model with the consumption utility in its general form and show that the allowable switching laws are of a more general applicability.

Consider a neo-classical economy with a constant returns production function $Q = F(K, L) = L\, f(k)$ where $k \equiv K/L$ = capital-labour ratio, with $f''(k) < 0 < f'(k)$, $\lim_{k \to 0} f'(k) = \infty$, $\lim_{k \to \infty} f'(k) = 0$ and constant labour growth $\dot{L}/L = n$ (see Cass (1965), Uzawa 1964). Let z be the capital per head invested abroad if $z > 0$ and borrowed from abroad if $z < 0$ and $g(z)$ be a concave increasing function, $g''(z) < 0 < g'(z)$ with $g(0) = 0$. Thus, per capita net income is $y = f(k) + g(z)$.

Let s_1, s_2 be the proposition of net income invested at home and abroad respectively ($s_2 < 0$ means borrowing, up to a limit β, see fig. 7.14) i.e.

$$s_1 \geq 0 \quad , \quad s_2 \geq -\beta$$

$$1 - s_1 - s_2 \geq 0$$

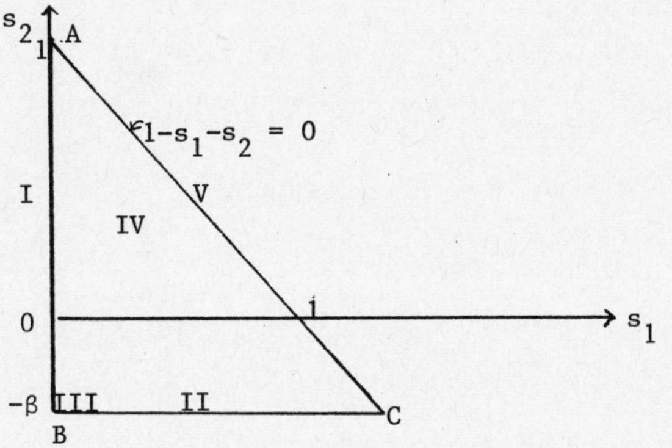

Fig. 7.14 The control space

The objective is to maximize the discounted utility functional

$$J = \int_0^\infty u(c_t)\, e^{-\rho t} \tag{39}$$

where

$$c_t \equiv (1 - s_1 - s_2)\, y \equiv (1 - s_1 - s_2)\,[f(k) + g(z)] \tag{40}$$

and

$$\dot{k} = s_1 y - nk \equiv s_1 [f(k) + g(z)] - nk \tag{41}$$

$$\dot{z} = s_2 y - nz \equiv s_2 [f(k) + g(z)] - nz \tag{42}$$

At $t = 0, [k(0), z(0)] = (k_0, z_0)$ is assumed to be such that

$y_0 = f(k_0) + g(z_0) > 0$ (see fig. 7.15) whose shape reflects the concavity of y (i.e. with $f'' < 0 < f'$, $g'' < 0 < g'$, $f(k) + g(z) = 0 \Rightarrow dz/dk = - f'/g' < 0$ and $d^2z/dk^2 > 0$).

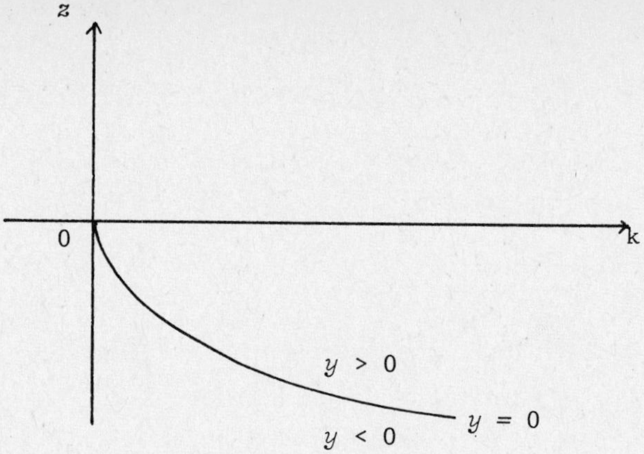

Fig. 7.15 The state space

The Hamiltonian is

$$H = e^{-\ t}\,[u(c) + p_1(s_1 y - nk) + p\,(s_2 y - nz)]$$

The Maximum Principle gives

$$\dot{k} = s_1\,[f(k) + g(z)] - nk \tag{43}$$

$$\dot{z} = s_2\,[f(k) + g(z)] - nz \tag{44}$$

$$\dot{p}_1 = (\rho + n)p_1 - q\ f'(k) \tag{45}$$

$$\dot{p}_2 = (\rho + n)p_2 - q\ g'(z) \tag{46}$$

where
$$q \equiv (1 - s_1 - s_2)u'(c) + p_1 s_1 + p_2 s_2 \tag{47}$$

$$\partial H/\partial s_1 = (p_1 - u')(f + g) \leq 0 \quad \begin{cases} = 0 \text{ when } s_1 > 0 \\ < 0 \text{ when } s_1 = 0 \end{cases} \tag{48}$$

$$\partial H/\partial s_2 = (p_2 - u')(f + g) \leq 0 \quad \begin{cases} = 0 \text{ when } s_2 \quad 0 \\ < 0 \text{ when } s_2 = 0 \end{cases} \tag{49}$$

The transversality conditions are $\lim_{t\to\infty} p_i\ e^{-\rho t} = 0\ (i = 1, 2)$ (50)

The solution is best examined in 5 phases.

Phase I $s_1 = 0$, $s_2 > -\beta$, $1-s_1-s_2 > 0$ (open segment AB in fig. 7.14)

$\partial H/\partial s_1 < 0 = \partial H/\partial s_2 \Rightarrow p_1 \leq p_2 = u'(c) = q$ and (57) - (60) give

$$\dot{k} = -nk$$

$$\dot{z} = s_2(f + g) - nz$$

$$\dot{p}_1 = (\rho + n)p_1 - p_2 f'(k)$$

$$\dot{p}_2 = (\rho + n)p_2 - p_2 g'(z)$$

The country decumulates domestic capital ($\dot{k} = -nk$) to the limit. This is not an equilibrium case since the transversality condition (50) is violated.

Phase II $s_1 > 0$; $s_2 = -\beta$; $1-s_1-s_2 > 0$ (open segment BC)

(48) and (50) imply

$$p_2 \leq p_1 = u'(c) \leq q \equiv (1 + \beta)p_1 - \beta p_2$$

(43) - (46) give

$$\dot{k} = s_1(f + g) - nk$$

$$\dot{z} = -\beta(f + g) - nz$$

$$\dot{p}_1 = (n + \rho)\, p_1 - qf'$$

$$\dot{p}_2 = (n + \rho)\, p_2 - qg'$$

The country borrows to the limit. This is not an equilibrium case since it violates (50).

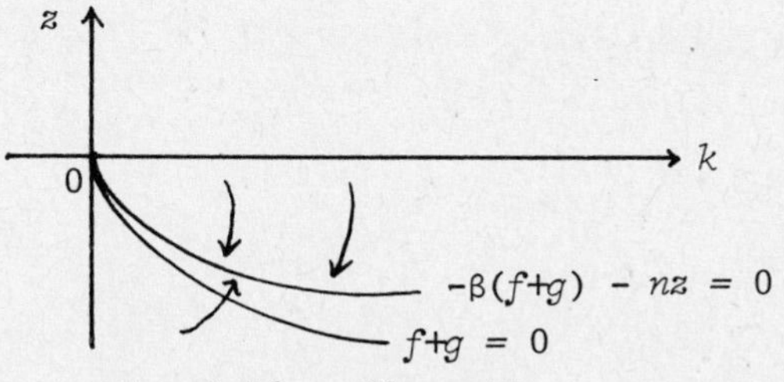

Fig. 7.16 : Phase II

<u>Phase III</u> $s_1 = 0$, $s_2 = -\beta$ at B $(0, -\beta)$

(48) and (49) imply

$$p_1 \leq u'(c) \; ; \; p_2 \leq u'(c)$$

$$u'(c) < q = (1 + \beta)\, u'(c)$$

(43) through (46) imply

$$\dot{k} = -nk$$

$$\dot{z} = -\beta(f + g) - nz$$

$$\dot{p}_1 = (\rho + n)\, p_1 - qf'$$

$$\dot{p}_2 = (\rho + n)\, p_2 - qg'$$

The country decumulates both domestic and foreign capital to the limit. Clearly this is not an equilibrium case.

<u>Phase V</u> $1-s_1-s_2 = 0$ (on the closed segment AC)

This implies

$$p_i \geq u'(0) \text{ and } \infty = u'(0) < q = s_1 p_1 + s_2 p_2$$

where $i = 1, 2$, $u'(0) \equiv \lim_{c \to 0} u'(c) = \infty$. Clearly this case need not be considered.

<u>Phase IV</u> $s_1 > 0$, $s_2 > -\beta$, $(1-s_1-s_2) > 0$ (interior of triangle ABC)

(48) through (50) imply $p_1 = p_2 = u'(c) = q$. This is the equilibrium case since it satisfies all conditions. (43) through (47) give

$$\dot{k} = s_1 (f + g) - nk$$

$$\dot{z} = s_2 (f + g) - nz$$

$$\dot{p}_1 = (\rho + n - f')\, p_1$$

$$\dot{p}_2 = (\rho + n - g')\, p_2$$

The equilibrium given by $(\dot{k}, \dot{z}, \dot{p}_1, \dot{p}_2) = (0, 0, 0, 0)$ is represented by the stationary solution $P(k^*, z^*)$ in fig. 7.17 where $f'(k^*) = g'(z^*) = n + \rho$ and $-\beta < s_2^* = nz^*/[f(k^*) + g(z^*)]$, $0 < s_1^* = nk^*/(f + g)$ and $s_1^* + s_2^* = (f' - \rho)/(f + g) < 1$.

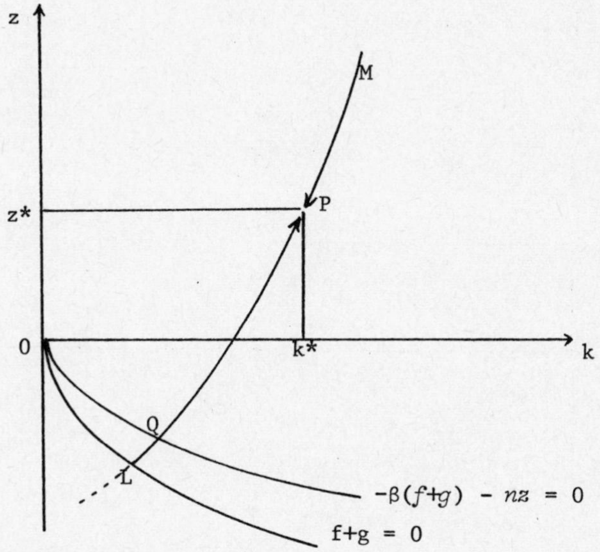

Fig. 7.17 The equilibrium of the economy

$P(k^*, z^*)$, situated to the NE of Q in view of $-\beta < s_2^*$, constitutes the unique interior equilibrium to which all other phases switch.

The Allowable Switching

It can be seen from the above analysis and fig. 7.17 that LM acts as an attractor. If the trajectory initially lies on LM, the system will stay on it and move along it to $P(k^*, z^*)$. If (k_0, z_0) is not on LM, the system will switch to it according to the allowable switching laws discussed in Example 7.3.1, and eventually terminate at $P(k^*, z^*)$.

A switch from phase I where $p_1 \leq p_2$ to phase IV where $p_1 = p_2$

is possible only if at the switching time t_s, $\dot{p}_1/p_1 \geq \dot{p}_2/p_2$. But in phase I, $p_1 \leq p_2 = q$ hence at t_s,

$$\frac{\dot{p}_1}{p_1} - \frac{\dot{p}_2}{p_2} = \frac{qf'}{p_1} + \frac{qg'}{p_2} \leq g' - f'$$

i.e. the switch is allowed only if $g' > f'$ i.e. in the SE of *LM*. See fig. 7.18.

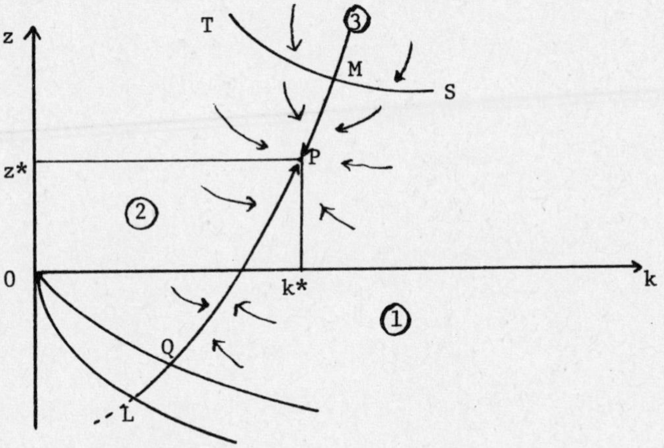

Fig. 7.18 Allowable and Forbidden Switchings

A switch from phase II where $p_2 \leq p_1 \leq q$ to phase IV where $p_1 = p_2 = q$ is possible only if at t_s

$$\frac{\dot{p}_1}{p_2} - \frac{\dot{p}_2}{p_2} = -\frac{qf'}{p_1} + \frac{qg'}{p_2} \geq (g' - f')\,\frac{q}{p_1} \geq 0$$

i.e. only if $g' \leq f'$ at t_s only in the NW of *LM* (see fig. 7.18).

A switch from phase I to II, however, is not allowed: at t_s, $\dot{p}_1 - \dot{p}_2 = -q(f' - g') \geq 0$ implies $f' \leq g'$ i.e. implies that the switching zone ① is to the SE of *LM*. Similarly, a switch from II to I implies $f' - g' \leq 0$ i.e. the region ②, SW of *LM*. However ① and ② are the regions in which II and I must switch to IV as shown above. Hence I and II cannot switch to each other.

Similarly, above the upper limit of phase IV at M, both domestic and foreign capital are excessive and must be decumulated, phase I and II cannot give meaningful solutions: the system must move to ① or ② and thence to LM and eventually to $P(k^*, z^*)$ along LM.

Thus LM acts as an attractor to which the optimal trajectory with given arbitrary initial conditions, switches either directly from phase I and II or indirectly from zone ③ via ① or ②. This completes the examination of allowable switchings.

7.7 Summary and Conclusion

Thus, it can be seen that when the Hamiltonian H is linear in the control variable (u), optimal control will not, in general, exist unless u is bounded. Thus bang-bang control is the optimal course: in maximization (minimization) problems, optimal control $u^* = u_{max}(u_{min})$ if the switching function $\sigma(t)$ is positive and $u^* = u_{min}(u_{max})$ if $\sigma(t) < 0$. A switch from u_{max} to u_{min} occurs when $\sigma(t)$ changes its sign from positive to negative in maximization problems and vice versa in minimisation problems. However when $\sigma(t) \equiv 0$ for some non-zero time interval, bang-bang control provides no information since H is then independent of u: the system is singular. Singular control sometimes turns out to be the most interesting case: it indicates the equilibrium of the system, as in the case of Resource Economics and Optimal Linear growth models above. Finally, transversality conditions are important: in some cases, they provide the only information concerning the choice between the bang-bang and singular control or their combinations as well as the mandatory switches from the one to the other.

CHAPTER 8

STABILIZATION CONTROL MODELS

8.1 Introduction

Fluctuations are inherent in the nature of free enterprise economies. National product, prices, interest rates and other exhibit ups and downs. Prosperity is followed by depression and vice versa. Stabilization aims at eliminating or at least minimizing these fluctuations.

In this chapter, two important types of such stabilizers are examined: the Linear Regulator (LR) and Linear Tracking (LT) models. These constitute an important class of control: they are very useful in analysing the various types of macro-economic policy and planning. Their solutions give linear feedback control laws which are highly suitable for the stabilization of the multiplier-accelerator type models and of market prices.

In these problems, the objective functional is of a quadratic form. More precisely, given the time-varying dynamic system

$$\dot{x}(t) = A(t)x(t) + B(t)\,u(t) \tag{1}$$

the objective is to minimize

LR $$J_R = \tfrac{1}{2}\, x(T)'S\,x(T) + \tfrac{1}{2}\int_0^T (x'Qx + u'Ru)\,dt \tag{2}$$

or LT $$J_T = \tfrac{1}{2}\, e(T)'S\,e(T) + \tfrac{1}{2}\int_0^T (e'Qe + u'Ru)\,dt \tag{3}$$

where $e(t) \equiv x(t) - \hat{x}(t)$ = deviation of the n-state variable vector $x(t)$ from some desired or reference level $\hat{x}(t)$

$u(t)$ = r-control vector

$R(t)$ = positive symmetric definite rxr matrix of relative weights

$Q(t)$ = positive symmetric semi-definite nxn matrix of relative weights

S = constant positive symmetric semi definite nxn matrix.

The objective of the LT problem is to minimize the deviations of the system from some desired level $x^*(t)$, while keeping the control expenditure $u'Ru$ to a minimum.

The LR is a special case of the LT in which x^* is a constant which, by a simple translation, could be set at zero. Since this problem is simpler, we shall deal with it first. This type of problem was solved by Kalman [1960, 1962, 1963a, 1963b] whose method is followed in our presentation.

8.2 Linear Regulator Problems

Consider the problem described in (1) and (2).

The Hamiltonian is

$$H = \tfrac{1}{2}(x'Qx + u'Ru) + p'Ax + p'Bu \tag{4}$$

where $x(t)$, $u(t)$, $Q(t)$, $R(t)$, $A(t)$ and $B(t)$ are written without t to alleviate notations and $'$ denotes transposition.

$$\dot{x} = Ax + Bu \tag{5}$$

$$-H_x = \dot{p} = -Qx - A'p \tag{6}$$

$$\text{with} \quad p(T) = Sx(T) \tag{7}$$

The conditions for the minimization of H are

$$H_u = Ru + B'p = 0$$

giving

$$u = -R^{-1}B'p \tag{8}$$

also $\partial^2 H/\partial u^2 = R$, positive definite by assumption. Substituting into (5) gives

$$\dot{x} = Ax - BR^{-1}B'p \tag{9}$$

With this, (6) and (9) give

$$\begin{bmatrix} \dot{x} \\ \dot{p} \end{bmatrix} = \begin{bmatrix} A & -BR^{-1}B' \\ -Q & -A' \end{bmatrix} \begin{bmatrix} x \\ p \end{bmatrix} \tag{10}$$

whose solution is

$$\begin{bmatrix} x(T) \\ p(T) \end{bmatrix} = \begin{bmatrix} \Phi_{11}\ (T,t) & \Phi_{12}\ (T,t) \\ \Phi_{21}\ (T,t) & \Phi_{22}\ (T,t) \end{bmatrix} \begin{bmatrix} x(t) \\ p(t) \end{bmatrix} \tag{11}$$

where $\left[\Phi_{ij}\ (T,t)\right]$ is the Transition matrix (see Appendix) written in its partionned form. In view of (7), multiplication of the first n equations of (11) by S gives $p(T) = Sx(t)$ which are the last n equations of (11) i.e.,

$$p(T) = S\,\Phi_{11}\ x(t) + S\,\Phi_{12}\ p(t) = \Phi_{21}\ x(t) + \Phi_{22}\ p(t) \tag{12}$$

giving

$$p(t) = \left[\Phi_{22} - S\,\Phi_{12}\right]^{-1} \left[S\,\Phi_{11} - \Phi_{21}\right] x(t) \tag{13}$$

i.e.,

$$\boxed{p(t) = K(t)\,x(t)} \tag{14}$$

where $\quad K(t) \equiv \left[\Phi_{22} - S\,\Phi_{12}\right]^{-1} \left[S\,\Phi_{11} - \Phi_{21}\right]$

and $\Phi_{ij}(T,t)$ are written without T and t to alleviate notations. Note that Kalman has shown that $\left(\Phi_{22} - S\,\Phi_{12}\right)$ is non-singular.

Thus $p(t)$ is a linear function of the state variables and substitution of (14) into (8) gives the feed-back control law

$$u^*(t) = -R^{-1}B'K\ x\ (t) \tag{15}$$

where $-R(t)^{-1}\ B'(t)\,K(t)$ is often referred to as the Kalman matrix.

Substituting (15) into (5) gives

$$\dot{x} = \left(A - BR^{-1} B'K\right)x \tag{16}$$

Differentiating (14) and substituting from (16) yield

$$\begin{aligned}\dot{p} &= \dot{K}x + K\dot{x} \\ &= \left[\dot{K} + K\left(A - BR^{-1}B'K\right)\right]x \end{aligned}\tag{17}$$

But (6) and (14) give

$$\dot{p} = -\left(Q + A'K\right) x \tag{18}$$

Equating (17) with (18) gives

$$[\dot{K} + KA + A'K - KBR^{-1}B'K + Q]\ x(t) = 0 \tag{19}$$

Since (19) holds for all arbitrary choice of the initial state $x(0)$ and $K(t)$ does not depend upon the initial state vector, (19) must hold for all arbitrary $x(t)$ and the matrix in the brackets of (19) must vanish, i.e., K must satisfy the Riccati equation matrix in the differential equations

$$\dot{K} = -KA - A'K + KBR^{-1}B'K - Q \tag{20}$$

with the boundary conditions at $t = T$ given by (7) and (14) as

$$K(T) = S \tag{21}$$

Thus we have shown that if $x(t)$ and $p(t)$ are the solutions of the canonical equations (10) and $p(t) = K(t)\,x(t)$ where $K(t)$ is an unknown matrix, then $K(t)$ must satisfy the differential equation system (19).

Equation (20) is called the Riccati equation matrix. It could be solved backward in time from T to 0 with the Kalman matrix $-R^{-1}B'K$ in (15) stored in order to obtain the feed-back control law

$$u(t) = -R^{-1}B'Kx(t)$$

It is easy to verify that (15) is a minimizer of H by noting the positive semi-definiteness of the following matrix

$$\begin{bmatrix} H_{xx} & H_{xu} \\ H_{ux} & H_{uu} \end{bmatrix} = \begin{bmatrix} Q & 0 \\ 0 & R \end{bmatrix} \tag{22}$$

It is also easy to verify the uniqueness of u^* in (15).

The results obtained could be summarised as follows

Given the Linear Regulator system

$$\dot{x}(t) = A(t)\, x(t) + B(t)\, u(t) \tag{23}$$

and the objective functional

$$J = \tfrac{1}{2}\, x(T)'\ S\, x(T) + \tfrac{1}{2}\int_0^T [x(t)'Q(t)x(t) + u(t)'R(t)\, u(t)]dt \tag{24}$$

where $u(t)$ is unbounded, T specified, S and Q are positive semi-definite and R positive definite matrices, there exists a unique optimal feedback control u^*t given by

$$u^*(t) = -\, R(t)^{-1}B'(t)\, K(t)\, x(t) \tag{25}$$

where $K(t)$ is the unique solution of the Riccati equation (20) *satisfying the boundary conditions* (21).

Note that $K(t)$ is symmetric. To see this, take the transpose of (20), and remembering that Q and $BR^{-1}B'$ are symmetric

$$\left(\frac{dK}{dt}\right)' = -A'K - K'A + K'BR^{-1}B'K - Q = \frac{d}{dt}K' \tag{26}$$

with $K(T) = S = K'(T)$ (21), positive symmetric positive definite. Thus K and K' are the solutions of the same differential equation and satisfy the same boundary conditions. Hence $K(t) = K'(t)$ by the uniqueness of solutions of differential equations.

$K(t)$ being a symmetric matrix, (20) consists of $n(n+1)/2$ different terms representing $n(n+1)/2$ first order non-linear time-varying differential equations.

Note that $K(T)$ is positive definite at $t = T$ by (21) but $K(t)$ is also positive definite for all $t \in [0,T]$ as shown by Kalman [1960], Athans and Falb [1966], Barnett [1975].

Finally, note that in the case A, B, R and Q are all constant matrices, Kalman (1960) has shown that if the system is controllable and $S = 0$ then $K(t)$ is still a time-varying matrix but $\lim_{t\to\infty} K(t) = K$ which is a constant matrix and the Riccati equation (20) becomes simply an algebraic, not a differential equation system

$$-KA - A'K + KBR^{-1}B'K - Q = 0 \tag{27}$$

with K being a time invariant positive definite matrix.

Example 8.1

As an example, let us examine the simplest case where $T = \infty$ $S = 0$ and all coefficients are constants. Consider a country with a foreign debt of $x(t)$ dollars at time t, and a repayment policy $u(t)$ (negative u for repayment and positive u for additional borrowing). In the absence of all repayment, debt increases at a constant rate a, due to the addition of interest payable to capital and other considerations such as population growth, developmental projects already started and the like. In other words, the dynamic system is, for a given $x(0) = x_0$,

$$\dot{x}(t) = ax(t) + u(t) \;, \quad x(0) = x_0$$

The objective functional is to minimize

$$J = \frac{1}{2}\int_0^\infty (qx^2 + ru^2)\, dt$$

where q, $r > 0$. In this case, $S = 0$, $A = a$, $B = b = 1$, $Q = q$ and $R = r$.

The Hamiltonian (with t omitted) is

$$H \equiv \frac{1}{2}(qx^2 + ru^2) + p\,ax + pu$$

The optimal policy, obtained from $\partial H/\partial u = 0$, is

$$u^* = -p/r$$

Applying $p(t) = kx(t)$ from (14) gives

$$u^* = -(k/r)\,x(t)$$

k being constant, the Riccati equation (27) gives

$$k^2 - 2\,rak - rq = 0$$

whose solution is

$$k = ar \pm r\sqrt{a^2r + q/r}$$

The ambiguity of signs is removed by recognizing the fact that $k > 0$ (K being a symmetric positive definite matrix in general), and writing $\beta \equiv \sqrt{a^2r + q/r}$ (> 0), we obtain

$$k = (a + \beta)r$$

The optimal policy is

$$u^*(t) = -(a + \beta)\,x(t)$$

$$\dot{x}(t) = ax(t) + u(t) = -\beta x(t)$$

i.e.,

$$u^*(t) = -(a + \beta)x_0e^{-\beta t}$$

$$x^*(t) = x_0e^{-\beta t}$$

which give the optimal repayment policy and the resulting debt which decreases at the exponential rate β over time.

8.3 Linear Tracking Problems

Linear tracking (LT) consists of devising a control mechanism to cause the system's output $y(t)$ (or state $x(t)$) to track a certain desired or reference output $\hat{y}(t)$ (or state $\hat{x}(t)$). A classical Engineering example is a missile intercepting some given rocket. In Economics, an example is the problem of finding some appropriate control such as Government expenditure, interest rate or money supply to cause the national output to track a certain optimal level determined by full employment or some other long run considerations, in such a way as to stabilize the economy.

More specifically, given the dynamic system

$$\dot{x}(t) = A(t)\, x(t) + B(t)\, u(t) \tag{28}$$

$$y(t) = C(t)\, x(t) \tag{29}$$

the objective is to minimize

$$J = \frac{1}{2}\, e(T)'Se(T) + \frac{1}{2}\int_0^T [\, e(t)'Q(t)e(t) + u(t)'R(t)u(t)]dt \tag{30}$$

where $e(t) \equiv y(t) - \hat{y}(t)$: the deviation of output $y(t)$ from some desired level $\hat{y}(t)$ in the output tracking problem, or

$e(t) = x(t) - \hat{x}(t)$: the deviation of the state vector $x(t)$ from some given optimal level $\hat{x}(t)$, in the state tracking.

Thus, $e(t)$ is the error vector whose weighted square $e'Qe$ is to be minimized without too much expenditure on control $u'Ru$.

The distinction between the state $x(t)$ and output $y(t)$ lies in the measurement of observations. Sometimes, only a few state variables are important or observable. Then output is some combination of state $x(t)$ and possibly control (or input) variables $u(t)$. For example $y(t) = C(t)\, x(t)$ or $y(t) = C(t)\, x(t) + D(t)\, u(t)$. For this reason, output $y(t)$ is also called measurement vector. In a missile intercepting problem, output $y(t)$ may represent the position at time t of a missile launched to intercept some rocket at position $\hat{y}(t)$. Output is a function of a state variable vector $x(t)$, representing say velocity and fuel and some control $u(t)$ representing say the thrust. In

Economics, output is sometimes simply an aggregation problem. For example, some data are recorded weekly, monthly or quarterly and others only annually. Aggregation consists in converting them all into annual data. (See for example, Peston 1973).

In practice, it is sometimes difficult or costly to measure all state variables directly but easier to observe output. If the measurement problem does not arise, the distinction between the state variable vector $x(t)$ and output vector $y(t)$ disappears, i.e., $C(t) = I$. Since our interest does not lie in the measurement problem, we shall not make such distinction and shall not mention output. The extension of the analysis to the case where $C \neq I$ is straightforward (see, for example, Athans & Falb (1966), Sage (1968) or any textbook on Optimal Control in Engineering).

Since LT is only a generalisation of the LR problem to the case where the desired vector $\hat{x}$ (or $\hat{y}$) is not set at the origin, we shall make use of the results derived in the LR case and be brief in our exposition.

Consider the problem of minimizing the functional

$$J = \frac{1}{2}\,[x(T) - \hat{x}(T)]'S\,[x(T) - \hat{x}(T)] + \frac{1}{2}\int_0^T [(x - \hat{x})'Q(x - \hat{x}) + u'Ru]dt \qquad (31)$$

given the dynamic system

$$\dot{x}(t) = A(t)\,x(t) + B(t)\,u(t) \qquad (32)$$

$$x(0) = x_0$$

where $Q = Q(t)$ = symmetric positive semi-definite nxn matrix

$R = R(t)$ = symmetric positive definite rxr matrix

S = constant positive semi-definite nxn matrix

$\hat{x}(t)$ = desired or reference non-zero n-state vector at time t

$u(t)$ = unbounded r-control vector

t is omitted whenever possible to alleviate notations.

The Hamiltonian is

$$H \equiv \frac{1}{2}(x - \hat{x})'Q(x - \hat{x}) + \frac{1}{2}u'Ru + p'Ax + p'Bu \tag{33}$$

The adjoint system is

$$\dot{p} = -H_x = -Qx - A'p + Q\hat{x} \tag{34}$$

$$\text{with } p(T) = Q(x_T - \hat{x}_T) \tag{35}$$

Optimal control, obtained from $H_u = 0$, is

$$Ru + B'p = 0$$

or

$$u^* = -R^{-1}B'p \tag{36}$$

Substitution into (32) gives, together with (34), the canonical system

$$\begin{pmatrix} \dot{x} \\ \dot{p} \end{pmatrix} = \begin{bmatrix} A & -BR^{-1}B' \\ -Q & -A \end{bmatrix} \begin{bmatrix} x \\ p \end{bmatrix} + \begin{pmatrix} 0 \\ Q\hat{x} \end{pmatrix} \tag{37}$$

Solving this, as in (11) gives

$$p(t) = K(t)\,x(t) + v(t) \tag{38}$$

where $K(t)$ is the same as $K(t)$ in (14), and $v(t)$ takes care of $Q\hat{x}$ in (37) which is due to the fact that unlike the LR case, $\hat{x} \neq 0$. (For the detailed derivation of $v(t)$ see, for example, Athans & Falb (1966)).

Substitution of (38) into (37) gives

$$\dot{x} = (A - BR^{-1}B'K)x - BR^{-1}B'v \tag{39}$$

$$\dot{p} = -(Q + A'K)x - A'v + Q\hat{x} \tag{40}$$

Differentiation of (38) and substitution from (39) and (40) give

$$\dot{p} = \dot{K}x + K\dot{x} + \dot{v}$$

$$= (\dot{K} + KA - KBR^{-1}B'K)x - KBR^{-1}B'v + \dot{v} \tag{41}$$

Equating (40) to (41) gives

$$(\dot{K} + KA + A'K - KBR^{-1}B'K + Q)x$$
$$+ (\dot{v} + A'v - KBR^{-1}B'v - Q\hat{x}) = 0 \qquad (42)$$

Since this must hold for all $x(t)$, $\hat{x}(t)$ and t, (42) implies

$$\dot{K} = -KA - A'K - Q + KBR^{-1}B'K \qquad (43)$$

$$\dot{v} = -(A' - KBR^{-1}B')v + Q\hat{x} \qquad (44)$$

with the boundary conditions from (31) and (38) as

$$p(T) = S[x(T) - \hat{x}(T)]$$
$$= K(T)\ x(T) + v(T)$$

But $\quad K(T) = S$

Hence $\quad v(T) = -S\ \hat{x}(T)$

Note that (43) is exactly the Riccati equation of the LR problem and $K(T)$ is a symmetric positive definite matrix involving only $n(n+1)/2$ different terms.

Thus, the LT problem is composed of two components: a LR part and the part associated with the forcing function $Q\hat{x}$. If $v = 0 = \hat{x}$, the LT problem is reduced to the LR model.

The optimal control law is

$$u^*(t) = -R^{-1}B'(Kx + v) \qquad (45)$$

It is left to the reader to verify that, like the LR case, u^* is unique.

The results obtained can be summarised as follows:

Given the linear dynamic system

$$\dot{x}(t) = A(t)\ x(t) + B(t)\ u(t)$$

the desired state vector $\hat{x}(t)$ and the error vector $e(t) \equiv x(t) - \hat{x}(t)$ and the functional

$$J = \frac{1}{2} e(T)'S\, e(T) + \frac{1}{2} \int_0^T [e(t)'Q(t)e(t) + u(t)'R(t)u(t)]dt$$

where T is specified, $R(t)$ is positive definite, S and $Q(t)$ positive semi-definite, there exists a unique optimal control $u^*(t)$ given by

$$u^*(t) = -R^{-1}B'(Kx + v)$$

The nxn real symmetric and positive definite matrix $K(t)$ is the solution of the Riccati equation (43) with the boundary condition $K(T) = S$. The n-vector $v(t)$ is the solution to the linear differential equation (44) with the boundary condition $v(T) = -S\hat{x}(T)$. The optimal trajectory is given by solving (39).

An illustration of the application to the U.S. economy of the LT model, in its discrete form, will be given in Chapter 10.

8.4 Controllability

As has been seen in Chapter 1, controllability refers to the existence of a control $u(t)$ which is capable of transferring a dynamic system from a given initial state x_0 at time $t = t_0$ to a given state x_T at $t = T$. There is no loss of generality in setting $x_T = 0$. Then the controllability condition simply requires the existence of a control $u(t)$ which is capable of driving the system home, to $x_T = 0$, in finite time. Stability is thus implied.

Consider the dynamic system

$$\dot{x}(t) = Ax(t) + Bu(t) \qquad (46)$$

$$x(0) = x_0$$

where A is nxn and B, nxn constant matrices. The case of time varying A and B involves more tedious and cumbersome calculations but the analysis is the same. For expository purposes, we shall only deal with the time invariant case.

It will be shown that the controllability conditions for the system (46) are the $nxrn$ matrix

$$M \equiv [B \quad AB \quad A^2B \quad \dots \quad A^{n-1} B] \tag{47}$$

be of maximal rank n.

To see this, note that the solution of (46) is

$$x(t) = e^{At} (x_0 + \int_0^T e^{-A\tau} Bu(\tau) d\tau) \tag{48}$$

At some final time T, $x(T) = 0$ and (48) becomes

$$-x_0 = \int_0^T e^{-At} Bu(t)\, dt \tag{49}$$

By Caley-Hamilton theorem and the fact that e^{At} is a polynomial in A of at most n-1 degree

$$e^{-At} = \sum_{i=0}^{n-1} (-1)^i A^i \frac{t^i}{i!} \equiv \sum_{i=0}^{n-1} c_i(t) A^i$$

where $c_i(t)$ are scalar functions of t.

Substituting into (49) gives, with $u(t)$ written as $u(t)I \equiv \Sigma u_j(t) e_j$ where e_j is the jth unit vector,

$$-x_0 = \int_0^T (c_0 I + c_1 A + c_2 A^2 + \dots + c_{n-1} A^{n-1}) Bu(t) dt$$

$$\equiv \sum_0^{n-1} \int_0^T c_i(t) A^i Bu(t) dt = \left[\sum_{j=1}^{r} \left(\int_0^T c_i(t) u(t) dt\right)\right] A^i Be_j$$

$$\equiv \left[\sum_{j=1}^{r} \left(\int_0^T c_i(t) u_j(t) dt\right)\right] M \tag{50}$$

It is easy to see from (50) that x_0 is in the linear space spanned by the columns of M, i.s., the system (46) is controllable if and only if M defined in (47) is of maximal rank n.

In the particular case in which $B = b$, a constant n-vector and $u(t)$ is a scalar, M becomes

$$M \equiv \begin{bmatrix} b & Ab & A\ b \ldots . & A^{n-1}b \end{bmatrix}$$

which is an nxn time-invariant matrix and controllability is reduced to the requirement that M be non-singular.

For example for the following system $x = Ax + bu$

$$\begin{pmatrix} \dot{x}_1 \\ \dot{x}_2 \end{pmatrix} = \begin{bmatrix} 1 & 2 \\ 0 & 1 \end{bmatrix} \begin{bmatrix} x_1 \\ x_2 \end{bmatrix} + \begin{pmatrix} 1 \\ -1 \end{pmatrix} u$$

the controllability matrix $M = \begin{bmatrix} 1 & -1 \\ -1 & -1 \end{bmatrix}$ is of rank 2, hence the system is controllable. If $b' = (1, 0)$ instead, M would be of rank 1 and hence the system would be uncontrollable. This is obvious since in this case, we would have $\dot{x}_2 = x_2$ which would remain unaffected by $u(t)$.

8.5 Observability

Observability refers to the possibility of determining the state of a system by measuring only its output. A system is said to be completely observable if, given any t_0 and x_0, knowledge of $u(t)$ and $y(t)$ ($t_0 \leq t \leq T$) is sufficient to determine x_0.

Observability and Controllability are closely linked with

each other: controllability refers to the possibility of controlling a dynamic system from the control $u(t)$ and observability examines the possibility of estimating and measuring the state $x(t)$ from the output $y(t)$. Consider the constant coefficient system

$$\dot{x}(t) = Ax(t) + Bu(t) \tag{51}$$

$$y(t) = Cx(t) \tag{52}$$

$$= Ce^{At}(x_0 + \int_0^t e^{-At} Bu(\tau)d\tau) \quad \text{by (48)} \tag{53}$$

It can be seen that $u(t)$ and $y(t) = g[x_0, u(t)]$ $\forall\ t \in [0, T]$ determines x_0.

For simplicity, let $u(t) = 0\ \forall\ t \in [0, T]$, (52) becomes simply

$$y(t) = Ce^{At}\ x_0 = C\ [c_0 I + c_1 A + \ldots + c_{n-1} A^{n-1}]x_0 \tag{54}$$

and the observability conditions are reduced to the requirement that

$$H \equiv [C' \vdots A'C' \vdots \ldots (A')^{n-1}\ C'] \tag{55}$$

be of maximal rank and in the case C is a row vector,

$$H = [c' \vdots A'c' \vdots \ldots \vdots (A')^{n-1}\ c'] \tag{56}$$

be non-singular.

8.6 Some Economic Applications

Example 8.6.1 The Multiplier-Accelerator Model

As an example of the application of LT in Economics, consider the Samuelson (1939) Hicks (1950) type of Multiplier – Accelerator model characterised by the following equations

$$I(t) = v\dot{Y}(t) + G(t)$$

$$S(t) = s_1 Y(t) + s_2 Y(t) \equiv sY(t)$$

$$\dot{Y}(t) = h\ [I(t) - S(t)]$$

Setting the constant adjustment speed h equal to 1 for simplicity

$$\dot{Y}(t) = -\frac{s}{1-v}\ Y(t) + \frac{1}{1-v}\,G(t)$$

or

$$\dot{y}(t) = ay(t) \qquad + bg(t)$$

the objective is to minimize

$$J = \frac{1}{2}\int_0^{\infty} (qy^2 + rg^2)\ dt$$

where $I(t)$ = total investment demand on the part of both the private sector $v\dot{Y}(t)$ (v = constant accelerator) and the Government sector $G(t)$

$S(t)$ = total saving = private saving $s_1Y(t)$ + Government saving $s_2Y(t)$ where s_1 = private marginal propensity to save and s_2 = Government's marginal propensity to tax

$y(t) \equiv Y(t)-Y^*$; $g(t) \equiv G(t)-G^*$ where Y^*, G^* are optimal (full employment) GNP and Government Expenditure respectively and $a \equiv -s/(1-v)$, $b \equiv 1/(1-v)$

q, r : positive weights assumed constant

The objective is to stabilize the economy by minimizing the deviations of income and public spending from their optimal levels, subject to the dynamic system $\dot{y} = ay + bg$.

The Hamiltonian of the system, omitting t, is

$$H \equiv \frac{1}{2}(qy^2 + rg^2) + p(ay + bg)$$

$\partial H/\partial g = 0$ implies (see (45))

$$g = -(b/r)p = -(b/r)(ky + v)$$

where $\lim_{t\to\infty} k(t) = k$, constant. Equation (27) gives

$$k^2 - 2(ar/b^2)k - rq/b^2 = 0$$

or $$k = (r/b^2)(a + \sqrt{a^2 + qb^2/r})$$

where the ambiguity of signs (±) is removed by recognizing the fact that $k > 0$ (K is a symmetric positive definite matrix in general). By (43), $v(t)$ is the solution of

$$\dot{v} = -(a - kb^2/r)v + qY^*$$

with $v(\infty) = 0$ (since $S = 0$). Assuming Y^* is a constant and $G^* = 0$,

$$\lim_{t\to\infty} v(t) = \frac{qY^*}{\alpha} + Ae^{-\alpha t} = 0$$

where A is a constant of integration and

$$\alpha \equiv a - kb^2/r = -\sqrt{a^2 + qb^2/r} < 0$$

thus $$\lim_{t\to\infty} g^*(t) = -(b/r)ky(t)$$

$$\equiv (s - \sqrt{s^2 + q/r})y \equiv \beta y(t)$$

where $$\beta \equiv s - \sqrt{s^2 + q/r} < 0 \quad \text{since } q,\ r > 0$$

This means that if there is a deficit of GNP from its optimal level Y^*, i.e., $y < 0$, the Government must correct it by increasing its spending, i.e., $g^* > 0$ in the long run. (See Turnovsky 1981).

The resulting optimal deviation of Y from Y^* is easily seen to be

$$\dot{y} = ay + bg^*$$

$$= (a - b^2k/r)y \equiv my$$

where $m \equiv a - b^2k/r \equiv -(s + q/r)^{\frac{1}{2}} /(1-v)^2 < 0$.

The solution obtained is

$$y^*(t) = y_0 e^{mt}$$

i.e., in an optimal programme, $Y(t)$ tends to $Y^*(t)$ over time.

Example 8.6.2 Production and Inventory Stabilization Model

The problem of minimizing the cost of holding inventories (x) and production (u) above or below some desired level $\hat{x}$ and $\hat{u}$ has been investigated by Simon (1952), Hwang et al. (1967), Holt et al. (1969) and lately by Bensoussan (1974). For a review, see Sethi (1978). We shall present Bensoussan's model to illustrate the Linear Tracking approach.

Given an exogenously given sale rate $s(t)$ where $s(t) > 0$ but not necessarily continuous, the accumulation of inventories $\dot{x}$ is the difference between the production rate $u(t)$ and the sale rate $s(t)$,

$$\dot{x}(t) = u(t) - s(t) \quad ; \quad x(0) = x_0 \tag{57}$$

The objective is to choose the production level $u(t)$ such as to minimize

$$J = \int_0^T [q\,(x - \hat{x})^2 + r(u - \hat{u})^2\,]\,dt \tag{58}$$

where q, r are given positive constant weights attached to the deviation from the desired stock and production levels respectively.

The Hamiltonian (in maximization form) is

$$H \equiv -\,q(x - \hat{x})^2 - r(u - \hat{u})^2 + p(u - s) \tag{59}$$

The adjoint equation is

$$-H_x = \dot{p} = 2q(x - \hat{x}) \quad ; \quad p(T) = 0 \tag{60}$$

The optimal production rate u^* obtained by

$$H_u = 0 = -\,2r\,(u - \hat{u}) + p$$

is

$$u^* = p/2r + \hat{u} \tag{61}$$

Substitution into the state equation gives

$$\dot{x} = p/2r + \hat{u} - s(t) \quad ; \quad x(0) = x_0 \tag{62}$$

(38) gives

$$p(t) = k(t)\ x(t) + v(t) \tag{63}$$

which, on differentiation, is

$$\dot{p} = k\dot{x} + \dot{k}\ x + \dot{v}$$

$$= k\ [p/2r + \hat{u} - s(t)] + \dot{k}\ x(t) + \dot{v} \tag{64}$$

$$= 2q\ (x - \hat{x}) \quad \text{by} \quad (60) \tag{65}$$

Equating (64) and (65) gives, for all $t \in [0, T]$ and $p(T) = 0$

$$(\dot{k} + \frac{k^2}{2r} - 2q)\ x + \dot{v} + \frac{k}{2r} v - k(\hat{u} - s) + 2q\hat{x} = 0 \tag{66}$$

(43) and (44) give

$$\dot{k} + \frac{k^2}{2r} - 2q = 0 \tag{67}$$

$$\dot{v} + \frac{k}{2r} v - k(\hat{u} - s) + 2q\hat{x} = 0 \tag{68}$$

with $k(T) = 0 = v(T)$ $(\ p(T) = 0)$.

The solution of the Riccati equation (67) gives

$$k(t) = -\ 2r\alpha\ tanh\ [\alpha\ (T - t)] \tag{69}$$

where $\alpha \equiv (q/r)^{\frac{1}{2}}$

Substituting into (68), assuming $\hat{u} = s$ and solving

$$v(t) = 2r\alpha\hat{x}\ \ tanh\ [\alpha\ (T - t)] \tag{70}$$

Substitution into (61) gives the optimal production rate

$$u^*(t) = \alpha(\hat{x} - x)\ tanh\ [\alpha\ (T - t)] + s(t) \tag{71}$$

and after some manipulation, we obtain the optimal inventory level $x^*(t)$ at each period as

$$x^*(t) = \hat{x} + cosh\ [\alpha\ (T - t)]\ (x_0 - \hat{x})/cosh\ (\alpha T) \tag{72}$$

Thus, the optimal production rate $u^*(t)$ is a feed-back control: in addition to the parameters of the system and the exogenously given sales rate $s(t)$, it depends on the current inventory level $x(t)$.

Equation (72) shows that if originally, inventory x_0 is at its desired level x, it will remain at that level all the time and if $x_0 \neq \hat{x}$, it will approach the desired level $\hat{x}$ asymptotically.

Example 8.6.3 Economic Stabilization: the Austrian case

Neck and Posch (1982) has revently investigated the optimality of Macro-economic Policies in Austria.

The variables and weights of his model are

Variables			Weights
PV	=	deflator	0
$PV\%$	=	inflation rate	1000
R	=	long term bond yield	1
M'	=	imports	5
V'	=	total real demand	0
IF'	=	real fixed investment	5
GDP'	=	real gross domestic product at market prices	10
GDP	=	nominal " " " " "	0
YD	=	nominal disposable income	0
C'	=	real private consumption	5
G'	=	real public consumption	5
T	=	net tax rate	1
$M1$	=	Money supply $M1$	1

The behavioural equations for Consumption, Investment, imports and money demand were estimated by OLS for the period 1955-1980. The objective is to minimize the welfare loss W_1 where

$$W_1 = \frac{1}{N}\left[\sum_{t=1}^{N} (y_t - a_t)'K(y_t - a_t) - \sum(y_t^* - a_t)'K(y_t^* - a_t)\right] \quad (43)$$

subject to the linearised system

$$y_t = A_t y_{t-1} + C_t u_t + b_t \quad (44)$$

where A and C are matrices; $b_t \equiv y_t^0 - A_t y_t^0 - C_t u_t^0$; y_t = historical values of the endogenous and control variables listed above, y_t^* = their values obtained by optimization, a = their target values and K is the matrix of the weights listed above: it is set at 5% annual growth for total demand and its components V', GDP', C', IF', M', G', GDP, YD, $M1$ and zero for R, PV, $PV\%$ and T.

An alternative welfare loss function W_2 was also used where

$$W_2 = \left[\sum_{t=1}^{N} (y_t - a_t)'K(y_t - a_t) \Big/ \sum_{t=1}^{N} (y_t^* - a_t)'K(y_t^* - a_t) \right] - 1 \quad (45)$$

Optimization gives

$$u_t = G_t\, y_{t-1} + g_t \qquad (t = 1, 2, \ldots, N) \quad (46)$$

where

$$G_t = -(C_t'\, H_t\, C_t)^{-1}\, (C_t'\, H_t\, A_t)$$

$$g_t = -(C_t'\, H_t\, C_t)^{-1}\, C_t'\, (H_t\, b_t - h_t)$$

$$H_{t-1} = K_{t-1} + A_t'\, H_t\, (A_t + C_t\, G_t)$$

$$h_{t-1} = K_{t-1}\, a_{t-1} - A_t'\, H_t\, (b_t + C_t\, g_t) + A_t'\, h_t$$

with boundary conditions $H_N = K_N$ and $h_N = K_N\, a_N$

For the W_1 where inflation rate gets a weight of 1000, the results are W_1 = 37.874 for the ÖVP government (1967-1976) and W_1 = 1128.8544 for the SPÖ government (1970-1975). Clearly the performance of the SPÖ was much worse than the ÖVP. However W_2 = .2935 for ÖVP and W_2 = 1.87325 for SPÖ i.e. although the SPÖ's performance is still worse, it was not to the extent indicated by W_1. It must be mentioned that the model is only tentative: it has its weaknesses among which prices are largely exogenous and the weight associated with inflation is perhaps excessive. Nevertheless, it provides valuable insights for the Austrian Economy.

Additional examples will be given in Chapter 9.

8.7 Conclusion

Thus, it can be seen that Linear Regulator and Linear Tracking are useful tools for economic stabilisation. By penalizing all undesirable deviations--the degree of severity being reflected by the relative weights in Q and R, assigned by the policy maker, it is hoped that the economy can be brought under control the same way as a guided missile. Theoretical as well as applied economic models have been devised (for a review of some major work, see, for example, D. Kendrick (1977)). With continued research in this direction, no doubt fluctuations in income and employment will be mitigated and inflation and unemployment will not deviate too far from some desired or tolerable paths.

CHAPTER 9

DISCRETE CONTROL SYSTEMS

9.1 Introduction

So far, the Calculus of Variations and Maximum Principle have been analysed in forms of differential equation system where changes are continuous or treated as continuous over time. In the real world, however, many observations are made and recorded only at discrete intervals, for example gross national product is measured only once a year, imports and exports figures are reported only quarterly. A discrete counterpart of the continuous Variational problems and Maximum Principle must be developed. This will be done in this chapter.

The discrete Maximum Principle and Calculus of Variations have been investigated by Rozonoer (1959), Fan and Wang (1964), Holtzman (1966) and others in the 1960's. There are many approaches to the problem and in this chapter, we choose to extend our analysis of the previous chapters to the discrete systems in a natural way, with a minimum of additional material.

9.2 Discrete Calculus of Variations

In the continuous variational problems, the objective functional is

$$J = \int_0^T f_0(x, \dot{x}, t)\, dt$$

In the discrete variational case, the counterpart of the above objective functional is a summation of scalar functions such as

$$J = \sum_{K=1}^{N-1} g\ (x_k, x_{k-1}, k) \equiv \sum_{K=1}^{N-1} g_k \tag{1}$$

where $x_k = n-$ state vector at period k $(1 \leq k \leq N-1)$

and $g(x_k, x_{k-1}, k)$ is a differentiable scalar function

g_k is the incremental value of the functional in period k, assuming observations are made at equal intervals

thus $g_k \equiv g(x_k, x_{k+1}, k)$

$$g_{k+1} \equiv g(x_{k+1}, x_{k+2}, k+1) \text{ etc...} \tag{2}$$

Using . to denote dot or inner products, we can write the first variations of (1) as

$$\partial J = \sum_{k=1}^{N-1} \left(\delta x_k \cdot \frac{\partial g_k}{\partial x_k} + \delta x_{k+1} \cdot \frac{\partial g_k}{\partial x_{k+1}}\right) \tag{3}$$

With the notation of (2), the last term of (3) is written as

$$\sum_{k=1}^{N-1} \delta x_{k+1} \cdot \frac{\partial g_k}{\partial x_{k+1}} \equiv \sum_{k=2}^{N} \delta x_k \cdot \frac{\partial g}{\partial x_k} (x_k, x_{k+1}, k)$$

$$\equiv \sum_{k=1}^{N-1} \delta x_{k+1} \cdot \frac{\partial g}{\partial x_{k+1}} (x_k, x_{k+1}, k)$$

$$\equiv \sum_{k=1}^{N-1} \delta x_k \cdot \frac{\partial g}{\partial x_k} (x_{k-1}, x_k, k-1) + \delta x_k \cdot \frac{\partial g}{\partial x_k} (x_{k-1}, x_k, k-1) \Bigg|_{k=1}^{k=N} \tag{4}$$

In this notation, the necessary conditions for an extremum of (1) obtained by requiring $\delta J = o$ in (3), can be written as

$$o = \sum_{k=1}^{N-1} \delta x_k \cdot \left[\frac{\partial g}{\partial x} (s_k, x_{k-1}, k) + \frac{\partial g}{\partial x_k} (x_{k-1}, x_k, k-1)\right] + \delta x_k \cdot \frac{\partial g}{\partial x_k} (x_{k-1}, x_k, k-1) \Bigg|_{k=1}^{k=N} \tag{5}$$

For arbitrary variations of $\delta x_k \; \forall k$, (5) implies

$$\frac{\partial g}{\partial x_k} (x_k, x_{k+1}, k) + \frac{\partial g}{\partial x_k} (x_{k-1}, x_k, k-1) = o \tag{6}$$

$$(1 \leq k \leq N - 1)$$

and

$$\delta x_k \cdot \frac{\partial g}{\partial x_k} (x_{k-1}, x_k, k-1) = o \text{ for } k=1 \text{ and } k=N \tag{7}$$

Clearly (6) is the discrete counterpart of the Euler-Lagrange equation and (7) is the Transversality condition.

Example 9.1 The discrete optimal growth model

The optimal economic growth model analysed in Chapter 5 (5.6.2) ignoring depreciation and assuming constant returns to scale, now appears, for each period t, in its discrete guise as

$$C_t + K_{t+1} - K_t = F(K_t, L_t) \equiv F(K_t, L_t) \equiv L_t f(k_t) \tag{9}$$

where $k_t \equiv K_t / L_t$ and t is discrete time (subscript t now corresponds to subscript k in the last section, k_t being used here to denote capital per worker at period t) .

Population is assumed to grow at a constant rate n i.e.,

$$L_{t+1}/L_t = 1 + n \tag{10}$$

Writing per capita consumption C_t/L_t as c_t, we have the neo-classical fundamental growth equation in its discrete form as

$$c_t = f(k_t) + k_t - (1+n)k_{t+1} \tag{11}$$

where $f\ (k_t)$ is the neo-classical production function and $k(o) = k_o$, $k(T\text{-}1) = k_{T-1}$ are prescribed.

The objective functional is to maximize the sum of discounted per-capita consumption

$$J = \underset{c_t}{Max.} \ \overset{T}{\underset{t=o}{\equiv}} \ \lambda^{-t}u(c_t)$$

$$= \underset{k_t}{Max.} \ \overset{T}{\underset{t=o}{\equiv}} \ \lambda^{-t}u\left[f(k_t) + k_t - (1+n)k_{t+1}\right] \tag{12}$$

where λ is a positive constant rate of future discount.

Using the notation of (2) to write g as

$$g(k_t, k_{t+1}, t) \equiv \lambda^{-t}u(c_t) \equiv \lambda^{-t}u\ \left[f(k_t) + k_t - (1+n)k_{t+1}\right]$$

$$g(k_{t-1}, k_t, t\text{-}1) \equiv \lambda^{-(t-1)}u\left[f(k_{t-1}) + k_{t-1} - (1+n)k_t\right]$$

we obtain from the Euler equation (6)

$$u'(c_t)\lambda^{-t}\left[1+f'(k_t)\right] - u'(c_{t-1})\lambda^{1-t}(1+n) = o$$

or divided by λ^{-t}

$$u'(c_t)\ \left[1+f'(k_t)\right] - u'(c_{t-1})\ \lambda(1+n) = o \tag{13}$$

and from the transversality conditions (7)

$$-\delta k_t(1+n)u'(c_t)\Big|_{t=o}^{t=T} = o \tag{14}$$

which are clearly satisfied in this two-fixed end point problem i.e., $\delta k_o = o = \delta k_T$.

The Euler equation (13) thus gives the familiar intertemporal optimality condition

$$\frac{u'(c_t)}{u'(c_{t-1})} = \frac{(1+n)\lambda}{1+f'(k_t)} = \text{discounted MRS}$$

i.e. an optimal investment (or consumption) programme should be carried out such that the ratio of the marginal utility of consumption in two periods is brought into equality with $(1+n)\lambda/(1+f'(k_t))$ which is the discounted marginal rate of substitution (MRS) of consumptions between two periods. In particular, when the consumption stream is stationary and the utility function remains unchanged, i.e. $u'(c_t) = u'(c_{t-1})$ for all t, we have

$$1 + f'(k_t) = (1+n)\lambda$$

In the absence of discounting rate i.e. $\lambda=1$, we obtain $f'(k_t) = n$, the Swan Phelps' golden rule that investment should be carried out to the point where the rate of return on capital is equal to the rate of population growth n. In the case of stationary population ($n=o$) and no future discount ($\lambda=1$), $f'(k_t) = o$, i.e. investment should be carried out to the point of capital saturation.

9.3 Discrete Maximum Principle

Similarly, the discrete Maximum Principle could be developed in the same way as the Continuous optimal control presented in Chapters 5 and 6. The analysis in these chapters will now be extended to the discrete case.

Consider the discrete dynamic system

$$x_{k+1} = f(x_k, u_k, k) \tag{15}$$

The objective is to maximize or minimize J where

$$J = S(x_k, k)\Big|_{k=1}^{k=N} + \sum_{k=1}^{N-1} g(x_k, u_k, k) \tag{16}$$

where x_k = n – state vector at period k

u_k = r – un bounded control vector at period k

f = a differentiable vector function

$S(x_k, k)$ = a differentiable "scrap" function

$g(x_k, u_k, k)$ = a differentiable scalar functions as in (1)

. = dot or inner products

As in chapter 5, we define the augmented functional J_a as

$$J_a \equiv S(x_k, k)\Big|_{k=1}^{k=N} + \sum_{k=1}^{N-1} \{g(x_k, u_k, k) - p_{k+1}\left[x_{k+1} - f(x_k, u_k, k)\right]\}$$

$$\equiv S(x_k, k)\Big|_{k=1}^{k=N} + \sum_{k=1}^{N-1} (H_k - p_{k+1} \cdot x_{k+1}) \tag{17}$$

where p_k = *co*- state or adjoint variable vector

$$H_k = g(x_k,u_k,k) + p_{k+1}\cdot f(x_k,u_k,k) \tag{18}$$

the discrete Hamiltonian

For an extremum, the first variations give

$$\frac{\partial S}{\partial x_k}\cdot\delta x_k\Bigg|_{k=1}^{k=N} + \sum_{k=1}^{N-1}\frac{\partial H_k}{\partial x_k}\cdot\delta x_k - \sum_{k-1}^{N-1} p_{k+1} + \sum_{k=1}^{N-1}\frac{\partial H_k}{\partial u_k}\cdot\delta u_k = 0 \tag{19}$$

The second last term of (19) can be written as

$$-\sum_{k=1}^{N-1} p_{k+1}\cdot\delta x_{k+1} \equiv -\sum_{k=2}^{N} p_k\cdot\delta x_k$$

$$\equiv -\sum_{k=1}^{N-1} p_k\cdot\delta x_k - p_N\cdot\delta x_N + p_1\delta x_1 \tag{20}$$

which is the discrete counterpart of the integration by parts.

Substitution into (19) yields

$$\left(\frac{\partial S}{\partial x} - p\right)\cdot\delta x_k\Bigg|_{k=1}^{k=N} + \sum_{k=1}^{N-1}\left(\frac{\partial H_k}{\partial x_k} - p_k\right)\cdot\delta x_k + \sum_{k=1}^{N-1}\frac{\partial H_k}{\partial u_k}\cdot\delta u_k = 0 \tag{21}$$

For (21) to hold, for arbitrary variations, it is plain that the following should hold true

$$\boxed{p_k = \frac{\partial H_k}{\partial x_k}} \quad \forall k \tag{21}$$

$$\boxed{\frac{\partial H_k}{\partial u_k} = 0} \quad \forall k \tag{22}$$

$$\left.\left(\frac{\partial S}{\partial x_k} - p_k\right) \cdot \delta x_k\right|_{k=1}^{k=N} = o \tag{23}$$

which means $\delta x_k = o$ $(k=1,N)$ for the case x_1, and x_N are specified; and for the case these are unspecified,

$$p_k = \frac{\partial S}{\partial x_k} \quad (k=1,N)$$

The last equation (23) gives the transversality conditions. They are the same as in the continuous case (Ch. 5, eq. 32).

Note that the Lagrangean multiplier vector or co-state vector p_k in (17) could be preceded by the negative or positive sign. The results are unaffected.

Note also that if the dynamic system is given as

$$x_{k+1} - x_k = f(x_k, u_k, k) \tag{24}$$

with $\qquad x(o) = x_o$ and $x(N) = x_N$

i.e. x_k on the LHS of (24) cannot be absorbed into the RHS to give the form (15) then (21) would be

$$p_k - p_{k-1} = - \delta H_k / \delta x_k \tag{25}$$

where the Hamiltonian $H_k \equiv g_k(x_k, u_k, k) + p_k \cdot f_k(x_k, u_k, k)$.

(Note the difference between this H_k and the Hamiltonian H_k in (18).

To see this, define the augmented objective functional J_a as follows

$$J_a \equiv \sum_{k=1}^{N} \{g_k(x_k, u_k, k) - p_k \left[x_{k+1} - x_k - f_k(x_k, u_k, k)\right]\} \tag{26}$$

The necessary conditions for (26) to be an extremum are

$$\frac{\partial J_a}{\partial x_k} = \frac{\partial g_k}{\partial x_k} + p_k + p_k \frac{\partial f_k}{\partial x_k} - p_{k-1} = 0 \qquad (2 \le k \le N-1)$$

or

$$\boxed{p_k - p_{k-1} = -\frac{\partial H_k}{\partial x_k}} \tag{21'}$$

$$\frac{\partial J_a}{\partial u_k} = \frac{\partial g_k}{\partial u_k} + p_k \frac{\partial f_k}{\partial u_k} \equiv \boxed{\frac{\partial H_k}{\partial u_k} = 0} \qquad (1 \le k \le N) \tag{22'}$$

$$\frac{\partial J_a}{\partial p_k} = x_{k+1} - x_k - f_k\,(x_k, u_k, k) = 0 \quad (1 \le k \le N-1)$$

or

$$\boxed{x_{k+1} - x_k = f_k\,(x_k, u_k, k)} \tag{27}$$

These give the exact counterpart of the continuous Maximum Principle. They contain $3N$-3 conditions for the determination of the $3N$-3 unknowns $x_2, \ldots, x_{N-1}$, $u_i, \ldots, u_N$ and $p_1, \ldots, p_{N-1}$.

Example 9.2 Optimal Management of Renewable Resources

Consider a model of optimal harvesting of a renewable resource, say fish (C. Clark 1976, Ch. 7). The fish population P_k at generation k gives birth to R_k some of which may be harvested H_k. The "stock-recruitment" relation is

$$R_k = F\,(P_k)$$

The dynamics of the fish population are

$$P_{k+1} = R_k - H_k = F(P_k) - H_k$$

or $R_{k+1} = F(R_k - H_k)$

or, to preserve the similarity with the continuous model,

$$R_{k+1} - R_k = F(R_k - H_k) - R_k$$

Subject to this, the objective is to maximize the present value of the profit functional

$$\sum_{k-1}^{\infty} \Pi\,(R_k, H_k)\alpha^{k-1}$$

where $\alpha = 1/(1+i)$ is the discount factor (i = interest rate)

Clearly H_k is the control and R_k, the state variable.

The Hamiltonian for each period k is

$$\mathcal{H}_k = \alpha^{k-1}\Pi(R_k, H_k) + \lambda_k \; [F(R_k - H_k) - R_k]$$

The optimal harvesting policy is obtained by $\partial\mathcal{H}_k/\partial H_k = 0$ as

$$\lambda_k = \alpha^{k-1}\Pi_H/F'$$

or $\lambda_k - \lambda_{k-1} = (\alpha^{k-1} - \alpha^{k-2})\; \Pi_H/F'$

where $F' = \partial F/\, \partial(R_k - H_k)$ and $\Pi_H \equiv \partial\Pi/\partial H_k$

But the adjoint system gives

$$\lambda_k - \lambda_{k-1} = -\,\partial\mathcal{H}_k/\partial R_k = -\,\alpha^{k-1}\;\Pi_R - \lambda_k(F'-1)$$

$$= -\,\alpha^{k-1}[\Pi_R + \Pi_H\;(1-1/F')$$

i.e. $F'\;(1+\Pi_R/\,\Pi_H) = \alpha^{-1} \equiv 1+i$

i.e., in an optimal harvesting programme, fish should be caught up to the point where the marginal product (or reproduction) of the fish population F', adjusted for the marginal stock effect $1+\Pi_R/\Pi_H$, is equal to the discount factor. If the fishing cost C is not affected by the size of the fish population, e.g., $\Pi(H) \equiv pH_k - C(H_k)$ then $F' = \alpha^{-1} = 1+i$ which is the well known equilibrium condition in the theory of capital. However, it is normally expected that the larger the fish population, the lower the cost is i.e., $\Pi(R,H) = pH_k - C(R_k,H_k)$ with $\partial C/\partial R_k < 0$, and the term $1+\Pi_R/\Pi_H$ reflects the effect of dwindling stock on cost and acts as a link between the present gain and the future loss resulting from overfishing. The above rule is the discrete counterpart of the equilibrium rule (Ch. 7., eq. 36) in the continuous case.

Example 9.3 Discrete Linear Regulator

Consider the discrete counterpart of the continuous Linear Regulator problem examined in the previous chapter. The objective is to minimize the functional.

$$J \equiv \tfrac{1}{2}x(N).Sx(N) + \tfrac{1}{2}\sum_{k=0}^{N-1}(x_k.Qx_k + u_k.Ru_k) \tag{28}$$

subject to the discrete dynamic system

$$x_{k+1} = Ax_k + Bu_k \quad (o \le k \le N) \tag{29}$$

where $x(o) = x_o$ but $x(N)$ is unspecified

x_k = n- state vector at period k

u_k = r- control vector at period k

S = Scalar differentiable "scrap" function, giving the value or cost of the terminal period.

Q, R = positive symmetric matrices

. = dot or inner products and $'$ denoted transposition.

The Hamiltonian is

$$H = \tfrac{1}{2}\, x_k . Qx_k + \tfrac{1}{2} u_k . Ru_k + p_{k+1} . (Ax_k + Bu_k) \tag{30}$$

The co-state vector equation, given by (21) is

$$p_k = Qx_k + A'p_{k+1} \tag{31}$$

At the terminal period $k=N$, the transversality condition (23)

gives

$$p(N) = Sx(N) \tag{32}$$

The optimal control given by (22) is

$$u_k = - R^{-1} B' p_{k+1} \tag{33}$$

Substitution into (29) gives

$$x_{k+1} = Ax_k - BR^{-1}B'p_{k+1} \text{ , } x(o) = x_o \tag{34}$$

$$p_k = Qx_k + A'p_{k+1} \text{ ,} p(N) = Sx(N)$$

From Kalman equation (8.14), we have

$$p_k = K_k x_k$$

Substitution into (34) to eliminate p_k and p_{k+1}, gives

$$x_{k+1} = Ax_k - BR^{-1}B'K_{k+1}x_{k+1}$$

$$K_k x_k = Qx_k + A'K_{k+1}\, x_{k+1}$$

Solving, as in chapter 8

$$K_k x_k = Qx_k + A'K_{k+1} \left[I+BR^{-1}B'K_{k+1}\right] Ax_k$$

which must hold for any arbitrary x_k i.e. the following Riccati equation systems must hold for each period k

$$K_k = Q + A'K_{k+1}\ (I+BR^{-1}B'K_{k+1})\ A$$

with the boundary condition for the final period N

$$K_N = S$$

Thus the parallel with the continuous Linear Regulator problem is complete.

Example 9.4 Linear Tracking and Economic Stabilization

Consider the discrete counterpart of the Linear Tracking model analysed in Chapter 8. The problem, it will be recalled, consists of tracking a certain nominal (optimal or otherwise such as the reference path in the missile intercepting problem) state $\hat{x}_k$ and control $\hat{u}_k$ vector at period k $(o \leq k \leq N)$. In the economic stabilization problem, $\hat{x}_k$ represents economic variables such as GNP, consumption, Investment, Labour, Wage, Price and others and $\hat{u}_k$, government expenditure, tax, money supply, etc. The objective is to minimize the deviation of x_k and u_k from their desired levels $\hat{x}_k$ and $\hat{u}_k$. More precisely, the objective is to minimize

$$J = \tfrac{1}{2}(x_N-\hat{x}_N).\ S(x_N-\hat{x}_N) + \tfrac{1}{2}\sum_{k=o}^{N} \left[(x_k-\hat{x}_k).Q(x_k-\hat{x}_k) + (u_k-\hat{u}_k).R(u_k-\hat{u}_k)\right] \tag{35}$$

subject to

$$x_{k+1} - x_k = Ax_k+Bu_k+Cz_k \tag{36}$$

where

x_k = $n-$ state variable vector at period $k(o\le k\le N)$ with $x_o = \bar{x}_o$

given but x_N unspecified

u_k = $r-$ control vector at period k

z_k = S. exogenous vector which is assumed known but is beyond

the control of the planner,

A,B,C are given constant matrices

Q = : symmetric positive semi-definite matrix

R = : positive definite matrix

S = : positive matrix

The model is the exact counterpart of the continuous Linear Tracking (L.T.) model in Chapter 8 except for the exogenous vector z_k which lends a sense of econometric realism and usefulness.

The discrete Hamiltonian H is

$$H\equiv\tfrac{1}{2}(x_k-\hat{x}_k).Q(x_k-\hat{x}_k) + \tfrac{1}{2}(u_k-\hat{u}_k).R(u_k-\hat{u}_k) + p_{k+1}(Ax_k+Bu_k+Cz_k) \tag{37}$$

$$p_{k+1}-p_k=-\frac{\partial H}{\partial x_k} = -Q(x_k-\hat{x}_k) - A'p_{k+1} \tag{38}$$

with the transversality condition

$$p_N = \tfrac{1}{2}\frac{\partial}{\partial x}(x_N-\hat{x}_N).S(x_N-\hat{x}_N) = S(x_N-\hat{x}_N) \tag{39}$$

Following Kalman (1960), assume p_k to be of a feedback form (See Ch.8,).

$$p_k = K_k x_k + v_k \tag{40}$$

where K_k and v_k are unknown.

The optimal control u^*_k obtained by $\partial H/\partial u_k = 0$, is given by

$$u^*_k = -R^{-1}B' p_{k+1} + \hat{u}_k$$

$$= -R^{-1}B'(K_{k+1}x_{k+1} + v_{k+1}) + \hat{u}_k \tag{41}$$

using (40). Substituting u^*_k in (41) into (36) rearranging, gives

$$x_{k+1} = M^{-1}(I+A)x_k - M^{-1}BR^{-1}B'v_{k+1} + M^{-1}(B\hat{u}_k + Cz_k) \tag{42}$$

where $M \equiv I+BR^{-1}B'K_{k+1}$, assumed non-singular

Substitution of p_k in (40) into (38) gives

$$(I+A')K_{k+1}x_{k+1} = (K_k - Q)x_k - (I+A')v_{k+1} + Q\hat{x}_k + v_k \tag{43}$$

Substitution of x_{k+1} in (42) into (43) and rearranging, gives

$$[Q+(I+A)'K_{k+1}M^{-1}(I+A)]x_k - (I+A^*)'(K_{k+1}M^{-1}BR^{-1}B'-I)v_{k+1}$$

$$+(I+A)'K_{k+1}M^{-1}(B\hat{u}_k + Cz_k) - Q\hat{x}_k \tag{44}$$

$$= \quad K_k x_k + v_k$$

Note that the RHS of (44) is p_k with any arbitrary conditions $\bar{x}_o$, we must have, by equating the coefficients and terms of the LHS with the RHS of (44)

$$K_k = Q + (I+A)'K_{k+1}M^{-1}(I+A) \tag{45}$$

and

$$v_k = (I+A)'(I-K_{k+1}M^{-1}BR^{-1}B')v_{k+1}$$

$$+(I+A)'K_{k+1}M^{-1}(B\hat{u}_k+Cz_k)-Q\hat{x}_k \quad (46)$$

(45) is the discrete Riccati equation (see Ch.8 eq.43) and (46) is the tracking equation, the discrete counterpart of eq. (44) in Ch. 8.

The problem is reduced to solving (45) and (46), account taken of the boundary conditions which are given by (39) and (40) as

$$p_N = S(x_N - \hat{x}_N) = K_N x_N + v_N \quad (47)$$

which must hold for any arbitrary x_N, i.e.

$$K_N = S \quad (48)$$

$$v_N = -S\hat{x}_N \quad (49)$$

Pindyck (1972 ,1973) applies this LT model to the U.S. economy.

The model involves 20 quarterly time periods, from 1957 to 1962, a 3-control vector $u(t)$ and a 28-state vector $x(t)$ of which the last 18 elements are lagged endogenous variables and the first 10, with their initial values, i.e. values as of the first quarter of 1957, are

$x_1(t)$	=	real consumption, with initial value	x_{10}	=	286.7
$x_2(t)$	=	non residential investment	x_{20}	=	47.7
$x_3(t)$	=	residential investment	x_{30}	=	20.7
$x_4(t)$	=	inventory investment	x_{40}	=	2.10
$x_5(t)$	=	short term interest rate	x_{50}	=	3.1
$x_6(t)$	=	long term interest rate	x_{60}	=	3.27

$x_7(t)$	=	price level	x_{70}	=	96.4
$x_8(t)$	=	unemployment rate	x_{80}	=	.04
$x_9(t)$	=	money wage rate	x_{90}	=	1.82
$x_{10}(t)$	=	disposable income	x_{100}	=	386.6

The control vector $u(t) = (u_1(t), u_2(t), u_3(t))$ = (tax surcharge, government expenditure, change of money supply).

The nominal state $\hat{x}(t)$ and control $\hat{u}(t)$ vectors are arrived at by assigning the different growth rates, deemed the most desirable, to the historically determined initial conditions. For example $\hat{x}_1(t) = x_{10}(1+.01)^t$ where x_{10} = consumption figure of the first quarter of 1957 which is 286.7. This corresponds to an annual growth rate of 4%. The desired control vector is $\hat{u}(t) = (o, 97(1+.01)^t, 1.4)$ i.e. the desired tax surcharge is zero, the desired government expenditure grows at the rate of 1% per quarter (4% per annum) from its initial level of \$97b and the change in money supply grows at the nominal quarterly rate of \$1.4b. The values of these desired state and control variables for each of the 20 quarters are given in Pindyck (1972) Tables I and II p. 291.

While A, B, and C in the dynamic system (36) are determined by the econometric relations (see Pindyck 1972, Appendix pp. 298-300, also Pyndyck 1973 ch. 4,5 and Appendix) of the U.S. economy, the weight matrices Q and R reflect the subjective judgement of the model builder. Pindyck, with some justifications, chose them to be the following diagonal matrices

$Q = \text{diag}\ (1, 6, 15, 0, 0, 0, 6, 4 \times 10^6, 0, 0)$

$R = \text{diag}\ (6, 3, 300)$

The results obtained are shown in fig. 1, 2,..., 10 in Pindyck (1972) and also in Pindyck (1973). Consumption, non-residential investment, disposable income, unemployment and prices all run higher than their nominal paths but unemployment drops to 3% in 20 periods (1% higher than its desired level of 2%) and prices show an inflation rate of about 5%. Wages rise at an annual rate of 8% to 12% as against their desired rate of 6%. The optimal policy called for a decrease in the tax surcharge, an increase in government spending and money supply.

It should be noted that given A. B, and C in the dynamic system (36) and the initial conditions for x_o and u_o, the results obtained depend on the weights used in Q and R. These weights are changed in the subsequent two runs and different policy recommendations emerged. For details, see Pindyck, op. cit. It should also be noted that the desired or nominal values of the state and control vectors are arbitrary: they reflect what the researcher thinks are the optimal paths for the economy to track. Clearly a different set of quantitative policy recommendations would be expected for different nominal trajectories.

Example 9.5 Optimal Wage Price Control

As another illustration, consider Shupp's model of Temporary Incomes Policy (1976). It is designed to contain the wage-price spiral within some acceptable level.

The model is a conventional two-equation system giving the behaviour of wage and price. Money wage increases (w_t) are related to the expected price increase (p_t^e), average productivity increase (w_t') and the deviation in the previous period of the actual unemployment rate (u_{t-1}) from the targeted rate (u_{t-1}^*) by the equation

$$w_t = p_t^e + w_t' + \eta(u_{t-1} - u_{t-1}^*) \tag{50}$$

The expected price increase (p_t^e) is related to past price changes p_{t-i} $(i = 1, 2, \ldots)$ by

$$p_t^e = \gamma(p_{t-1} + \delta p_{t-2} + \delta^2 p_{t-3} + \ldots) \tag{51}$$

where $0 < \gamma < 1$ and $\delta \equiv 1 - \gamma^2$

The wage formation equation obtained by using a Koyck transformation is

$$w_{t+1} = \delta w_t + \gamma p_t + w_{t+1}' - \delta w_t' + \eta(u_t - u_t^*) - \eta\delta(u_{t-1} - u_{t-1}^*) \tag{52}$$

Price increases are assumed to be related to increases in average unit labour costs, $\tilde{w}_t \equiv w_t - w_t'$, excess demand $y_t - y_t^*$ where y_t^* is the output obtained at the targeted unemployment rate u_t^*. Assuming price and wage changes obey a Koyck lag structure, we have

$$p_t = \beta[\tilde{w}_{t-1} + \alpha\tilde{w}_{t-2} + \alpha^2\tilde{w}_{t-3} + \ldots] + v(y_{t-1} - y_{t-1}^*) \tag{53}$$

where $0 < \beta < 1$ and $\alpha = 1 - \beta$

This implies the price formation equation

$$p_{t+1} = \alpha p_t + \beta w_t - w_t' + v(y_t - y_t^*) - v\alpha(y_{t-1} - y_{t-1}^*) \tag{54}$$

Defining the optimal wage increase w_t^* to be equal to productivity increase (w_t') and inflation rate i.e., $w_t^* \equiv w_t' + p_t$ to protect wage earners, clearly the objective is to minimize the deviation of the actual wage and price inflation rates from these optimal levels of wage increase w_t^* and price increase p^*, i.e.,

$$Min\ J = \tfrac{1}{2} \sum_{1}^{T} \mu_t\ (p_t - p^*)^2 + (w_t - w_t^*)^2 \tag{55}$$

subject to (52) or (54) which are constrained such that $u_t - u_t^* = 0$ and $y_t - y_t^* = 0$, i.e., Minimize (55) subject to

$$p_{t+1} = \alpha p_t + \beta(w - w_t') \tag{56}$$

or

$$w_{t+1} = \delta w_t + \gamma p_t + w'_{t+1} - \delta w_t' \tag{57}$$

depending on which rate is to be controlled.

The Hamiltonian is

$$H \equiv \tfrac{1}{2}[\mu_t(p_t - p^*)^2 + (w_t - w_t^*)^2] + \lambda_{t+1}[\alpha p_t + \beta(w_t - w_t')] \tag{58}$$

With w_t as the control and p_t as the state variable, optimality implies

$$\partial H/\partial w_t = 0 \quad => w_t = -\beta\lambda_{t+1} + p_t + w_t' \tag{59}$$

$$\partial H/\partial \lambda_{t+1} = \quad p_{t+1} = \alpha p_t \quad + \beta(w_t - w_t') \tag{60}$$

$$\lambda_t = \partial H/\partial p_t = (1 + \mu_t)p_t - w_t + \alpha\lambda_{t+1} + w_t' - \mu_t p^* \tag{61}$$

These last three equations combined give

$$p_{t+1} = \frac{1}{\alpha+\beta} \left\{[(\alpha + \beta)^2 + \beta^2\mu_t]p_t - \beta^2\lambda_t - \beta^2\mu_t p^*\right\} \tag{62}$$

$$\lambda_{t+1} = \frac{1}{\alpha+\beta} \left(- \mu_t p_t + \lambda_t + \mu_t p^*\right) \tag{63}$$

with the boundary conditions $p(0) = p_0$ and $\lambda_{T+1} = \mu_{T+1}\ p_{T+1}$.

Remembering that $\alpha + \beta = 1$, we have

$$p_{t+1} = (1 + \beta^2\mu_t)p_t - \beta^2\lambda_t - \beta^2\mu_t p^* \tag{62'}$$

$$\lambda_{t+1} = \quad \lambda_t \quad - \mu_t\ (p_t - p^*) \tag{63'}$$

The shadow price λ_t plays an important part: it measures the marginal disutility and in an optimal programme, is expected to decrease over time. Indeed (63') clearly shows that it does, whenever $p_t > p^*$ (and p^* could be set at zero, i.e., zero inflation is ideal). With $\lambda_{t+1} - \lambda_t < 0$, i.e., $\lim_{t\to\infty} \lambda_t = 0$, $\beta\lambda_{t+1} \to 0$ in (59) and as $t \to \infty$

$$w_t \to p + w_t' \equiv w_t^*$$

The resultant p_t obtained by substituting w_t in (59) into (60), remembering that $\alpha + \beta = 1$, is

$$p_{t+1} - p_t = -\beta^2\lambda_{t+1} \to 0 \text{ as } t \to \infty.$$

This shows that the program is optimal: both wage and price increases will eventually be brought under control.

Alternatively we can impose price control by treating p_t as the control variable and w_t as the state variable. In this case, the problem is one of minimizing (55) subject to (57). The procedure is exactly the same.

Shupp then used the quarterly data for the U.S. economy for the period 1967-1971 to illustrate the solutions obtained. The results, presented in Tables I and II (op. cit.) show that in both wage and price control models, λ_t falls from 1.911 at $t = 1$ to .001 at $t = 13$ and the resultant price increase p_t falls from 1.1 in the wage model and .834 in the price model, to .6 in both models. Thus, by the 13th quarter, wage and price inflation are reasonably brought under control.

CHAPTER 10

SENSITIVITY ANALYSIS

10.1 Introduction

We have seen that Optimal Control theory is concerned with finding a control vector $u(t)$ to steer the dynamic system in an optimal manner, i.e., in order to maximize or minimize an objective functional. State variables $x(t)$ are thus controlled. All this, however, takes place as of a certain environment characterised by some parameter vector $\alpha(t)$. For a given $u^*(t)$, a change in one or more parameters causes a corresponding change in the dynamic system and hence in the objective achieved. Sensitivity analysis studies the displacement of the dynamic system or objective functional in response to such a change in parameters.

Sensitivity was first introduced by Bode (1945) and after some twenty years, further developed by Lindorff (1963), Tomovic (1963) and 1972) . In Economics, the concept had been used much earlier, among others, by Marshall (1890), Slutsky (1915), Hicks (1932, 1939), Allen (1934), unified and further developed by Samuelson (1947) under the name of Comparative Statics. This has recently been extended to dynamic systems and called Comparative Dynamics by Oniki (1973) and others.

This chapter deals with Sensitivity theory or Comparative Dynamics in general and particularly in the Calculus of Variations and Optimal Control.

10.2 Sensitivity Theory

Consider a dynamic system

$$F(x^{m}, x^{m-1}, \ldots, \dot{x}, x, \alpha, t) = 0 \tag{1}$$

where x is an n-state vector and x^{i} its ith time derivative,

α is an r-parameter vector and t is time.

For the expository purpose of this chapter, our presentation will be restricted to the case where $m \leq 2$ and write (1) as

$$F(\ddot{x}, \dot{x}, x, \alpha, t) = 0 \tag{2}$$

The state vector $x(t)$ is assumed to vary continuously with any parameter change and the relationship between $x(t)$ and $\alpha(t)$ is assumed unique.

The sensitivity of $x(t)$ with respect to $\alpha_i (1 \leq i \leq r)$ is defined as

$$\frac{\partial x}{\partial \alpha_i} \frac{\alpha_i}{x} \equiv v_i \frac{\alpha_i}{x} \tag{3}$$

where the sensitivity coefficient v_i is defined as

$$v_i \equiv \frac{\partial x}{\partial \alpha_i} \equiv \lim_{\epsilon \to 0} \frac{1}{\epsilon} \left[x(\alpha_1, \ldots, \alpha_{i-1}, \alpha_i + \epsilon, \ldots \alpha_r, t) - x(\alpha_1 \ldots, \alpha_r, t) \right] \tag{4}$$

This is the concept of elasticity in Economics.

With the usual assumption that the solution

$$x(t) = x(\alpha, t) \tag{5}$$

of (1) or (2) are analytically dependent on, and continuous in the parameters, the sensitivity equation is obtained by partially differentiating (2) with respect to any parameter k, all α_h $(h \neq k)$ remaining unchanged and denoting time derivatives by dots,

$$F'_{\ddot{x}}\, \ddot{v}_k + F'_{\dot{x}}\, \dot{v}_k + F'_x\, v_k = -F_{\alpha_k} \tag{6}$$

where $F'_x = \left(\frac{\partial F}{\partial \ddot{x}_1}, \frac{\partial F}{\partial \ddot{x}_2}, \ldots, \frac{\partial F}{\partial x_n}\right)$ (′ denotes transposition)

and similarly for $F_{\dot{x}}$ and F_x

$$v_k \equiv \partial x/\partial \alpha_k = (\partial x_1/\partial \alpha_k, \ldots, \partial x_n/\partial \alpha_k) \quad (1 \le k \le r)$$

$$\dot{v}_k \equiv \frac{d}{dt}\frac{\partial x}{\partial \alpha_k} = \frac{\partial}{\partial \alpha_k}\frac{dx}{dt} \equiv \frac{\partial \dot{x}}{\partial \alpha_k}$$

$$\ddot{v}_k \equiv \frac{d}{dt}\dot{v}_k$$

$$F_{\alpha_k} \equiv \frac{\partial F}{\partial \alpha_k}$$, a known function once F is given

This is a linear differential equation as the coefficients $F_{\ddot{x}}$, $F_{\dot{x}}$ and F_x are not functions of v or their derivatives.

For example, when $m = 1$, and (1) is written in explicit form, this is

$$\dot{x} = f(x, \alpha, t) \tag{7}$$

with $x(0) = x_0$, where x = n-state and α = r-parameter vectors. The sensitivity function is, for each i and j $(1 \le i \le n,\ 1 \le j \le r)$

$$\frac{\partial x_i}{\partial \alpha_j} \equiv \frac{\partial}{\partial \alpha_j}\frac{dx_i}{dt} \equiv \frac{d}{dt}\frac{\partial x_i}{\partial \alpha_j} \equiv \frac{dv_{ij}}{dt} = \sum_{k=1}^{n} \frac{\partial f_i}{\partial x_k} v_{kj} + \frac{\partial f_i}{\partial \alpha_j} \tag{8}$$

where $v_{k_j} \equiv \frac{\partial x_k}{\partial \alpha_j}$, or, in matrix form,

$$\dot{V}(\alpha) = A(x^*, \alpha, t)\, V(\alpha) + B(x^*, \alpha, t) \tag{9}$$

$$V(\alpha, t_0) = [0]$$

where $V(\alpha) \equiv [\partial x_i/\partial \alpha_k] \quad (1 \le i \le n\ ;\ 1 \le k \le r)$

$$A(x^*,\alpha,t) \equiv [\partial f_i(x,\alpha,t)/\partial x_h] \quad (1 \leq i \leq n \;;\; 1 \leq h \leq n)$$

$$B(x^*,\alpha,t) \equiv [\partial f_i/\partial \alpha_k] \quad (1 \leq i \leq n \;;\; 1 \leq k \leq r)$$

x^* is the unperturbed solution of (7)

If only α_i changes, all other a_j $(j \neq i)$ remaining unperturbed, (9) is simply

$$v = A(x^*,\alpha,t)v + b(x^*,\alpha,t) \tag{10}$$

where $v \equiv \partial x/\partial \alpha_i$, $b \equiv \partial f/\partial \alpha_i$, written without subscripts for simplicity.

A simultaneous solution of (7) and (10) gives $x^*(\alpha,t)$ and $v^*(\alpha,t)$. The solution of (10) is

$$v(t) = \Phi\,(t,t_0)v_0 + \int_{t_0}^{t} \Phi\,(t,\tau)\, b(\tau)\, d\tau \tag{11}$$

where $\Phi\,(t,\tau)$ is the fundamental or transition matrix of A (see Appendix)

i.e., $$\frac{d}{dt}\Phi\,(t,t_0) = A(x^*,\alpha,t)\,\Phi\,(t,t_0)$$

$$\text{with} \quad \Phi\,(t_0,t_0) = I$$

If A in (10) is time-invariant and $t_0 = 0$, (11) is simply

$$v(t) = e^{At}v_0 + \int_0^t e^{A(t-\tau)}b(\tau)\, d\tau \tag{12}$$

where $$\Phi\,(t) = e^{At}$$

and $$\Phi\,(t)\,\Phi^{-1}\,(t) = \Phi(t-\tau) = e^{A(t-\tau)}$$

Note that (10) holds true for any α_i $(1 \leq i \leq r)$, i.e., the Jacobian matrix $A(x^*,\alpha,t)$ is the same regardless of which parameter is made to change.

If x_0 is independent of the changing parameter, say α_i.

the initial condition for (10) is $v(0) = 0$. In other cases, $v(0) = v_0 \neq 0$. For example if the perturbed and unperturbed systems have different initial conditions and the effect of a change in initial conditions is to be examined, (7) gives

$$x(t) = x_0 + \int_0^t f(x,\alpha,\tau)\, d\tau \tag{13}$$

and $$v \equiv \partial x/\partial x_0 = (1,1,\ldots,1)$$

Clearly (12) shows how a dynamic system is displaced by a parameter variation: corresponding to each value of the parameter vector, there is a different trajectory of the state vector. This is the gist of Comparative Dynamics.

If both A and b are time invariant and A is non-singular, the solution of (10) is simply

$$v(t) = e^{At}c - A^{-1}b \tag{14}$$

where $$c \equiv v_0 + A^{-1}b = A^{-1}b \text{ if } v_0 = 0.$$

10.3 Cross Sensitivity

It may be necessary, in some cases, to investigate higher and cross-sensitivity coefficients by taking higher or cross-derivatives. Taking the second derivative of (6) and writing $z \equiv \partial v/\partial\alpha$, $\dot{z} \equiv dz/dt$ etc...

$$F'_{\ddot{x}}\,\ddot{z} + F'_{\dot{x}}\,\dot{z} + F'_{x}\,z = -(F'_{\alpha\ddot{x}}\,\ddot{v} + F'_{\alpha\dot{x}}\,\dot{v} + F'_{\alpha x}\,v + F_{\alpha\alpha}) \tag{15}$$

Note that the coefficients of the homogeneous part of (15) and (6) are identical and the non-homogeneous part of (15) is a function of the derivatives of lower order, (15) could be solved in a recursive manner. For example, the sensitivity functions of (7) for $\alpha = (\alpha_1,\alpha_2)$

is

$$\frac{\partial \dot{x}}{\partial \alpha_1} \equiv \dot{v} = A(x^*,\alpha,t)v + b_1(x^*,\alpha,t) \tag{16}$$

$$\frac{\partial \dot{x}}{\partial \alpha_2} \equiv \dot{w} = A(x^*,\alpha,t)w + b_2(x^*,\alpha,t) \tag{17}$$

$$\frac{\partial^2 \dot{x}}{\partial \alpha_2 \partial \alpha_1} \equiv \dot{z} = A(x^*,\alpha,t)z + c(x^*,\alpha,t) \tag{18}$$

where $b_i \equiv \partial f/\partial \alpha_i \qquad (i=1,2)$

$$c \equiv f_{\alpha_1 \alpha_2} \equiv \partial^2 f/\partial \alpha_1 \partial \alpha_2$$

The solution is obtained by solving (7), (16), (17) and (18) simultaneously for $x(t)$, $v(t)$, $w(t)$ and $z(t)$. This is the essence of Comparative Dynamics. Comparative Statics emerges as a special case in which $\dot{x}$, $\ddot{x}$ are absent. In this case, (16) or (17) is simply

$$Av + b_1 = 0$$

$$Aw + b_2 = 0$$

whose solutions, provided A is non singular, are

$$v = A^{-1} b_1 \quad \text{and} \quad w = -A^{-1} b_2$$

10.4 Objective Functional Sensitivity

Consider the problem of maximizing or minimizing

$$J = \int_0^T f_0(x,u,\alpha,t)\, dt$$

subject to

$$\dot{x} = f(x,u,\alpha,t)$$

where x,u,α are respectively the n-state, m-control and r-parameter vectors.

The variations in J due to small variations of α, are given by

$$\Delta J \approx dJ = \sum_{i=1}^{r} \frac{\partial J}{\partial \alpha_i} \Delta\alpha_i \equiv J'_\alpha \cdot \Delta\alpha$$

where $J'_\alpha \equiv \partial J/\partial\alpha$ is the objective functional sensitivity row vector. But

$$\frac{\partial J}{\partial \alpha_j} = \int_0^T \sum_{i=1}^{n} \frac{\partial x_i}{\partial \alpha_j} dt$$

$$\equiv \int_0^T \left(\frac{\partial x}{\partial \alpha}\right)' \frac{\partial F}{\partial x} dt$$

$$\equiv \int_0^T V' \quad F_x \, dt \tag{19}$$

where $V \equiv [\partial x_i/\partial\alpha_j]$ is the solution matrix of (9). This gives the variations in the performance index or objective functional caused by a change in parameter.

10.5 Stability and Sensitivity

It has been seen that a change in parameter causes a displacement of the dynamic system from one trajectory to another. The question of stability arises as to whether such displacement bends to a certain limit for sufficiently long periods of time. In other words, a sensitivity function $v(t)$ is said to be stable if

$$\lim_{t\to\infty} v(t) = c$$

for some finite constant c. If $c = \pm\infty$, $v(t)$ is said to be unstable.

As noted earlier, the Jacobian matrix A is the same for all parameters and furthermore, is independent of the various sensitivity coefficients v, w, z. This simplifies the problem greatly: the same stability analysis applies to the dynamic and sensitivity systems.

The stability of differential equation systems has been exhaustively studied in Mathematics, Engineering and Economics, it will not be dealt with here. Basically, stability requires the real part of the eigen values of the Jacobian A to be negative. For a linear differential equation of order n with constant coefficients, for example

$$L(D)\ x(t) = \left(a_0 D^{n-1} + \ldots + a_n\right) x(t) - g(t) = 0 \tag{20}$$

with initial conditions $D^i\ x(0) = b_i\ (0 \leq i \leq n-1)$

where $D^i \equiv d^i/dt^i\ (0 \leq i \leq n-1)$. The solution of (20) is of the form

$$x(t) = \sum_1^n c_i e^{\lambda_i t} + x_p(t) \tag{21}$$

where λ_i are characteristic roots or eigen values, c_i arbitrary constants to be determined by the given initial conditions and $x_p(t)$ is the particular solution.

The sensitivity function of (20) for an arbitrary change in parameter α_k for example, is

$$L(D)v \equiv \left(a_0 D^n + a\ D^{n-} \quad + \ldots + a_n\right) v = -D^{n-k} x(t) \tag{22}$$

Substituting (21) into (22) yields

$$L(D)v \equiv \left(a_0 D^n + a_1 D^{n-1} + \ldots + a_n\right) v = -D^{n-k}\left(\sum_1^n c_i e^{\lambda_i t} + x_p\right) \tag{23}$$

It can be seen that the equation of motion (20) and of sensitivity (23) have the same characteristic equation: i.e., if either is stable so is the other. This, however, does not mean they have identical time paths because of the different non-homogeneous parts.

Non-linear systems are more complicated, because all depends on which parameter is perturbed and to what extent. For the Van de Pol equation, for example,

$$\ddot{x} - a(1-x^2)\dot{x} + x = 0$$

The sensitivity equation is

$$\ddot{v} - a(1-x^2)\dot{v} + (1 + 2ax\ \dot{x})v = (1-x^2)\dot{x}$$

The dynamic and sensitivity systems are not too different when a is small and its variations are small: both have a limit cycle. But when a is large, the two systems exhibit vastly different behaviours.

10.6 Some Economic Applications

Let us illustrate Sensitivity theory with some economic examples, starting with the most familiar supply. Demand model with tax and moving into the optimal growth model in the Calculus of Variations and Optimal Control models.

Example 1. Tax and Taste Sensitivity and Cross-sensitivity

Consider the following demand (D) and supply (S) model with price (p) as the variable, tax rate (τ) and constant coefficients a_0, a_1, a, b_0, b_1 and b as parameters (24)

$$D(p) = a_0 + a_1 p$$

$$S(p) = b_0 + b_1(1-\tau)\ p \tag{25}$$

$$\dot{p} = k(D-S) = ap + b \tag{26}$$

where $a \equiv a_1 - b_1(1-\tau)$ and $b \equiv a_0 - b_0$

The solution of (26) is

$$p(t) = ce^{at} - b/a \tag{27}$$

where $\qquad c \equiv p_0 + b/a$

The tax sensitivity function $v(t) \equiv \partial p/\partial\tau$ is

$$\dot{v} = av + b_1 p$$

$$= av + b_1(ce^{at} - b/a) \tag{28}$$

with $\qquad v(0) = 0$, the solution of which is

$$v(t) = b_1 e^{at}(ct - b/a^2) + b_1 b/d^2 \tag{29}$$

The sensitivity of the system with respect to a change in taste a_0, (i.e., $w(t) \equiv \partial p/\partial a_0$), causing a shift in the demand function, is

$$\dot{w} = aw + 1$$

the solution of which, for $w(0) = 0$, gives

$$w(t) = (e^{at} - 1)/a \tag{30}$$

It is plain that the stability of the sensitivity functions (29) and (30) is intimately tied to the stability of the price system (27): if the dynamic system is stable, so is the sensitivity system and vice versa. Stability conditions in this case require that the slope (with respect to the p-axis) of the supply function, net of tax, is greater than the slope of the demand curve, i.e., $a<0$, for then

$$\lim_{t\to\infty} p(t) = -\frac{b}{a}$$

$$\lim_{t\to\infty} v(t) = b_1\frac{b}{a^2}$$

$$\lim_{t\to\infty} w(t) = -\frac{1}{a}$$

These conditions are always realised in a model with downward sloping demand and upward sloping supply curves.

Note that if $\dot{p}$ is absent, so are $\dot{v}$ and $\dot{w}$ and the static sensitivity functions simply give

$$v = b_1 b/a^2$$

$$w = -a^{-1}$$

The sign of w is now ambiguous as it stands but will cease being so when the corresponding dynamic sensitivity system is examined. This would allow statements to be made such as "$w > 0$ (i.e., an increase in tax would lead to an increase in price) if the system is stable, i.e., if $a < 0$." This is Samuelson's (1947) Correspondence Principle.

Similar remarks could be made about the cross-sensitivity $z(t) \equiv \partial w/\partial\tau \equiv \partial^2 p/\partial\tau\partial a$ which is

$$z(t) = b_1 e^{at}(at - 1)/a^2 + b_1/a^2$$

with $$\lim_{t\to\infty} z(t) = b_1/a^2 \quad \text{if } a < 0$$

i.e., a change in taste, coupled with an increase in tax rate, will cause price to increase if the system is stable.

Example 2 Optimal Growth Model: Sensitivity and Comparative Dynamics

As a further illustration of sensitivity analysis, let us examine the neo-classical model of Optimal Growth in the context of the Calculus of Variations and Optimal Control.

The optimal growth problem consists of maximizing, for a given neo-classical dynamic system of investment $\dot{k} = f(k) - \lambda k - c$, the integral of discounted utility of per capita consumption (c), i.e.,

$$Max\ J = \int_0^T u(c)\ e^{-\delta t}\ dt \tag{31}$$

where $$c = f(k) - \lambda k - \dot{k} \tag{32}$$

$\lambda \equiv n + u$: constant rates of population growth and capital decay

δ = positive constant discount rate

$f(k)$ = per capita neo-classical production function as a function of capital per worker $(\equiv K/L)$

$\dot{k}$ = investment per head of worker, $k(0) = k_0$; $k(T) = k_T$

The Euler equation of this Calculus of Variations problem implies

$$E \equiv \ddot{k} - (f' - \lambda)\dot{k} - \frac{u'}{u''}\ (f' - \lambda - \delta) = 0 \tag{33}$$

where $d^2u/dc^2 \equiv u'' < 0 < u'(c) \equiv du/dc$

and $d^2f/dk^2 \equiv f'' < 0 < f' \equiv df/dk$

For illustrative purposes, let $u(c) \equiv \dfrac{c^{1-\sigma}}{1-\sigma}$ $(0<\sigma<1)$ and $f(k) = ak\ (a>0)$,

E becomes $$E \equiv \ddot{k} - \beta\dot{k} + \gamma k = 0 \tag{34}$$

where $\beta \equiv a-\lambda + (a-\lambda-\delta)/\sigma$

$\gamma \equiv (a-\lambda-\delta)(a-\lambda)/\sigma$

The solution of (34) is

$$k(t) = Ae^{m_1 t} + Be^{m_2 t} \tag{35}$$

where $(m_1, m_2) = \frac{1}{2}(\beta \pm \sqrt{\beta^2 - 4\gamma})$ and A, B are arbitrary constants to be determined by k_0 and k_T.

The discount sensitivity $v \equiv \partial k/\partial\delta$ is obtained by differentiating $\partial E/\partial\delta = 0$, putting $\dot{v} = dv/dt$ etc..., as

$$\ddot{v} - \beta\dot{v} + \gamma v = -\frac{\dot{k}}{\sigma} + \frac{(a-\lambda)}{\sigma}k \tag{36}$$

The population sensitivity is similarly obtained by $\partial E/\partial n = 0$ putting $w \equiv \partial k/\partial n$, $\dot{w} \equiv dw/dt$ etc..., as

$$\ddot{w} - \beta\dot{w} + \gamma w = -\frac{(1+\sigma)}{\sigma}\dot{k} + \frac{(2a-2\lambda-\sigma)}{\sigma}k \tag{37}$$

Finally the cross-sensitivity is obtained by $\partial^2 E/\partial n\partial\delta = 0$, putting $z \equiv \partial v/\partial n \equiv \partial^2 k/\partial n\partial\delta$, and $z \equiv dz/dt$ etc....

$$\ddot{z} - \beta\dot{z} + \gamma z = -\frac{(1+\sigma)}{\sigma}\dot{v} + \frac{(2a-2\lambda-\delta)}{\sigma}v - \frac{\dot{w}}{\sigma} + \frac{(a-\lambda)}{\sigma}w - \frac{k}{\sigma} \tag{38}$$

It can be seen from (34) through (38), that the dynamic and sensitivity systems all have the same characteristic equation to their homogeneous part: the same stability analysis applies to each. The non-homogeneous parts of (36) and (37) are known functions of k and $\dot{k}$ which have been found in (35) and the non-homogeneous part of the cross-sensitivity function (38) is a function of $\dot{w}$, w, $\dot{v}$, v and k which are the solutions of (34), (36) and (37).

The objective functional sensitivity is given by

$$\delta J = \int_0^T \left(\frac{\partial F}{\partial c}\frac{\partial c}{\partial k}\frac{\partial k}{\partial \alpha}\right) dt \tag{39}$$

where $F \equiv u(c)e^{-\delta t}$ and α is the relevant parameter such as n and δ in the above analysis.

As a further illustration, the same optimal growth model can be set up as an optimal control problem and examined as follows (see Oniki 1973).

$$Max\ J \equiv \int_0^T u(c)e^{-\delta t}dt$$

$$\text{given} \quad \dot{k} = f(k) - \lambda k - c \tag{40}$$

$$k(0) = k_0 \quad \text{and} \quad k(T) = k_T$$

The Hamiltonian of the system is

$$H \equiv e^{-\delta t}\left\{u(c) + q\left[f(k) - \lambda k - c\right]\right\} \tag{41}$$

$\partial H/\partial c = 0$ implies

$$q = u'(c)$$

which, in view of the assumption $u''<0<u'$, could be expressed as

$$c = c(q)$$

with $dc/dq < 0$. With this (40) becomes $\dot{k} = f(k) - \lambda k - c(q)$.

$\frac{d}{dt} qe^{-\delta t} = -\frac{\partial H}{\partial k}$ gives

$$\dot{q} = \left[\lambda + \delta - f'(k)\right] q \tag{42}$$

Writing $v' \equiv (v_1, v_2)$ where $v_1 \equiv \partial k/\partial\delta$, $v_2 \equiv \partial q/\partial\delta$ and $x' \equiv (0,q)$ (′ denoting transposition), the discount parameter sensitivity system is obtained by differentiating (40)

$$\begin{bmatrix} \dot{v}_1 \\ \dot{v}_2 \end{bmatrix} = \begin{bmatrix} f'-\lambda & -c' \\ -qf'' & \lambda+\delta-f' \end{bmatrix} \begin{bmatrix} v_1 \\ v_2 \end{bmatrix} - \begin{bmatrix} 0 \\ q \end{bmatrix} \tag{43}$$

or, in matrix notation, denoting the matrix of (43) as A,

$$v(t) = Av(t) + x(t) \tag{44}$$

The population sensitivity system, writing $w' \equiv (w_1, w_2)$ where

$w_1 \equiv \partial k/\partial n$; $w_2 \equiv \partial q/\partial n$; $y' \equiv (k, -q)$, becomes

$$\dot{w} = Aw - y \tag{45}$$

Finally, writing $z' \equiv (z_1, z_2) \equiv (\partial w_1/\partial\delta, \partial w_2/\partial\delta)$ where, of course,

$$z_1 \equiv \partial w_1/\partial\delta \equiv \partial^2 k/\partial\delta\partial n \text{ and } z_2 \equiv \partial w_2/\partial\delta \equiv \partial^2 q/\partial\delta\partial n$$

the cross-sensitivity system is

$$\dot{z} = Az + Bw - \xi \tag{46}$$

where

$$B \equiv \begin{bmatrix} f''v_1 & -c''v_2 \\ -qf'''v_1 + f''v_2 & 1-f''v_1 \end{bmatrix} \quad \text{and } \xi \equiv \begin{pmatrix} v_1 \\ -v_2 \end{pmatrix}$$

and the objective functional sensitivity is the same as (39).

It can be seen that matrix A is the same in (44), (45) and (46) as noted earlier.

The explicit solution of the dynamic system and its various sensitivity systems is not simple unless the specific forms of the various functions are given and heroic assumptions are made. These simultaneous solutions are usually obtained by use of high speed computers. For a phase diagram of some of the above, see Oniki (1973).

10.7 Conclusion

Sensitivity analysis, under the name of Comparative Statics and recently extended to Comparative Dynamics, is perhaps one of the oldest tools of economic analysis. Parameters play an important part. From some given initial position, the course of a dynamic system is steered in different directions by a choice of control variables and as of a given environment, represented by some parameters. When this environment changes, the optimal trajectory is affected. Although all this is expected, what is often neglected is how sensitive some changes are as a result of a parameter change and whether the system remains stable after the change. Since the outcome could be widely different, policy makers should pay particular attention to those parameters which are the most sensitive.

This chapter has drawn attention to this sensitive issue in an introductory manner and provided some economic examples. An investigation in depth would cover unstable situations such as bifurcations and catastrophes and which range of parameter change will cause a dynamic system to change its stability, but this is clearly beyond the scope of this chapter.

CHAPTER 11

SOME ECONOMIC AND MANAGEMENT APPLICATIONS

11.1 Introduction

Optimal Control (OC) has been extensively applied to the various fields of economic theory and policy, from Micro to Macro-economics, from International Trade to Economic Development, from Natural Resource to Urban Economics, from leisure allocation, education and training to Population Economics. Books like Shell (1967), Arrow & Kurz (1970), Pitchford & Turnovsky (1977), Kemp & Long (1980), Cass & Shell (1976) contain nothing but economic applications of OC. At periodical level, the use of OC is commonplace. Choosing among these, even some major contributions to present in this chapter is virtually an impossible task. And perhaps a futile one too: all it does would be to convince readers that OC has useful economic applications, which is obvious. Besides, applications have been provided in each chapter.

Nevertheless, our discussion of the Calculus of Variations and Optimal Control would be somewhat deficient if the beauty of some final products built with these tools is not displayed. In view of the difficulty of choice, only some representative works can be discussed and only for illustrative purposes. No attempts will be made to survey the literature in each field: such surveys have been done and would be out of place here, our purpose being simply to show how the Calculus of Variations and the Maximum Principle have proved valuable dynamic optimisation tools in Economic theory. We shall restrict our selective presentation to some economic and Management Science applications.

11.2 Some Economic Applications

The Calculus of Variations and Optimal Control have been applied especially in the following fields

(i) Optimal growth and Economic Planning

(ii) Economic Stabilisation

(iii) Dynamic Theory of the Firms

(iv) International Trade

(v) Regional Economics

(vi) Urban Economics

(vii) Education, Training and Human Capital

(viii) Natural Resources Economics

(ix) Pollution Control

(x) Optimal Population Control

(xi) Arms Race Control

Among these applications, Optimal Growth models--one-sector, two-sector and multi-sector models--stand out as the most important ones both for their intrinsic beauty and their inspiration for other "growth related" models. In fact it will be seen that (iv), and (vii) through (xi) may arguably be called "growth related" models in that the objective functions are almost always the discounted utility and the dynamic system, a variation of the fundamental neo-classical growth equation in which capital per head, k, is replaced by the relevant state variables. For these reasons, Optimal Growth models will be presented in greater details than the others.

11.2.1 Optimal Economic Growth

Economic growth models were first examined by Harrod (1939) and Domar (1946). In these models, input coefficients are fixed and full employment equilibrium growth is a rare coincidence. This undesirable feature led Solow (1956) and Swan (1956) to the formulation of the so-called neo-classical model which allows for factor substitution and hence full employment equilibrium. The optimality aspect of the model, however, was not investigated until 1965 when Cass (1965) combined it with Ramsey's (1928) optimal saving theory and cast it in the framework of Optimal Control theory. We shall present this aggregate model first, then its extensions to the two-sector and multi-sector models.

11.2.1.1 The One-Sector Optimal Growth Model

Consider the one-sector model discussed in sections 3.5.2 and 5.6.2.

The assumptions are

(i) Constant returns to scale, i.e., the production function $F(K, L)$ is homogeneous of degree one in both capital (K) and labour (L), with $F(K, L) = Lf(K/L) \equiv Lf(k)$ where $k \equiv K/L$, and $f''(k) < 0 < f'(k)$ with $\lim_{k\to 0} f'(k) = \infty$, $\lim_{k\to\infty} f'(k) = 0$;

(ii) Diminishing but non vanishing marginal utility of per-capita consumption (c), i.e., $u''(c) < 0 < u'(c)$ with $\lim_{c\to 0} u'(c) = \infty$, $\lim_{c\to\infty} u'(c) = 0$

(iii) Labour grows at the constant positive rate n, i.e., $L = e^{nt}$

With these three assumptions, the stylised economy, expressed in per-capita terms, is represented by

$$f(k) = \dot{k}(t) + \lambda k(t) + c(t)$$

or

$$\dot{k}(t) = f(k) - \lambda k(t) - c(t), \; k(0) = k_0 \tag{1}$$

which gives the allocation of per-capita output $f(k)$ between investment requirements $\dot{k} + \lambda k(t)$ (where $\lambda \equiv \mu + n$ = constant depreciation rate μ and population constant growth rate n) and consumption $c(t)$. Thus, net capital formation $\dot{k}$ ($\equiv dk/dt$) absorbs all output left over after replacement investment $\lambda k(t)$ and consumption requirements $c(t)$ have been met. This is the fundamental differential equation of the neo-classical growth model.

Subject to this fundamental dynamic law (1), the objective is to maximize the present value of per-capita utility, i.e., maximize

$$J = \int_0^\infty u(c)\ e^{-\delta t}\ dt$$

The Hamiltonian is

$$H = e^{-\delta t} \left\{u(c) + q[f(k) - \lambda k - c]\right\} \tag{2}$$

where $q(t)\ e^{-\delta t}$ is the co-state variable where δ is a positive constant discount rate.

Clearly this Hamiltonian H is a present valuation of per capita GNP: $u(c)\ e^{-\delta t}$ is the present valuation of consumption utility and $qe^{-\delta t}\ (f - \lambda k - c) = qe^{-\delta t}\ \dot{k}$ is the current value of gross per capita investment. Maximizing H amounts to maximizing the present value of per capita GNP.

The Maximum Principle gives

$$\partial H/\partial c \equiv H_c = 0 \quad \text{implying}$$

$$u'(c) = q(t) \quad \text{or} \quad c = c(q) \tag{3}$$

in view of assumption (ii), by the Implicit Function theorem.

Substituting (3) into (1) gives

$$\dot{k} = f(k) - \lambda k - c(q) \equiv \psi(q,\ k) \tag{4}$$

$\frac{d}{dt}(qe^{-\delta t}) = -H_k$ gives

$$\dot{q}/q = [a - f'(k)] \equiv \phi(q, k) \tag{5}$$

where $a \equiv \lambda + \delta \equiv \mu + n + \delta > 0$.

Equations (3), (4) and (5) together with $k(0) = k_0$ and $\lim_{t\to\infty} qe^{-\delta t}[k(t) - k^*] = 0$ provide all the information required. The solution can be presented in a phase diagram in the qk or ck plane. (See fig. 11.2 and 11.3)

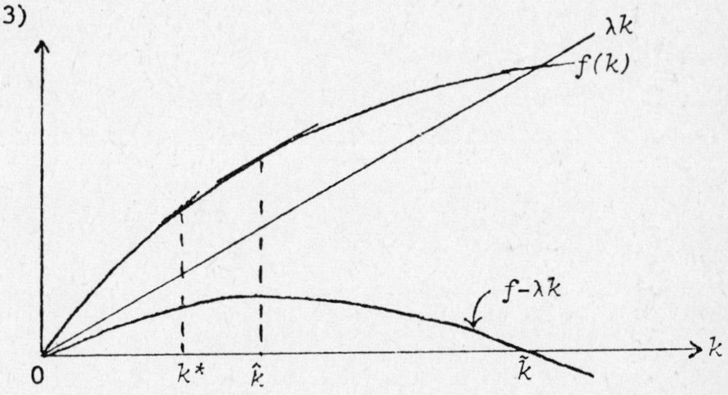

Fig. 11.1 The neo-classical growth model

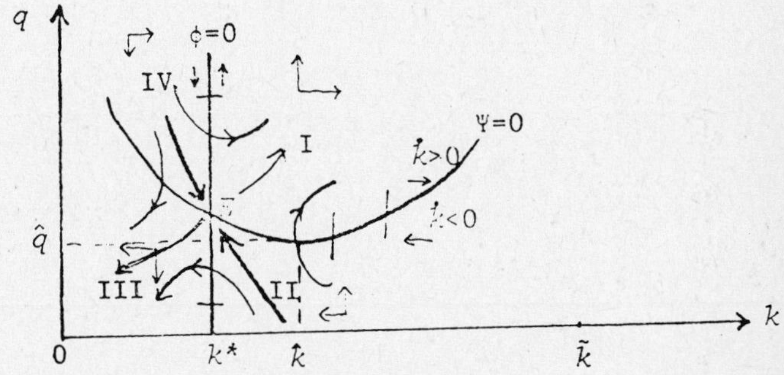

Fig. 11.2 Phase diagram in the qk plane

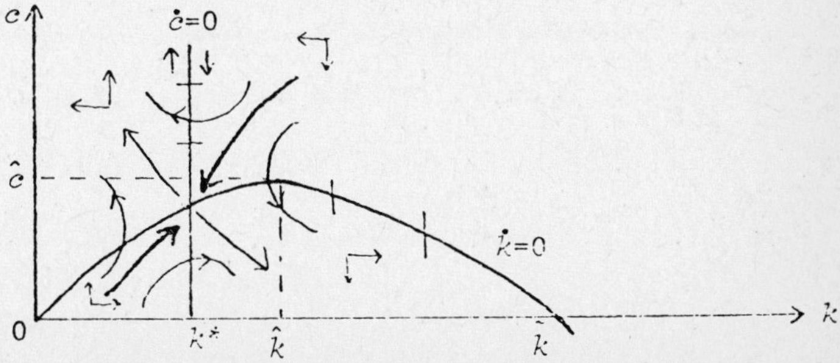

Fig. 11.3 Phase diagram in the ck plane

Let us examine the structure of the solution in the qk plane first. For this, consider the equilibrium of (4) and (5) characterised by $\psi = 0 = \phi$ in fig. 11.2.

Since $\phi = 0$ implies $f'(k) = a$ and $f''(k) < 0$, there is a unique k^* for which $f'(k) = a$. Hence the curve $\phi = 0$ is a vertical line crossing the k-axis at k^*. By the law of diminishing returns $f'' < 0 < f'$ (Assumption (i)), it is clear that $\phi > 0$, (i.e., $\dot{q} > 0$, q rising) whenever $k > k^*$ and $\phi < 0$, i.e., q is falling for all $k < k^*$. Of course q is unchanged on the curve $\phi = 0$. Note that q in ϕ is always positive by the assumption (ii) i.e., $q = u'(c) > 0$ in (3).

Now consider the curve $\psi\ (q, k) = 0$. Implicit differentiation gives

$$\psi_q\ dq/dk + \psi_k = -c'(q)\ dq/dk + f' - \lambda = 0$$

i.e.,

$$\left.\frac{dq}{dk}\right|_{\psi=0} = -\frac{\psi_k}{\psi_q} = \frac{f'(k) - \lambda}{c'(q)}$$

Clearly the curve $\psi(q, k) = 0$ reaches a stationary point at $\hat{k}$ for which $f'(\hat{k}) = \lambda$. In view of Assumption (i) and the fact that $c'(q) = 1/u''\ (c) < 0$ in (3),

$$\left.\frac{d^2q}{dk^2}\right|_{k=\hat{k}} = \frac{f''(k)}{c'(q)} > 0$$

i.e., this stationary point $(\hat{k}, \hat{q})$ where $\hat{q}$ is the value of q corresponding to $\hat{k}$ is the minimum of the curve $\psi = 0$. Above this curve, $q > \hat{q}$, $\psi > 0$, i.e., $\dot{k} > 0$ and below it, $q < \hat{q}$, i.e., $\dot{k} < 0$. The arrows in fig. 11.2 give the direction of the movement of the system. When $k \to 0$, $c(q) = 0$ and $u'(0) = \infty$ by assumption (iii), i.e., the curve $\psi = 0$ is U-shaped with a unique minimum $(\hat{k}, \hat{q})$. This is called the golden age point at which per-capita consumption is highest (see fig. 11.1).

The curves $\psi = 0$ and $\phi = 0$ thus divide the qk plane into four quadrants: On $\psi = 0$, $\dot{k} = 0$ and above (below) it, $\dot{k} > 0$ (< 0). On the curve $\phi = 0$, $\dot{q} = 0$ and to the right (left) of it, $\dot{q} > 0$ (< 0). Their intersection at (k^*, q^*) gives the equilibrium of the system. It remains to show that this is a saddle point: not every arbitrary point in the qk plane approaches this equilibrium and hence, given any arbitrary initial capital stock k_0, the correct p_0 must be chosen.

Consider quadrant I. Here, both q and k are increasing and the system moves away from the equilibrium point (k^*, q^*) in the N.E. direction. It can never cross the curve $\phi = 0$ because $\psi > 0$, k is increasing, but may cross the curve $\psi = 0$. But then $\dot{k} = 0$ while $\dot{q} > 0$, i.e., the system cannot leave I.

Similarly, in quadrant III, both q and k are decreasing and the system moves away from the equilibrium point (k^*, q^*) in the S.W. direction. It cannot cross the boundary curves $\phi = 0$, $\psi = 0$ and hence any path starting in quadrant III will remain there for ever, continually moving away from the equilibrium point (k^*, q^*) until it becomes infeasible $(k < 0)$.

Clearly neither path is optimal. This leaves quadrants II and IV as the only remaining candidates. Consider quadrant II. Here $\dot{q} > 0$ and $\dot{k} < 0$ and the resultant motion is in the N.W. direction. In the limit, it can either cross the curve $\psi = 0$ and enter I or the curve $\phi = 0$ and enter III, as shown by the arrows in fig. 11.2. Neither is optimal. Thus, only the path approaching (k^*, q^*) is optimal. Similarly any path originating in IV must either cross the curve $\phi = 0$ and end up in I or the curve $\psi = 0$ and terminate in III. Neither is optimal. Hence the only optimal path is the one approaching (k^*, q^*) in the limit.

Thus, we have shown that the equilibrium $E(k^*, q^*)$ is a saddle point with stable branches in quadrants II and IV and unstable branches in the remaining two quadrants. The conclusion is clearly that for any given initial capital stock k_0, there exists a unique optimal trajectory which brings the system to its long run equilibrium $E(k^*, q^*)$. Any other path will end up in the wrong quadrant which is infeasible and non-optimal. The initial imputed price q_0 must therefore be chosen on this optimal trajectory.

Note that in the above analysis, an implicit assumption has been made that $k^* < \hat{k}$. If $k^* > \hat{k}$, the optimal trajectory does not exist. For a further analysis of this point, see Arrow and Kurz (1970) ch. III.

Similarly, a phase diagram can be shown in the (c, k) plane.

From (3) we have

$$u''(c)\dot{c} = \dot{q} = (a - f')q = (a - f')u'$$

using (5) and (3) again. With the definition $\sigma \equiv - cu''/u' > 0$, this can be written as

$$\frac{\dot{c}}{c} = \frac{1}{\sigma}(f' - a)$$

Combining this with (4), we have two differential equations in k and c whose phase diagram is given in fig. 11.3. It can be seen that the equilibrium is a unique saddle point to which the stable paths converge.

11.2.1.2 Technical Progress in the aggregate model

the disembodied Harrod neutral technical progress, the above well behaved neo-classical production function with constant returns to scale, is

$$Y(t) = F(K, L, t) = F[K, A(t)L] \qquad (6)$$

Assuming $A(t) = e^{\tau t}$, this can be written in terms of output, capital and consumption per head of effective labour, $y(t) = Y(t)/e^{\tau t} L(t)$, $k(t) \equiv K(t)/e^{\tau t} L(t)$, $c(t) \equiv C(t)\, e^{\tau t} L(t)$ and assuming constant returns as before, output per head of effective labour is

$$y(t) \equiv Y(t)\, e^{\tau t} L(t) = F[K(t)\, e^{\tau t} L(t) \equiv f[k(t)] \qquad (7)$$

Assuming $\dot{L}/L = n$ as before, the fundamental neo-classical growth function is

$$\dot{k}/k = \dot{K}/K - (n + \tau)$$

Remembering that investment ($\dot{K}$) is income not consumed (Y-C) the above becomes

$$\dot{k}(t) = f[k(t)] - (n + \tau)k(t) - c(t) \qquad (8)$$

But this is precisely (1) with λ replaced by $n + \tau$ and the analysis proceeds exactly as in the basic model.

11.2.1.3 Two-sector models

The above aggregate model has been extended to a large number of two-sector models. These are Investment and Consumption sectors (Uzawa 1961, 1963, 1964, 1966), public and private investment sectors (Arrow and Kurz 1970), physical and human capital sectors (Uzawa 1965, Tu 1966, 1969, Dobell and Ho 1967, Razin 1972, Manning 1975, 1976) production and pollution abatement sectors and various others. For reason of space, only Uzawa's Investment-Consumption two-sector model will be discussed since this constitutes an important landmark around which a number of other two-sector models have been built.

Uzawa's Two-Sector Model

The best known generalisation of the aggregate model to incorporate two sectors producing respectively consumption good (Y_c) and investment good (Y_I) is Uzawa's (1961, 1963, 1964, 1966).

The assumptions made are

(i) Labour grows at a constant rate n, i.e., $L(t) = e^{nt}$.

(ii) The production function in each sector, $Y_i(t) = F_i(K_i, L_i)$ $(i=C,I)$ is differentiable, homogeneous of degree one with all the neo-classical properties.

(iii) Factors are paid their marginal product in each sector.

The production functions are

$$Y_i(t) = F_i(K_i, L_i) \qquad i = C, I$$

$$= L_i(t)\, f_i(k_i) \text{ by assumption (ii)}$$

or

$$y_i \equiv Y_i/L_i = f_i(k_i) \tag{9}$$

where K_i, L_i = capital, labour in sector i $(i = C, I$ with t omitted)

$k_i \equiv K_i/L_i$, $y_i \equiv Y_i/L_i$ and $y \equiv Y/L$

The accounting identities of output are

$$Y(t) = Y_C(t) + pY_I(t)$$

where $p \equiv p_I/p_C$ = price of the investment good in terms of the consumption good. In per capita terms, this is

$$y(t) = y_C(t) + py_I(t) \tag{10}$$

The full employment implies

$$K_C(t) + K_I(t) = K(t)$$

$$L_C(t) + L_I(t) = L(t) = e^{nt}$$

which, with the definitions $\ell_i \equiv L_i/L$, $k_i \equiv K_i/L_i$ $(i = C, I)$ and $k \equiv K/L$, can be written as

$$\ell_C + \ell_I = 1$$

$$k_C\ell_C + k_I\ell_I = k \tag{11}$$

Solving (11) gives

$$\ell_C = (k - k_I)/(k_C - k_I)$$

$$\ell_I = (k_C - k)/(k_C - k_I) \tag{12}$$

Consumption and investment per head thus become

$$y_C \equiv Y_C/L = \ell_C f_C(k_C) = f_C(k - k_I)/(k_C - k_I)$$

$$y_I \equiv Y_I/L = \ell_I f_I(k_I) = f_I(k_C - k)/(k_C - k_I) \tag{13}$$

The total capital accumulation ($\dot{K}$) is the production of new machines (Y_I) less depreciation (μ), i.e.,

$$\dot{K}(t) = Y_I(t) - \mu K(t)$$

or in per capita terms, putting $\lambda \equiv \mu + n$,

$$\dot{k}(t) = y_I(t) - \lambda k(t) \tag{14}$$

Assumption (iii) implies that capital rental (r) is equal to the marginal product of capital and wage (w), to the marginal labour product in each sector, i.e.,

$$r = \partial Y_C/\partial K_C = f'_C = pf'_I = p\,\partial Y_I/\partial K_I$$
$$w = \partial Y_C/\partial L_C = f_C - k_C f'_C = p(f_I - k_I f'_I) = p\partial Y_I/\partial L_I \tag{15}$$

Define wage rental ratio $\omega \equiv w/r$. It is clear that ω determines the capital intensity in each sector (see fig. 11.4)

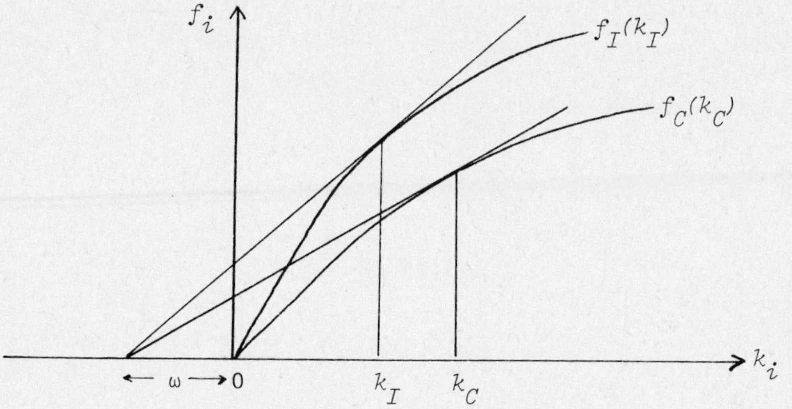

Fig. 11.4 Wage-rental ratio ω and capital intensity k_i

But from (15)

$$\omega(k_i) \equiv \frac{f_i(k_i)}{f'_i(k_i)} - k_i \qquad (i = C, I) \tag{16}$$

$\omega(k_i)$ is clearly a monotonically increasing function of k_i, by assumptions (i) and (ii), $k_i = k_i(\omega)$ by the Implicit function theorem.

Relative output price $p \equiv p_I/p_C$ is also dependent on ω: from (14)

$$p(\omega) = f'_C(\omega)/f'_I(\omega) \tag{17}$$

whose logarithmic differentiation gives

$$\frac{p'(\omega)}{p(\omega)} = \frac{1}{k_I(\omega) + \omega} - \frac{1}{k_C(\omega) + \omega} \tag{18}$$

which is positive (negative) for $k_C > k_I$ $(k_C < k_I)$.

This completes the specification of the model. Clearly ω plays an important part: it will be used as the control variable. Note that ω is bounded between $\omega_I(k) \equiv f_I/f'_I - k$ where only investment good is produced and $\omega_C(k) \equiv f_C/f'_C - k$ where only consumption good is produced.

The Maximum Principle can now be used to solve this problem. It consists of maximizing the consumption functional

$$J \equiv \int_0^\infty e^{-\delta t} \ell_C f_C(k_C)\, dt \tag{19}$$

subject to

$$\dot{k} = \ell_I f_I(k_I) - \lambda k \text{ , } k(0) = k_0$$

where ℓ_C and ℓ_I are given in (12) and δ is a positive constant discount rate.

The Hamiltonian is

$$H \equiv e^{-\delta t} [\ell_C f_C + q(\ell_I f_I - \lambda k)] \tag{20}$$

The canonical equations are, remembering (12)

$$\dot{k} = \ell_I f_I - \lambda k \tag{21}$$

$$\dot{q} = (\lambda + \delta)q - f_C/(k_C - k_I) + q f_I/(k_C - k_I) \tag{22}$$

Differentiation $\partial H/\partial \omega \equiv H_\omega$ gives

$$H_\omega = (qf'_I - f'_C) \left[\frac{k - k_I}{k_C - k_I} \cdot \frac{\omega + k_I}{k_C - k_I} + \frac{k_C - k}{k_C - k_I} \cdot \frac{\omega + k_C}{k_C - k_I} \cdot \frac{dk_C}{d\omega}\right] e^{-\delta t} \tag{23}$$

Assuming first that the consumption sector is more capital intensive, i.e., $k_C > k_I$. The second term on the RHS of (23) being positive, $H_\omega \gtrless 0$ depending on whether $\sigma \equiv qf'_I - f'_C \gtrless 0$.

Thus σ plays the role of the switching function in this non-linear (in ω) problem. If $H_\omega > 0$, i.e., $\sigma > 0$, $q > f'_C / f'_I$, only consumption good is produced and if $H_\omega < 0$, $\sigma < 0$, only investment good is produced. Equilibrium takes place when $\sigma \equiv 0$ and $p = f'_C / f'_I$, i.e., both goods are produced. Then the canonical system, with appropriate substitutions and manipulations (for detail, see Uzawa 1964), becomes

$$\dot{k} = \frac{k_C - k}{k_C - k_I} f_I - \lambda k \tag{24}$$

$$\frac{\dot{p}(\omega)}{p(\omega)} = \lambda + \delta - f'_I$$

But $\dot{p}/p = \dot{\omega}\, p'(\omega)/p$ and substitution from (18) gives

$$\dot{\omega} = \frac{\lambda + \delta - f'_I}{\dfrac{1}{k_I + \omega} - \dfrac{1}{k_C + \omega}} \tag{25}$$

Balanced growth equilibrium given by the intersection of the curves $\dot{k} = 0$ in (24) and $\dot{w} = 0$ in (25) is $w = w^*$ and $k = k(w^*) \equiv k^*$ where

$$f'_I\,[k_I\,(\omega^*)] = \lambda + \delta \tag{26}$$

$$k(\omega^*) \equiv k^* = \frac{k_C(\omega^*) + f_I\,[k_I(\omega^*)]}{f_I[k_I(\omega^*)] + \lambda[k_C(\omega^*) - k_I(\omega^*)]} \tag{27}$$

The movements of the optimal paths for the various initial values (k_0, ω_0) are given by the arrows in fig. 11.5 for $k_C > k_I$ (and fig. 11.6 for $k_C < k_I$). If the initial capital stock k_0 is too low, i.e., $k_0 < k_I$ in fig. 11.5, the economy specialises in the production of the capital good until k reaches k_I when both goods are produced. On the other hand, if k_0 is too high, i.e., $k_0 > k_C$, the consumption good alone will be produced until the capital stock

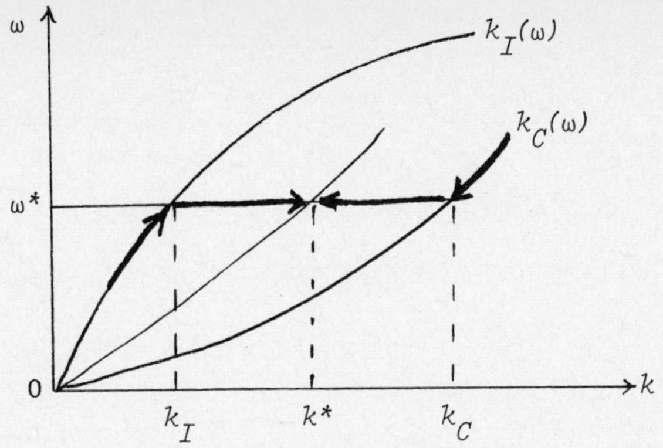

Fig. 11.5 Solution of k and ω for the case $k_C > k_I$

declines to k_C when both goods are produced. The economy thus moves to its equilibrium at (k^*, ω^*) in the long run.

Similarly, it can be shown that if the investment good sector is more capital intensive, i.e., $k_C < k_I$, the economy with $k_0 < k_I$ $(k_0 > k_I)$ will specialise in the production of the investment (consumption) good at first then switch to the production of both when $k \geq k_I$ $(k \leq k_C)$ and the balanced growth will be approached assymptotically and the long run equilibrium (k^*, ω^*) is a saddle point (see fig. 11.6).

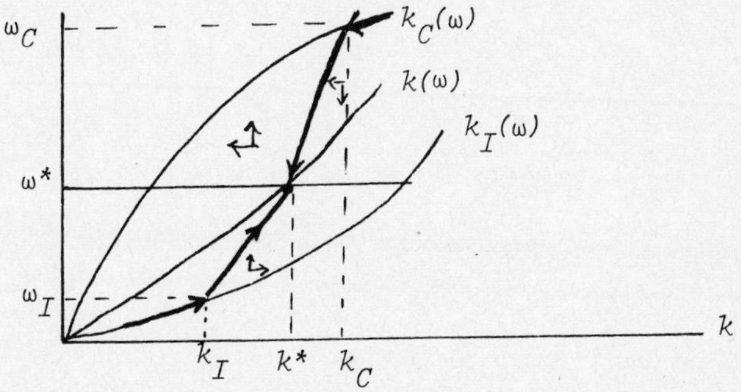

Fig. 11.6 Solution of k, ω when $k_C < k_I$

This completes the solution of the two-sector model.

11.2.1.4 The Multisectoral Optimal Growth Model

Uzawa's two-sector model can be generalised to incorporate many sectors. This has been done by Burmeister and Dobell (1970), Adams & Burmeister (1973).

The economy has $n + 1$ goods $Y \equiv (Y_0, Y_1, \ldots, Y_n)$ where Y_0 is the output of the consumption good and Y_i $(1 \leq i \leq n)$ the output of the ith capital good. The stock vector is $K \equiv (K_0, K_1, \ldots, K_n)$ where K_0 is the labour stock assumed to grow at a constant positive rate n and K_i $(1 \leq i \leq n)$ is the stock of the ith capital good. The transformation function may be written as

$$Y_0 = G(Y_1, \ldots, Y_n ; K_0, K_1, \ldots, K_n) \tag{28}$$

where G is assumed to be continuously differentiable and homogeneous of degree one. These assumptions allow us to write (28) in per capita terms, with $y_i \equiv Y_i/K_0$ $(0 \leq i \leq n)$ and $k_i \equiv K_i/K_0$ $(1 \leq i \leq n)$ as

$$\begin{aligned} y_0 &= G(y_1, \ldots, y_n ; 1, k_1, \ldots, k_n) \\ &\equiv T(y, k) \end{aligned} \tag{29}$$

where $y \equiv (y_1, \ldots, y_n)$ and $k \equiv (k_1, \ldots, k_n)$

The prices of y are $p \equiv (p_1, \ldots, p_n)$ and the rental rates of k is $w \equiv (w_1, \ldots, w_n)$ all in terms of the consumption good used as a numéraire. Factors are assumed to be paid their marginal product, i.e., $w = \partial T/\partial k$. Finally each k_i depreciates at a constant rate δ_i. The fundamental equation is thus

$$\dot{k} = y - (n + \delta) , \; k(0) = k_0 \tag{30}$$

The objective is to maximize per capita consumption y_0 over time, discounted at rate $\gamma > 0$, i.e., subject to (30)

$$Max. \int_0^\infty T(y, k)\, e^{-\gamma t}\, dt \tag{31}$$

The Hamiltonian is

$$\begin{aligned} H &\equiv T(y, k)\, e^{-\gamma t} + q\, [y - (n + \delta)\, k] \\ &\equiv \left\{ y_0 + p\, [y - (n + \delta) k] \right\} e^{-\gamma t} \end{aligned} \tag{32}$$

where $p(t) \equiv q(t)\, e^{\gamma t}$

Application of the Maximum Principle gives

$$\frac{\partial H}{\partial y} = \left(\frac{\partial T}{\partial y} + p \right) e^{-\gamma t} = 0$$

i.e., $\partial T/\partial y \equiv T_y = - p$ (33)

and

$$\frac{d}{dt}\, pe^{-\gamma t} = - \partial H/\partial k \text{ gives}$$

$$\dot{p} = - T_k + p\, (n + \delta + \gamma)$$

where $T_k = \partial T/\partial k = w$, i.e.,

$$\dot{p} = - w + p\, (n + \delta + \gamma) \tag{33}$$

This, together with (30) and (33) and the transversality conditions, provides all the information required. The solution is a saddle point in the (k, p) space

$$\lim_{t \to \infty} [k(t), p(t)] = (k^*, p^*)$$

This equilibrium (k^*, p^*) is unique: for a given initial k_0, the choice of a unique corresponding initial price $p(0)$ must be made in such a way that $\{k(0), p(0)\}$ lies on the stable branches which approach the equilibrium (k^*, p^*). The analogy with the one-sector model is thus complete.

11.2.1.5 Numerical methods for Optimal Growth Models

Numerical methods can be used to solve the finite horizon optimal growth models discussed in the previous sections. These have been carried out by Kendrick and Taylor (1971).

The production function (f) is a function of capital (k)

$$f(k) = e^{\tau t}\ \gamma k(t)^{\beta}(L_0 e^{rt})^{1-\beta}$$

$$= e^{gt}\ ak^{\beta} \tag{35}$$

where $a \equiv \gamma L_0^{\ 1-\beta}$; $g \equiv r(1-\beta) + \tau$

which can be approximated in discrete time with a one-period lag by

$$y_i = f(k_i) = (L + g)^i\ ak_i^{\beta} \tag{36}$$

where τ = rate of neutral technical progress; γ = efficiency parameter;

L = labour force growing at rate r and $g = r(1-\beta) + \tau$

The objective is to maximise the present value of consumption u utility

$$J = \sum_{i=0}^{T-1} (1+\rho)^{-i}\ (1 - v)^{-1}\ c_i^{\ 1-v} \tag{37}$$

subject to

$$k_{i+1} = f(k_i) - c_i + (1-\delta)k_i;\ k(0) = \bar{k}\ ;\ y(T) = \bar{y} \tag{38}$$

The consumption utility function $u(c_i) = c_i^{\ 1-v}/(1-v)$ exhibits diminishing marginal utility, i.e. $u'' < 0 < u'$ and $\lim_{v\to 0} u(c_i) = c_i$, $1-v$ = constant elasticity of consumption utility.

The Hamiltonian is

$$H = (1+\rho)^{-i} (1-v)^{-1} c_i^{1-v} + \lambda_{i+1}[f(k_i) - c_i + (1-\delta)k_i] \quad (39)$$

$H_{c_i} = 0$ gives

$$c_i = [(1 + \rho)^i \lambda_{i+1}]^{-1/v} \quad (40)$$

$\lambda_i = -H_{k_i}$ gives

$$\lambda_{i+1} = [(1+g)^i \beta a k_i^{\beta-1} + 1 - \delta]^{-1} \lambda_i \; ; \quad \lambda(T) = \bar{\lambda} \quad (41)$$

$$k_{i+1} = (1+g)^i a k_i^{\beta} - c_i + (1 - \delta)k_i \quad ; \quad k(0) = \bar{k} \quad (42)$$

The numerical method proceeds as follows:

1. Choose an arbitrary λ_0
2. Use k_0 and λ_0 in (41) to obtain λ_1
3. Use λ_1 in (40) to obtain c_0
4. Use c_0 and k_0 in (42) to obtain k_1
5. Repeat steps 2 through 4, increasing the index by one on each iteration until k_T is obtained.
6. Compare y_T to $\bar{y}$; stop if they are sufficiently close to each other, otherwise choose a new λ_0 and return to step 2.

The authors obtained convergence with only a few iterations. With the various values assigned to the parameters, the paths for optimal savings rate were obtained. These were presented graphically for the various values of the parameters, both for the case of a closed economy without foreign trade and an open economy with foreign trade, for a period of 50 years.

Thus, the optimal growth models are no longer an abstraction but are computable and useable for economic planning.

11.2.2 Economic Stabilization Models

The Linear Regulator and Linear Tracking methods discussed in Chapter 8 have been widely used to minimize economic fluctuations. Some applications have been discussed in Chapters 8 and 9 where the multiplier-accelerator type models as well as the production and inventory stabilization models were examined. Shupp's (1976) wage-price control model was discussed. Neck and Posch's (1982) work on the Austrian Economy was briefly presented. Additional examples and applications can be found in Chow (1975), Pitchford and Turnovsky (1977), Aoki (1976), Turnovsky (1981) and Myoken (1980), for example. It may be interesting to note that this stabilization approach sometimes proves very useful (Simaan and Cruz 1975). This will be discussed in section 11.2.11.

For lack of space, we shall not make any further review here.

11.2.3 Dynamic Theory of the Firms

The static Theory of the Firms consists of allocating resources such as to maximize profit. The Dynamic counterpart consists of maximizing the discounted profit functional, i.e. the present value of the profit made during the planning period, subject to a certain dynamic demand function (which incorporates some speculative element) $D = D(p, \dot{p}, t)$ or dynamic cost (which includes adjustment cost associated with the change in output level, $\dot{x}$). Clearly this is an area to which the first applications of the Calculus of Variations were made (Evans 1924, Hotelling 1931) for both the cases of Perfect Competition and Monopoly. Dynamic Duopoly and Oligopoly have also been investigated, mainly in the context of Differential games (see for example Levine and Thepot 1982, Feichtinger 1982). The theory has also been extended to the Mining Firms (Kemp & Long 1980).

Since the early applications have been examined in earlier chapters, the Mining Firm theory has been analysed in the context of the Theory of Natural Resources, and Differential Games theory goes beyond the scope of this book, we shall not go into these here.

11.2.4 International Trade

Although International Trade is static in nature: Hecksher-Ohlin, Stolper-Samuelson and Rybczynski's theorems, Comparative Advantage, Balance of Payments and others are all cast in the framework of static equilibrium, Optimal Control has been used since its earliest days and proved a valuable tool. The early applications focused on the international movement of factors such as capital and labour (Bardhan 1965, 1967, Hamada 1966) and commodities (Bardhan 1965, Ryder 1967, 1969). Extensions of the basic models have been made to optimal tax and tarrifs, multinational firms and transfer of technology (Koizumi and Kopecky 1980), devaluation (Calvo 1981), to mention a few more recent ones. Since international capital movement has been discussed in some detail in Chapter 7 mainly to illustrate allowable switchings, we shall restrict our presentation to Ryder's model (1967) of commodity movement. This is an extension of Uzawa's two-sector model presented in 11.2.1 with the same notations and assumptions such as Investment (I) and Consumption (C) good sectors, constant labour growth rate n, constant returns to scale production function $Y_i = F_i(K_i, L_i) = L_i F_i(K_i, 1) \equiv L_i f_i(k_i)$ per-capita income in terms of the consumption good

$$y(t) = y_C(t) + q(t)\, y_I(t)$$

where

$$y_I(t) = (k_C - k)\, f_I \,/(k_C - k_I) \quad \geq 0$$

$$y_C\, t = (k - k_I)\, f_C \,/(k_C - k_I) \quad \geq 0$$

$$x_i(t) = y_i(t) + z_i(t) \geq 0 \qquad (i = C, I)$$

$$\dot{k}(t) = x_I(t) - \lambda k(t) \equiv y_I(t) + z_I(t) - \lambda k(t)$$

where $x_i(t)$, $z_i(t)$ = per-capita consumption and import of good i ($i = C, I$)

$\lambda \equiv \mu + n$ = depreciation + labour growth rate

δ = discount rate

Let the offer curve be

$$z_C(t) = \theta[z_I(t)] \; , \; (0 < z_{I\,min} \leq z_I \leq z_{I\,max})$$

The objective is to maximize the present value of per-capita consumption utility $u(x_C)$, assumed linear for simplicity i.e. $u(x_C) = x_C$; $u'' = 0$,

$$Max.\; J = \int_0^\infty e^{-\delta t} x_C \, dt$$

subject to

$$\dot{k} = x_I(t) - \lambda k(t) \; , \; k(0) = k_0$$

The Hamiltonian is

$$H = y_C + z_C + q(y_I + z_I - \lambda k) = x - q\lambda k$$

The Maximum Principle gives

$$\dot{k} = H_q = x_I(k, q) - \lambda k \; ; \; k(0) = k_0$$

$$\dot{q} = \delta q - H_k = (\lambda + \delta)q - \partial x/\partial k$$

$$\lim_{t\to\infty} e^{-\delta t} q(t) = 0$$

$$\partial H/\partial z_I = \theta'(z_I) + q = 0$$

$$\frac{\partial H}{\partial k_I} = \left(\frac{k_C - k}{k_C - k_I}\right)^2 \left\{-f_C + q[f_I + (k_C - k_I)f_I']\right\} = 0$$

$$\frac{\partial H}{\partial k_C} = \left(\frac{k = k_I}{k_C - k_I}\right)^2 \left\{-f_C + (k_C - k_I)\; f'_C + qf_I\right\} = 0$$

The solution of this system is very similar to Uzawa's. The results are summarized in the following table. (For detailed computations, see Ryder 1967)

Table 11.1 Patterns of Specialization

	Production		
	I	II	III
Absorption	Consumption Good	Nonspecialized	Investment Good
A. Consumption	$y_1=0$ $y_C>0$ $z_1=0$ $z_C=0$ $x_1=0$ $x_C=y_C$	$y_1>0$ $y_C>0$ $z_1=-y_1$ $z_C>0$ $x_1=0$ $x_C>0$	$y_1>0$ $y_C=0$ $z_1=-y_1$ $z_C>0$ $x_1=0$ $x_C=z_C$
B. Non-specialized	$y_1=0$ $y_C>0$ $z_1>0$ $-y_C<z_C<0$ $x_1=z_1$ $x_C>0$	$y_1>0$ $y_C>0$ $z_1>-y_1$ $z_C>-y_C$ $x_1>0$ $x_C>0$	$y_1>0$ $y_C=0$ $-y_1<z_1<0$ $z_C>0$ $x_1>0$ $x_C=z_C$
C. Investment	$y_1=0$ $y_C>0$ $z_1>0$ $z_C=-y_C$ $x_1=z_1$ $x_C=0$	$y_1>0$ $y_C>0$ $z_1>0$ $z_C=-y_C$ $x_1>0$ $x_C=0$	$y_1>0$ $y_C=0$ $z_1=0$ $z_C=0$ $x_1=y_1$ $x_C=0$

The general conclusions are as follows:

If the initial capital stock k_0 is very low, consume nothing and invest everything. With these initial austerity measures, capital will be accumulated at the fastest possible rate and will soon reach some critical level at which consumption will be increased until the stationary values (k^*, q^*) are reached. On the other hand, if k_0 is high, the economy should "specialise" on consumption i.e. consume all and invest nothing, until (k^*, q^*) is reached.

11.2.5 Regional Economics

Regional Economics is concerned with the problem of optimal allocation of resources and investment among the various regions of the country.

Given two regions with different income levels Y_i $(i = 1, 2)$ labour growth rates $\dot{L}_i/L_i = n_i$, savings S_i, Datta-Chaudhuri (1967) seeks to allocate national saving between them in such a way that the country acquires a desired capital stock as quickly as possible. This is a time optimal problem whose objective is to minimize

$$J = \int_0^T dt$$

subject to

$$Y_i(t) = F_i(K_i, L_i) = L_i(t) f_i\,(k_i) \quad \text{where } k_i \equiv K_i/L_i \quad (i = 1, 2)$$

$$L_i(t) = L_{i0}\, e^{n_i t}$$

$$S_i(t) = s_i\,[Y_i(t) - w_i L_i(t)]$$

$$\dot{K}_i(t) = S_i(t) - T_i(t) + (1 - \delta)T_j(t) = [1 - u_i(t)\; S_i(t)] + (1 - \delta)u_j S_j(t)$$

$$T_i(t) = u_i(t)\; S_i(t)\;, \quad 0 \le u_i \le 1 \quad , \quad (i = 1, 2)$$

$$K_i(0) = K_{i0}\,,$$

where

$Y_i\ t \;= L_i\ F_i(k_i)$ is the neo-classical production function

T_i = transfer of resources from the ith region (δ = transportation cost)

u_i = proportion of ith's surplus to be transported, a control variable

w_i = constant wage rate in the ith region (with unlimited labour supply)

The Hamiltonian is

$$H = 1 + p_1 [(1-u_1)S_1 + (1-\delta)u_2 S_2] + p_2[(1-u_2)S_2 + (1-\delta)u_1 S_1]$$

The adjoint system is

$$\dot{p}_1 = -\left\{p_1 - [p_1 - (1-\delta)p_2]\, u_1\right\} \partial S_1/\partial K_1$$

$$\dot{p}_2 = -\left\{p_2 - [p_2 - (1-\delta)p_1]\, u_2\right\} \partial S_2/\partial K_2$$

The Hamiltonian being linear in u_i, the switching functions

$$\sigma_1 \equiv [-p_1 + (1-\delta)p_2]\, S_1 \;;\; \sigma_2 \equiv [-p_2 + (1-\delta)p_1]\, S_2 \quad (S_i > 0)$$

give the optimal control u_i as

$$u_1^* = \begin{cases} 0 & \text{for } p_1 > (1-\delta)\, p_2 \\ 1 & \text{for } p_1 < (1-\delta)\, p_2 \end{cases}$$

$$u_2^* = \begin{cases} 0 & \text{for } p_2 > (1-\delta)\, p_1 \\ 0 & \text{for } p_2 < (1-\delta)\, p_1 \end{cases}$$

This is a bang bang singular control model. If the price ratio $p_1/p_2 < 1 - \delta$, all investment takes place in zone II, if $p_1/p_2 > 1/(1-\delta)$, $u_2 = 1$, $u_1 = 0$ i.e. all investment takes place in zone I. For $1/(1-\delta) < p_1/p_2 < 1 - \delta$, two regions grow autarkically. The singular arc I means $\sigma_1 \equiv 0$ i.e. $p_1/p_2 \equiv 1 - \delta$ and the singular arc II implies $\sigma_2 \equiv 0$ i.e. $p_1/p_2 \equiv 1/(1 - \delta)$; $\sigma_1 \equiv 0 \equiv \sigma_2$ implies $p_1/p_2 = 1/(1-\delta)$ i.e. $\delta = 0$ which means complete autarky, $p_1 = p_2$, the 45° line. (See fig. 11.7.)

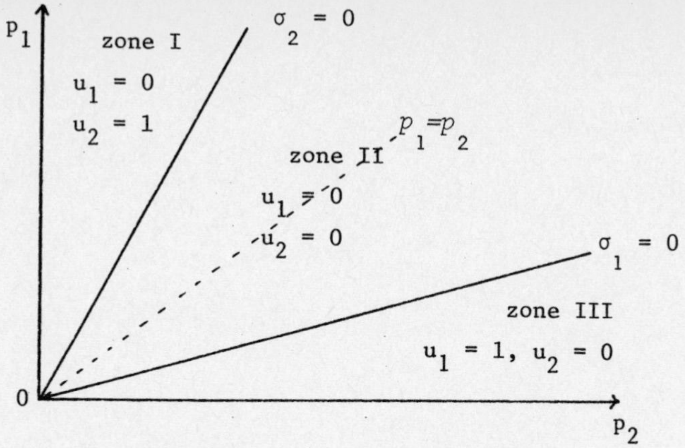

Fig. 11.7 Zones of specialisation

An alternative formulation of the problem (Rahman 1963, Takayama 1967) is to maximise national income $Y_1(T) + Y_2(T) = b_1K_1(T) + b_2K_2(T)$ where the production functions $Y_i = b_iK_i$ $(i = 1, 2)$, $\dot{K}_i = s_iY_i$, $\dot{K}_1 + \dot{K}_2 = s_1b_1K_1 + s_2b_2K_2$. Let u be the proportion of investment allocated to region 1 and $1-u$, to region 2, $0 \le u \le 1$ and define $g_i \equiv b_is_i$ $(i = 1, 2)$, we have the dynamic system.

$$\dot{K}_1 = u(g_1K_1 + g_2K_2) \quad ; \; K_1(0) = K_{10} > 0$$

$$\dot{K}_2 = (1-u)(g_1K_1 + g_2K_2); \; K_2(0) = K_{20} > 0$$

The Hamiltonian H is

$$H \equiv p_1u(g_1K_1 + g_2K_2) + (1-u)p_2(g_1K_1 + g_2K_2)$$

$$\equiv (g_1K_1 + g_2K_2)\,[p_2 + (p_1 - p_2)\,u]$$

The switching function σ is $\sigma \equiv (g_1K_1 + g_2K_2)(p_1 - p_2)$.

Clearly

$$u^* = \begin{cases} 1 & \text{if } p_1 > p_2 \\ 0 & \text{if } p_1 < p_2 \end{cases}$$

The adjoint system is

$$\dot{p}_1 = -[u(p_1 - p_2) + p_2]g_1 \quad ; \quad p_1(T) = b_1$$

$$\dot{p}_2 = -[u(p_1 - p_2) + p_2]g_2 \quad ; \quad p_2(T) = b_2$$

giving simply $\dot{p}_1/\dot{p}_2 = g_1/g_2$ which gives on integration, $p_1(t) = (g_1/g_2)p_2(t)+C_1$ where $C_1 = b_1 - g_1 b_2/g_2 \equiv (s_2 - s_1)\, b_1 b_2/g_2$ in view of $p_1(T) = b_1$. *Thus*

$$p_1(t) = (g_1/g_2)\, p_2 + (s_2 - s_1) b_1 b_2/g_2$$

$$\equiv \frac{b_1 s_1}{b_2 s_2} p_2 + \frac{b_1}{b_2}(s_2 - s_1)$$

is important since it determines the optimal policy. The various cases are based on the various combinations of regional productivities b_i and saving propensities s_i. For example if $g_1 > g_2$ and $s_2 > s_1$ then $p_1 > p_2$ and $u = 1$. If $s_1 = s_2$, and $b_1 > b_2$ then $p_1 > p_2$, $u = 1$, i.e. investment should be made in the region having higher productivity. The switching time t_s at which $\sigma(t_s) = 0$ can also be calculated.

11.2.6 Optimal Urban Economics

In the last few years, Optimal Control theory has been applied to Urban Economics to the point that Mills and MacKinnon (1973, p. 597) complained that

> ....We are concerned that the profession will quickly see that Pontryagin's Principle is as applicable to space as to time and the journals will be flooded with technically sophisticated but economically uninteresting urban models.

Urban Economics is a relatively new field. It is about twenty years old. Alonso (1964), Mills (1967) and Muth (1969) are among the pioneers. They were joined later by Oron, Pines and Sheshinski (1973),

Solow (1973), Mirrlees (1972), Dixit (1973) and Riley (1973). Our interest in this chapter is not to review the theory but rather to show how O.C. has been successfully applied to this new field. As ingredients of optimality, we shall concentrate on the issues of space and congestion as dealt with by Oron et al. (1973), Solow (1973), Mirrlees (1972) and Dixit (1973) and reviewed by Muth (1977) whose exposition is followed in this section.

Consider a city centre called Central Business District (CBD), surrounded by land used for residence (q) and transportation (p). There are $z(x)$ households living x miles from the CBD, each enjoying an identical utility function $u(c, q, x)$ where c is a composite consumption good. Each household earns an identical income (y) which is spent on consumption (c) and transportation cost $t(x)\ g(t, p)$ which is the number of trips $t(x)$ multiplied by the average cost per trip per unit distance $g(t, p)$, the latter being a decreasing function of t and p.

The objective is to maximize utility

$$\underset{c,q,z,p}{J} = \int_0^{\bar{x}} z(x)\ u(x)\ dx \quad - \tfrac{1}{2}\, v r \theta \bar{x}^2 \tag{43}$$

subject to

$$\int_0^{\bar{x}} [z(x)\ c(x) + t(x)\ g(t, p) - r\theta x]\ dx \leq Zy \tag{44}$$

$$t(x) = \int_x^{\bar{x}} z(x')\ dx'$$

or

$$t'(x) = -\ z(x) \tag{45}$$

$$t(\bar{x}) = 0, \ t(0) = Z$$

$$z(x)\ q(x) + p(x) \leq \theta\ x \tag{46}$$

where

$\bar{x}$ = distance from the CBD to the edge of the city

r = exogeneously given non-urban land rental

θ = angular measure of land surrounding the CBD

This is a typical O.C. problem where the dynamic system is (45), and the inequality constraints are (44) and (46). The state variable is $t(x)$ and control variables, c, q, z and p. If urban land is available in unlimited quantity, $\bar{x} = \infty$ and we have a problem of infinite space horizon. If it is strictly limited in quantity, say $\bar{x} = x_1$, we have a two fixed end points problem. If the city boundary is finite but unspecified, $\bar{x}$ is free and we have a free end point problem. These must be taken into account in the determination of transversality conditions.

The above formulation is equivalent to the problem of maximizing the following Hamiltonian H, account taken of the "Scrap" function $S(\bar{x}) = \frac{1}{2} \cdot vr\theta\bar{x}^2$

$$H \equiv zu(c, q, x) - v[zc + tg(t, p)] - \mu(zp + p - \theta x) - \lambda z \tag{47}$$

The necessary conditions for optimality, writing $u_c \equiv \partial u/\partial c$ etc..., are

$$H_c = 0 \Rightarrow zu_c - vz = 0 \tag{48}$$

$$H_q = 0 \Rightarrow zu_q - \mu z = 0 \tag{49}$$

$$H_z = 0 \Rightarrow u - vc - \mu q - \lambda = 0 \tag{50}$$

$$H_p = 0 \Rightarrow -vtg_p - \mu = 0 \tag{51}$$

$$\lambda'(x) + H_t = 0 \Rightarrow -v(g + tg_t) + \lambda'(x) = 0 \tag{52}$$

and the transversality conditions $[H(x) + S_x]\ \delta x|_{x=\bar{x}} = 0$ imply

$$vr\theta\bar{x} = H(\bar{x}) \tag{53}$$

(48), (49) and (51) imply the equi-marginal rule

$$\frac{u_q}{u_c} = -\ tg_p = \frac{\mu}{v} \tag{54}$$

i.e., in an optimal programme, allocation should be made such that the relative marginal utility of space and consumption good u_q/u_c is equal to their relative prices μ/v which in turn should be equal to the saving in commuter costs resulting from an additional unit of land allocated to transportation $-\ tg_p$.

Also, from (51) we have

$$\frac{d}{dx}(-tg_p) = \frac{\mu'(x)}{v} \tag{55}$$

which says that the decline in the relative price of land depends upon marginal transportation costs (MTC) relative to space consumption, MTC including the private cost per trip g and congestion cost tg_p.

Finally (53), (45) and (50) imply

$$\mu(\bar{x})\ /\ v\ =\ r$$

i.e., at the city's edge, the rental of urban land is equal to the rental of non-urban land.

11.2.7 Education, Labour training and Human Capital

The objective function is to maximize consumption utility subject to physical and human capital formation. This is an optimal growth related two-sector model, first investigated by Uzawa (1965), Ben Porath (1967), Dobell and Ho (1967), Razin (1972), Tu (1969) and

Manning (1976, 1978, 1979). Dobell and Ho's model has been discussed at length in Chapter 6 and Tu's model in Ch. 2. For lack of space, no further discussion will be carried out.

11.2.8 Natural Resources

The application of the Calculus of Variations and Optimal Control to the management of exhaustible and renewable resources has been discussed in great detail in Ch. 7 and will be omitted here.

11.2.9 Optimal Control of Pollution

This is a problem of optimal resource allocation between consumption and pollution control. Consumption goods (C) contribute to increasing happiness but the pollution (P) generated in the process of their production causes misery. The central issue is the consumption-pollution trade-off.

The problem has been investigated by Keeler et al. (1972), Plourde (1972) (given in example 2.6.8, Ch. 2 to illustrate the Variational techniques), Smith (1972), Forster (1977), Luptáčik and Schubert (1982), all using Optimal Control theory. We shall briefly analyse the problem, using Forster's model.

Forster (1977) considers an economy which produces in each period a fixed output Y^0 to be allocated to consumption (C) and expenditure (E) on pollution abatement, i.e. $Y^0 = C + E$. The pollution accumulation function is

$$\dot{P} = Z(C) - \alpha P \tag{56}$$

where $Z(C) \equiv g(C) - h(E) \equiv g(C) - h(Y^0 - C)$ indicates the pollution generated by the production of output, $g(C)$ (with $g(0) = g'(0) = 0$

and $g'(C)$, $g''(C) > 0 \; \forall \; C > 0$) less the quantity abated $h(E)$ where $h(0) = 0$ and $h''(E) < 0 < h'(E) \; \forall \; E > 0$. Clearly $Z''(C) < 0 < Z'(C)$; $\lim_{C \to Y^0} Z'(C) = \infty$ is assumed. With no production and pollution abatement, the waste (P) biodecomposes at the exponential rate $\alpha(\alpha > 0)$.

The objective is to maximize the present value of social utility $U(C, P)$. This function is assumed to be concave with diminishing marginal utility of consumption ($U_{CC} < 0 < U_C$) and increasing undesirability of pollution (U_P, $U_{PP} < 0$), also $U_{CP} < 0$ and $U_C(0, P) = \infty$, $U_P(C, 0) = 0$.

The optimal control problem is to choose a consumption level such as to maximize

$$\int_0^\infty e^{-t} \, U(C, P) \, dt \tag{57}$$

subject to

$$\dot{P} = Z(C) - \alpha P, \; P(0) = P_0 \; ; \; P(\infty) \text{ free} \tag{58}$$

$$E = Y^\circ - C \geq 0, \quad P \geq 0 \tag{59}$$

The current value Hamiltonian is

$$H = U(C, P) + \lambda(Z - \alpha P) + r(Z - \alpha P) + q(Y^\circ - C) \tag{60}$$

The Maximum Principle gives

$$H_C = 0 \Rightarrow \quad \lambda + r = (q - U_C) \, / Z'(C) \tag{61}$$

which gives, by the Implicit Function theorem, $C = C(\lambda, P)$ with $\partial C / \partial P < 0 < \partial C / \partial \lambda$.

$$\dot{\lambda} = p\lambda - H_P = (p + \alpha)\lambda - U_P + r\alpha \tag{62}$$

$$q \geq 0, \; q(Y^\circ - C) = 0 \tag{63}$$

$$r \geq 0, \; rP = 0 = r\dot{P} \tag{64}$$

The properties of the solution are obtained by a phase diagram examination of the equilibrium values (P^*, λ^*) for which $\dot{P} = 0$, $\dot{\lambda} = 0$ in (56) and (62) above. Clearly this pair of equations gives

$$\left.\frac{d\lambda}{dP}\right|_{\dot{P}=0} = \frac{\alpha - Z'\partial C/\partial P}{Z' \, \partial C/\partial\lambda} > 0$$

and

$$\left.\frac{d}{dP}\right|_{\dot{\lambda}=0} = \frac{-U_{CC} \, U_{PP} - U^2{}_{CP} - \lambda U_{PP} \, Z''}{- U_{CC} + \lambda Z'' \quad \rho + \alpha - U_{CP} \, \partial C/\partial\lambda} < 0$$

Thus, the curve $\dot{P} = 0$ is downward sloping and $\dot{\lambda} = 0$ is upward sloping (see fig. 11.8).

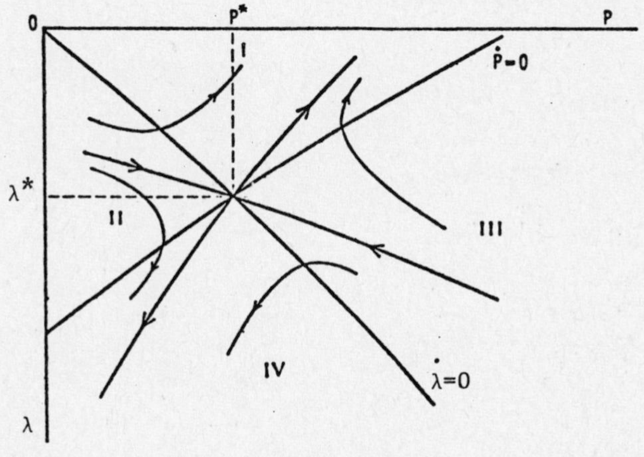

Fig. 11.8

Simple calculations show that the determinant of the Jacobian of system (56) and (62) is negative, i.e. the equilibrium is a unique saddle point at (P^*, λ^*), which is approached asymptotically.

11.2.10 Optimal Population Control

The Maximum Principle has been successfully applied in the theory of optimal population by Dasgupta (1969), Sato and Davis (1971),

Pitchford (1974), Lane (1975) and Arthur & McNicoll (1977). Since this application has been discussed in detail in Chapter 5 (section 5.6.4), we shall not go into further elaboration here.

11.2.11 Optimal Control of the Armament Build-up

The Maximum Principle has also been applied (Brito 1972 , Simaan and Cruz 1975) to Richardson's (1960) classical problem of Arms Race.

Richardson's model is

$$\dot{x} = Ax + b \tag{65}$$

where b = aggressiveness ($b > 0$) or goodwill ($b < 0$) n-vector,

$x = (x_1, \ldots, x_n)$ = armament level of n countries

$A = n\text{x}n$ constant armament coefficient matrix whose negative diagonal terms represent the war burden, called fatigue coefficient and the positive off-diagonal terms, the threat or defence coefficients.

In general, the model is non linear, $\dot{x} = F(x)$. Brito (1972) examined the problem in the context of resource allocation. The dynamics of the armament build-up is

$$\dot{x}_i = -\beta_i x_i + Z_i \qquad (i = 1,\ 2) \tag{66}$$

where $Z_i(t)$ = expenditure on weapons and $\beta_i x_i$ the resources necessary to maintain and operate these weapons. The objective is to maximize the present value of utility which is a function of consumption (C) and defense $D_i(x_1, x_2)$ i.e.

$$J_i = \int_0^\infty e^{-rt}\, U_i\ [C_i,\ D_i(x_i\ x_j)]\ dt \tag{67}$$

where $D_i(x_1, x_2)$ is the ith country's defence index and

U_i its utility function.

This is a differential game problem (see for example, Pontryagin 1962 Ch. 4, Intrilligator 1971, Isaacs 1965) which is beyond the scope of this elementary text. However the Maximum Principle could be used to solve the problem for the simple Cournot-Nash case where the assumption of a myopic behaviour of $x_j = \bar{x}_j$ (fixed) is made. The current value Hamiltonian for country i is

$$H_i \equiv U_i[C_i, D_i(x_i, x_j)] + p_i\ Z_i - \beta_i x_i) + \lambda_i(Y_i - Z_i - C_i) \quad (68)$$

Application of the Maximum Principle gives

$$\dot{p}_i = rp_i - H_{x_i} = - \frac{\partial U_i}{\partial D_i} \frac{\partial D_i}{\partial x_i} + (r + \beta_i)\, p_i \quad (69)$$

with the transversality conditions

$$\lim_{t\to\infty} e^{-rt} p_i(t) x_i(t) = 0$$

$$\partial U_i/\partial C_i - \lambda_i \leq 0 \ ; \quad C_i\ (\partial U_i/\partial C_i - \lambda_i) = 0$$

$$p_i \quad - \lambda_i \leq 0 \ ; \quad Z_i\ (p_i - \lambda_i) \quad = 0 \quad (70)$$

Since $Z_i \neq 0$, $C_i \neq 0$ (otherwise marginal utility is infinite), we have

$$\partial U_i/\partial C_i = \lambda_i = p_i$$

Substituting into (66) and (69) it can be shown that under certain assumptions concerning the utility and defense functions (see Brito 1972 for details), equilibrium exists. The existence proofs make use of the Fixed Point theorem and Contraction mapping. Furthermore, if myopic behaviour is assumed, this equilibrium is stable. This is expected: myopic behaviour means that country i takes the other's armament stock x_j as fixed at $x_j = \overline{x}_j$ and pursues an armament policy J_i. Similarly country j assumes $x_i = \overline{x}_i$ and chooses Z_j to maximize J_j. The equilibrium point is defined by (x_i^*, x_j^*) such that $\dot{x}_i = 0 = \dot{x}_j$ simultaneously. Thus, Richardson's equilibrium is obtained.

Brito's Nash-Cournot solution based on the assumption of myopic behaviour is rather unrealistically simplistic. Simaan and Cruz (1975) reformulated the Brito (1972) problem above as one of differential game and uses closed loop control. The Hamiltonian for country i (i = 1, 2) is

$$H_i = e^{-rt}\left\{U_i[Z_i,\ D_i(x_1,x_2)] + q_{i1}(-\beta_1 x_1 + Z_1) + q_{i2}(-\beta_2 x_2 + Z_2)\right\} \quad (71)$$

The necessary conditions for optimality are

$$\dot{q}_{ij} = -\frac{\partial U_i}{\partial D_i}\frac{\partial D_i}{\partial x_j} + (\beta_j + r)q_{ij} - \frac{\partial Z_i}{\partial x_j} q_{ij}\ ;\ q_{ij}(T) = 0\ ,\quad (i,\ j = 1,2) \quad (72)$$

$$\partial H_i/\partial Z_i = 0 \Rightarrow q_{ii} = -\ \partial U_i/\partial Z_i \quad (i = 1,2) \quad (73)$$

The closed-loop nature is given by $\partial Z_i/\partial x_j$. The solution leads to a system of non-linear non-autonomous differential equations, which is difficult to analyse. An alternative was then used. This consists of the linear tracking (see Ch. 8) problem of minimizing

$$J_1 = \tfrac{1}{2}\int_0^T e^{-rt}\ [R_1(W_1 - Z_1)^2 + Q_1(x_1 - a_1 x_2 - v_1)^2]dt \quad (74)$$

$$J_2 = \tfrac{1}{2}\int_0^T e^{-rt}\ [R_2(W_2 - Z_2)^2 + Q_2(x_2 - a_2 x_1 - v_2)^2]dt$$

where the defence indices are $D_1 \equiv x_1 - a_1 x_2 - v_1$, $D_2 = x_2 - a_2 x_1 - v_2$.

Using Kalman's equation $q = Kx$ (see Ch. 8, eq. 14), a system of time variant linear differential equations of the Richardson type $\dot{x} = Ax + b$ was obtained which is easy to solve and has the advantage of closed loop control.

Deger and Sen (1981) have recently developed the above models by introducing security (S) and threat (θ) into the utility function and focusing on a less developed country whose national income (Y) is allocated between civilian (C) income and military (M) expenditure

$$Y = C + M$$

of

$$1 = C/Y + M/Y \equiv c + m$$

Security, depending on the armament stock S as a proportion of Y i.e. $s \equiv S/Y$, enters the utility function $u(c, s, \theta)$

The armament build-up is given by

$$\dot{S} = M - \delta S$$

or

$$\frac{\dot{s}}{s} = \frac{\dot{S}}{S} - \frac{\dot{Y}}{Y}$$

or

$$\dot{s} = m - \alpha s \qquad (75)$$

where δ = depreciation,

$\dot{Y}/Y = \beta$, assumed constant and $\alpha \equiv \delta + \beta$.

The objective is to maximize, subject to (75),

$$J = \int_0^\infty e^{-\rho t} u(1 - m, s, \theta)\, dt \qquad (76)$$

The Hamiltonian is

$$H \equiv e^{-\rho t} u(1 - m, s, \theta) + p(m - \alpha s) \qquad (77)$$

The necessary conditions for optimality are

$$\partial H/\partial m = 0 = -e^{-\rho t}u + p$$

$$\Rightarrow p = e^{-\rho t}\, u \qquad (78)$$

where $u_1 \equiv \partial u/\partial c \equiv \partial u/\partial(1-m)$ = marginal utility of consumption.

Similarly for other partial derivatives.

$$\dot{p} = -\partial H/\partial s = - \; e^{-\rho t}\, u_2 - p\alpha)$$

$$= \; e^{-\rho t}(\alpha u_1 - u_2) \qquad (79)$$

(78) gives

$$\dot{p} \; = \; e^{-\rho t}(-\rho u_1 - u_{11}\dot{m} + u_{12}\dot{s}) \qquad (80)$$

Equating (79) and (80) and assuming for simplicity constant cross derivatives u_{ij} $(i, j = 1, 2)$, we have

$$\dot{m} = [-(\rho + \alpha)\, u_1 + u_2 + u_{11}\, m - u_{12}\, \alpha\, s]/u_{11} \qquad (81)$$

The steady state equilibrium obtained by $\dot{m} = 0 = \dot{s}$ (i.e. by differentiating (80) and (81) totally, with $\dot{m} = 0 = \dot{s}$) gives

$$\left.\frac{dm}{ds}\right|_{\dot{m} = 0} < 0 < \left.\frac{dm}{ds}\right|_{\dot{s} = 0} = \alpha \qquad (82)$$

i.e. the solution obtained is a saddle point equilibrium. See fig. 11.9

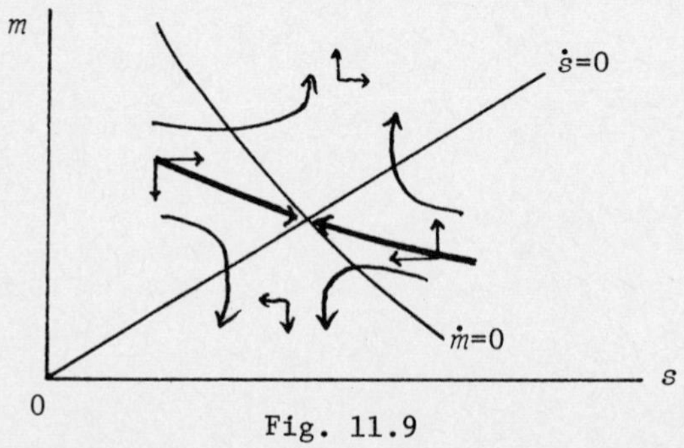

Fig. 11.9

It could be shown that for a given m, (i.e. $\dot{m} = 0$), differentiating (81) totally gives

$$\frac{\partial s}{\partial \theta} > 0$$

i.e. an increase in threat will cause the $\dot{m} = 0$ curve to shift to the right, causing a displacement of the equilibrium armament stock to a higher level.

This model has been developed into a differential game model but we shall not go into this, our purpose being only to show the application of optimal control in this field.

11.3 Some Management Science Applications: A Dynamic Theory of the Managerial Firm

Optimal Control (O.C.) has been extensively applied in Management Science in the last few years. (See, for example, Bensoussan et al. 1974, 1978, Kamien & Schwartz 1981, Sethi & Thompson 1981, Feichtinger 1982). The managerial firm faces the problem of choosing a financial structure, a production and inventory plan, a marketing strategy and a maintenance policy in order to maximize profit and the value of the firm. O.C. has proved a valuable tool in all these areas. For obvious reasons, we shall not survey the literature in this section: such surveys have been carried out (see, for example, Sethi 1977, 1978, Sethi and Thompson 1981). Rather we shall briefly show how O.C. has been applied in the dynamic theory of the Managerial Firm (Ekman 1982, van Loon 1982).

11.3.1 Optimal Financing Model

The optimal capital structure, dealing with an optimal combination of internal and external financing, equity and debt financing has been investigated by Miller and Modigliani (1961), Gordon (1962), Linter (1963), Magill (1970), Krouse and Lee (1973), Wong (1975), Sethi (1978), Ekman (1982) and van Loon (1982). The objective is

$$J = \int_0^T (1 - u - v)\, x(t)\, e^{-\rho t}\, dt$$

subject to

$$\dot{x} = r(cu + v)x\,, \quad x(0) = x_0$$

where $x(t)$ = current earnings

$u(t)$ = external equity

$v(t)$ = fraction of current earnings retained

r = rate of return on investment ($r > \rho$)

c = fraction left over after floatation costs ($0 \leq c < 1$)

$\dot{x}/x \leq g$ where g is some maximum allowable growth rate.

This is basically Krouse & Lee (1973) and Sethi's (1978) model. An alternative to the above objective function is to maximize the present value of dividends (Wong 1975, Magill 1970, van Loon (1982) for example).

The current value Hamiltonian is

$$\begin{aligned} H &= (1 - u - v)x + pr(cu + v)x \\ &\equiv [(cpr - 1)u + (pr - 1)\, v + 1]x \\ &\equiv (\sigma_1 u + \sigma_2 v + 1)x \end{aligned}$$

where $\sigma_1 \equiv cpr - 1$, $\sigma_2 \equiv pr - 1$ are the switching functions.

The Maximum Principle gives

$$\dot{p} = [\rho - r\,(cu + v)]\, p - (1 - u - v)$$

with $p(T) = 0$ ($x(T)$ being unspecified).

In view of $0 < c < 1$, the only cases of optimal decision rules of interest are

(i) $\sigma_1, \sigma_2 < 0$; $u^* = 0 = v^*$

(ii) $\sigma_1 < c\sigma_2$; $\sigma_2 > 0$; $u^* = 0$, $v^* = g/r$

(iii) $\sigma_1 < 0 = \sigma_2$; $u^* = 0$, $0 \leq v^* \leq min\ (1, g/r)$

(iv) $\sigma_2 < 0 = \sigma_1$; $0 \leq u^* \leq (g - r)/rc$, $v^* = 1$

(v) $\sigma_1 < 0 < \sigma_2$; $u^* = 0$, $v^* = 1$

The solution is similar to Dobell and Ho (1967, discussed in Ch. 6) except it is carried out in terms of "time to go" $\tau \equiv T - t$ as used by Krouse and Lee (1973). For example $\overset{\circ}{x} \equiv dx/d\tau = (dx/dt)(dt/d\tau) = -\dot{x}$. Similarly, $\overset{\circ}{p} = -\dot{p}$. With this notation, we have

$$\overset{\circ}{x} = -r(cu + v)x \quad , \quad x(0) \text{ free}$$

$$\overset{\circ}{p} = 1 - u - v - [\rho - r\ (cu + v)] \quad , p(0)(\equiv p(T)) = 0$$

The optimal evolution patterns can be summarised as follows. With $p(0) = 0$, $\sigma_1(0) = -1 = \sigma_2(0)$ hence $u^* = 0 = v^*$. But then $\overset{\circ}{x} = 0$, $\overset{\circ}{p} = 1 - \rho p$ gives $p(\tau) = 1/\rho - c_1 e^{-\rho\tau} = (1/\rho)(1 - e^{-\rho\tau})$ in view of $p(0) = 0$. Starting from $\sigma_2(0) = -1$, $\sigma_2 \equiv rp - 1$ will increase with τ until $\sigma_2(t_s) = 0$ i.e. until $t_s = (1/\rho)\ \ell n\ (r/(r-\rho))$. At this time, $\sigma_1(t_s) \equiv crp - 1 < 0$ and $u^* = 0$ while v^* is undetermined : this is a singular case. So long as $\sigma_2 \equiv 0$, $\overset{\circ}{\sigma}_2 = 0 \Rightarrow \overset{\circ}{p} = 0 \Rightarrow 1 = \rho p = \rho/r$ $\Rightarrow r = \rho$ i.e. the rate of return r is equal to the rate of discount ρ: it does not pay to invest. This is not a sustainable case: the dynamic evolution of the system will eventually bring v^* to its upper bound g/r and further switchings will take place. For further details, see

Krouse and Lee (1973).

11.3.2 Optimal production and inventory models

In these models, the objective is to minimize the costs involved in over producing and storing. This problem has been investigated by Holt et al. (1969), Hwang et al. (1967), Jammernegg (1982).

The objective is to minimize

$$J = \int_0^T [g_1(x) + g_2(u)]\, dt$$

subject to

$$\dot{x} = u - s\ , \quad x(0) = x_0$$

where x = inventories, u = production rate and s = sale rates. A particular form of $g_1(x)$ and $g_2(u)$ is the quadratic : $h(x-\hat{x})^2 + g(u-\hat{u})^2$ where h and g are constant.

This problem has been examined in Ch. 8 (Example 8.3) in some detail and will not be elaborated further.

11.3.3 Marketing Models

A central problem in marketing is advertising on which we shall concentrate. The pioneering work is done by Nerlove and Arrow (1962), extended by Tsurumi and Tsurumi (1971), Gould (1970), Jacquemin (1973) and others. For a survey, see Sethi (1977). We shall present Gould's (1970) version of the Nerlove-Arrow's diffusion model. The objective is to maximize the present value of the operating profit $\pi(x)$ net of advertisement cost $C(u)$ where π is concave, C is convex, x is the number of people who know of the product and u is the contact coefficient, i.e.,

$$Max. \int_0^\infty e^{-rt} [\pi(x) - C(u)]\, dt \tag{83}$$

subject to

$$\dot{x} = - bx + ux\ (1 - x/N), \quad x(0) = x_0 < N \tag{84}$$

i.e. x people inform, through contact, ux people whose $1-x/N$ are newly informed customers, N being the whole customer population and b being the positive rate of forgetfulness.

The current value Hamiltonian is

$$H \equiv \pi(x) - C(u) + px[-b + u(1 - x/N)] \tag{85}$$

The Maximum Principle gives

$$H_u = 0 \Rightarrow C'(u) = px\ (1 - x/N) \tag{86}$$

$$\dot{p} = (r + b - u)p + 2px\ u/N - \pi'(x) \tag{87}$$

Differentiating (86) totally, substituting and manipulating, gives

$$\dot{u} = \frac{1}{C''} \left[C'\left(r + \frac{bx}{N-x}\right) - \left(x - \frac{x}{N}\right)^2 \pi'\right]$$

$$\dot{x} = (u - b)x - ux^2/N \tag{89}$$

For $\dot{x} = 0$ there are two singular arcs $x = 0$ (i.e. the u-axis) and $x = N(1 - b/u)$ and for $\dot{u} = 0$,

$$C'(u) = \frac{(x - x^2/N}{r + bx/(N-x)} \equiv h(x) \tag{90}$$

See figs. 11.10, 11.11, 11.12.

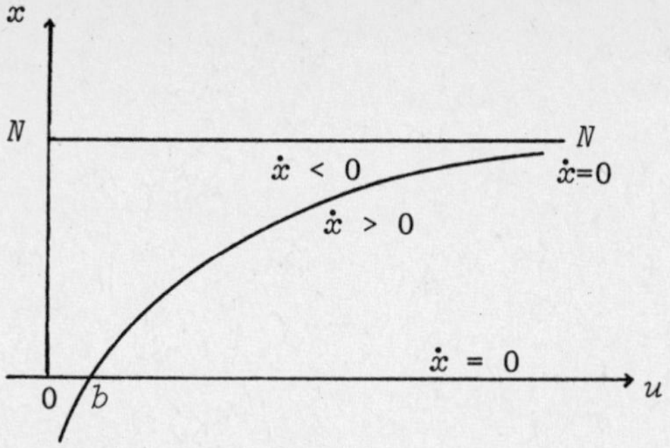

Fig. 11.10

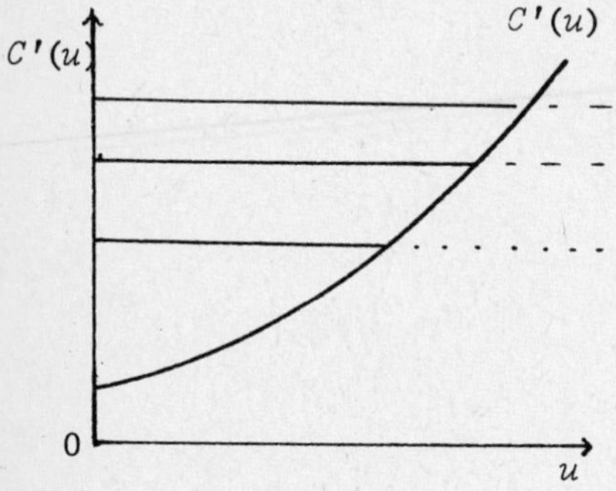

Fig. 11.11

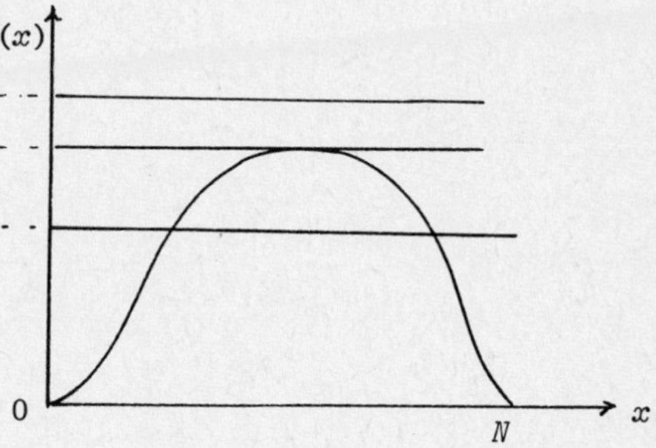

Fig. 11.12

Clearly there are two equilibrium x for small u, one for some intermediate u and none for "large" u as can be seen from figs. 11.11 and 11.12.

Linearisation of $\dot{u}$ and $\dot{x}$ in (88) and (89) about an equilibrium point (x^*, u^*) gives

$$\begin{pmatrix} \dot{x} \\ \dot{u} \end{pmatrix} = A \begin{pmatrix} x - x^* \\ u - u^* \end{pmatrix} \tag{91}$$

where

$$A \equiv \begin{bmatrix} a_{11} & a_{12} \\ a_{21} & a_{22} \end{bmatrix} \equiv \begin{bmatrix} \partial\dot{x}/\partial x & \partial\dot{x}/\partial u \\ \partial\dot{u}/\partial x & \partial\dot{u}/\partial u \end{bmatrix}$$

The result is that there are two, one or no equilibrium points depending on whether the curve $\dot{u} = 0$ and $\dot{x} = 0$ intersect twice, once or not at all. Where there are 2 equilibrium points, one is a saddle point where

$$\frac{du}{dx}\Big|_{\dot{u}=0} = -\frac{a_{21}}{a_{22}} < -\frac{a_{11}}{a_{22}} = \frac{du}{dx}\Big|_{\dot{x}=0}$$

the other an unstable node if the eigen values of A are real and positive, unstable focus (see Appendix) if these are complex with positive real part (see fig. 11.13).

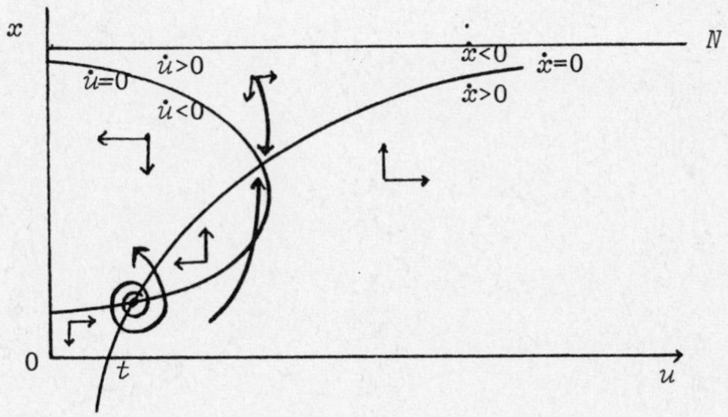

Fig. 11.13

Since $tr\ A = r > 0$, it is clear these are unstable (see Appendix).

When these two curves are tangent to each other, the stable (saddle) and unstable equilibrium points merge together, causing a "catastrophic" jump either into the no-equilibrium zone or back into the two-equilibrium zone, depending on the critical values of the parameters involved at this point (see Thom 1972). The jump is often so abrupt that it is hardly observable in practice (cf. the Euler buckling or Zeeman's Catastrophe machine, Zeeman 1977).

11.3.4 Maintenance Models

How long should we keep a car, a house or a machine, knowing that without maintenance expenditure $u(t)$, the asset in question will deteriorate at some rate δ and maintenance will only retard this deterioration process and furthermore beyond a certain age, it may not be worth spending any money on maintenance and repair. The problem has been examined among others, by Massé (1962), Rapp (1974), Thompson (1968), Luenberger (1975) Varaiya (1976), Sethi and Thompson (1981).

Basically the dynamic system is

$$\dot{x} = -\ \delta + u(t)\ g(t)$$

where $x(t)$ = quality or value of an asset

$g(t)$ = maintenance effectiveness function

Note that with $\dot{x} = -\delta x + ug$, the asset quality or value will never fall to zero even without maintenance, but with $\dot{x} = -\ \delta + ug$ it would.

The objective is to maximize

$$J = \int_0^T [wx(t) - u(t)]\ e^{-rt}\ dt \quad x(T)\ e^{-rT}$$

i.e. to maximize the present value of the rental income $wx(t)$ (w = rental rate) or production rate (w is then the constant production coefficient). Note that w could be made a concave increasing function of quality $w(x)$ with $w''(x) < 0 < w'(x)\ \forall\ x > 0$ and $w(0) = 0$. However, we shall restrict our illustration to the case of constant w.

The Hamiltonian is

$$H = e^{-rt}\ (wx - u) + p(-\delta + gu) \qquad (0 \le u \le U)$$

$$\equiv (pg - e^{-rt})\ u + e^{-rt}\ wx - p\delta$$

The optimal control u^* is given by

$$u^* = \begin{cases} U & \text{if} \quad \sigma > 0 \\ 0 & \text{if} \quad \sigma < 0 \end{cases}$$

where the switching function $\sigma \equiv pg - e^{-rt}$. It is unlikely that $\sigma \equiv 0$, hence singular control is ruled out.

$$-H_x = \dot{p} = - e^{-rt} w \quad \text{with} \quad p(T) = e^{-rT}$$

whose solution is

$$p(t) = e^{-rt} w/r + (1 - w/r) e^{-rT}$$

i.e. the marginal value of the money spent on maintenance and repair decreases exponentially with the asset's age t.

Clearly if $p(0)\ g(0) > 1$, $u^* = U$ until the switching time t_s at which $p(t_s)\ g(t_s) = e^{-rt_s}$ when it is best not to spend any money on maintenance ($u^* = 0$) until re-sale time T. This time T is given by the Transversality Condition

$$H(T) = rx(T)\ e^{-rT}$$

Note that many alternative ways of modelling maintenance are possible. For example $\dot{x} = -\delta x + g(x, u)$ $\quad 0 \le u \le U$, with $g(x, 0) = 0$, $g_{uu} < 0 < g_u$ i.e. diminishing returns to maintenance. Luenberger (1975) uses $g = u - cu^2/(X - x)$ where $0 \le u \le U = (X - x)/c$ i.e. when quality $x = X$ (= 100%, top quality), no maintenance is required, and with age (i.e. $x < X$), quality x in the absence of maintenance, will deteriorate at an exponential rate δ.

11.4 Conclusion

In this chapter, some Economics and Management applications of OC theory were provided. These are by no means exhaustive: only a few simple and arguably representative works were presented. Although applications were classified by major fields, this chapter is not a survey of the literature in each field. Clearly this could do justice neither to the contributions which were discussed nor to those which were omitted. Finally, only some deterministic models were examined: stochastic models are beyond the scope of these introductory lecture notes.

MATHEMATICAL APPENDIX

REVIEW OF DIFFERENTIAL AND DIFFERENCE EQUATIONS

A.1 Introduction

Differential and Difference equations play an important part in dynamic optimisation. Since there are excellent treatments of these at various levels of difficulty (see, for example Pontryagin 1962, Coddington & Levinson 1955, Hurewicz 1958, Arnold 1973, Hirsch & Smale 1974, Goldberg 1958, Yamane 1962, Chiang 1974), we shall focus our brief review on some selected aspects which were used in the text and with which Economics students may not be very familiar.

A. 2 Differential Equations

Differential Equations are equations involving derivatives. For example,

$$F[x(t),\ \dot{x}(t),\ \ddot{x}(t),\ \ldots,\ x^{n}(t),\ t] = 0 \tag{1}$$

The order of a differential equation is the order of the highest derivative in the equation and the degree of the equation is determined by the highest exponent of the derivatives. If the equation does not involve time t explicitly (i.e. if t is absent in (1)), it is called autonomous, otherwise it is non-autonomous. An example of a first order differential equation is

$$\dot{x} + P(t)\ x(t) = Q(t) \tag{2}$$

If $Q(t) = 0$, (2) is a homogeneous differential equation, if $Q(t) \neq 0$ it is non-homogeneous.

Solving a differential equation consists in recovering the primitive which gave rise to the equation, i.e. finding the integral curves of the differential equation. Such integral or primitive is called the general solution and the particular curve which fits the initial conditions, i.e. which is obtained by assigning definite values to the constants of integration such as to satisfy the initial conditions on $x(0)$ is called the particular solution. The complete solution is a combination of both. For example, (2) can be written as

$$\frac{d}{dt}\left(xe^{\int Pdt}\right) \equiv e^{\int Pdt}\left(\dot{x} + P(t)x\right) = e^{\int Pdt}\, Q(t)$$

Integrating both sides

$$xe^{\int Pdt} = \int e^{\int Pdt}\, Q(t)dt + c$$

$$x(t) = e^{-\int Pdt}\int e^{\int Pdt}\, Q(t)dt + ce^{-\int Pdt} \qquad (3)$$

This is the general solution of (2). A particular solution is obtained by specifying the value of the constant c such as to fit the initial condition $x(0) = x_0$. For example, if $P = 2$, $Q = 6$, $x(0) = 10$, the "integrating factor" $e^{\int Pdt}$ is $e^{\int 2dt} = e^{2t}$ and

$$x(t) = e^{-2t}\int e^{2t}\, 6dt + ce^{-2t}$$

$$= 3 + ce^{-2t} = 3 + 7e^{-2t}$$

where $c = x_0 - 3 = 7$ for $x_0 = 10$. This gives the complete solution of (2). An example of the second order constant coefficient differential equation if

$$\ddot{x} + 2b\dot{x} + cx = d, \quad x(0) = \alpha\ ; \quad x'(0) = \beta \qquad (4)$$

In the light of example 1, it could be shown that the solution is

$$x(t) = c_1 e^{\lambda_1 t} + c_2 e^{\lambda_2 t} + d/c$$

where d/c is the particular solution, the rest is the general solution and λ_1, λ_2 are the roots of the characteristic equation, i.e. $\lambda_1, \lambda_2 = -b \pm \sqrt{b^2 - c}$.

These are real and distinct if $b^2 > c$, repeated i.e. $\lambda_1 = \lambda_2 = \lambda$ if $b^2 = c$ and complex if $b^2 < c$. In the repeated root case, the above solution is

$$x(t) = (c_1 + tc_2)\, e^{\lambda t} + d/c$$

and in the complex case, $\lambda = u \pm iv$ where $i \equiv \sqrt{-1}$ and the solution is

$$x(t) = e^{ut}(c_1 \cos vt + c_2 \sin vt)$$

The above give the complete solution, with c_1, c_2 to be determined by the initial conditions $(x(0), x'(0)) = (\alpha, \beta)$ in each case.

Since higher order differential equations can be reduced to a system of first order equations, (see A.2.3), we shall restrict our review to the latter.

A.2.1 First Order Linear Differential Equation Systems

Consider first the case of time invariant coefficient systems $\dot{x}(t) = Ax(t)$, A constant and $x(o) = x_0$.

The solution to this is

$$x(t) = e^{At} x_0 \; .$$

A Taylor expansion about $t = o$ gives

$$x(t) = x_0 + t\dot{x}_0 + t^2\ddot{x}_0/2! + \ldots$$

Substituting $\dot{x} = Ax$, $\ddot{x} = A\dot{x} = A^2 x$ etc. ... gives

$$\begin{aligned} x(t) &= x_0 + tAx_0 + \frac{t^2}{2!} A^2 x_0 + \ldots \\ &= (I + tA + \frac{t^2}{2!} A^2 + \ldots)x_0 \\ &\equiv e^{At} x_0 \end{aligned} \tag{5}$$

where $e^{At} = I + tA + (t^2/2!)A^2 + \ldots$ by definition.

If all eigen values of A are real and distinct, there exists a non-singular matrix P called modal matrix whose columns are the eigen vectors of A such that $P^{-1}AP = \Lambda = \text{diag}(\lambda_i)$ i.e. $A = P\,\Lambda\,P^{-1}$ and

$$A^k = (P\,\Lambda\,P^{-1})(P\,\Lambda\,P^{-1})\ldots(P\,\Lambda\,P^{-1}) = P\,\Lambda^k\,P^{-1} \tag{6}$$

Substitution gives

$$e^{At} = I + P\,\Lambda P^{-1}t + P\,\Lambda^2\,P^{-1}\frac{t^2}{2!} + P\,\Lambda^3\,P^{-1}\frac{t^3}{3!} + \ldots$$

$$= P(I + \Lambda t + \Lambda^2\frac{t^2}{2!} + \ldots)P^{-1} = Pe^{\Lambda t}P^{-1} \tag{7}$$

The solution above is thus

$$x(t) = e^{At}x_0 = Pe^{\Lambda t}\,P^{-1}\,x_0 \tag{8}$$

which is indeed very simple to compute.

For example, the system

$$\begin{pmatrix} \dot{x}_1 \\ \dot{x}_2 \end{pmatrix} = \begin{bmatrix} 1 & 2 \\ 2 & 1 \end{bmatrix}\begin{bmatrix} x_1 \\ x_2 \end{bmatrix} ; \begin{pmatrix} x_{10} \\ x_{20} \end{pmatrix} = \begin{pmatrix} 1 \\ 3 \end{pmatrix}$$

$$\left|A - \lambda I\right| = \begin{vmatrix} 1-\lambda & 2 \\ 2 & 1-\lambda \end{vmatrix} = (\lambda-3)(\lambda+1) = 0$$

the eigen values are $\lambda_1 = 3$, $\lambda_2 = -1$ and the corresponding eigen vectors are $[v_1, v_2] = \begin{bmatrix} 1 & 1 \\ 1 & -1 \end{bmatrix} \equiv P$. Clearly $P^{-1} = \frac{1}{2}\begin{bmatrix} 1 & 1 \\ 1 & -1 \end{bmatrix}$,

$P^{-1}x_0 = \begin{bmatrix} 2 \\ -1 \end{bmatrix}$ and the complete solution is

$$x(t) = Pe^{\Lambda t}\,P^{-1}x_0 = \begin{bmatrix} 1 & 1 \\ 1 & -1 \end{bmatrix}\begin{bmatrix} e^{3t} & 0 \\ 0 & e^{-t} \end{bmatrix}\begin{bmatrix} 2 \\ -1 \end{bmatrix} = \begin{bmatrix} 2e^{3t} - e^{-t} \\ 2e^{3t} + e^{-t} \end{bmatrix}$$

It could be proved that the n solutions $\phi_1(t),\ldots,\phi_n(t)$ of the n differential equations $\dot{x} - Ax$ are linearly independent if and only if its WronsKian matrix $W(\phi)$ is non-singular where

$$W(\phi) \equiv \begin{bmatrix} \phi_1(t) & \cdots & \phi_n(t) \\ \phi_1'(t) & & \phi_n'(t) \\ \cdot & & \cdot \\ \cdot & & \cdot \\ \cdot & & \cdot \\ \phi_1^{n-1}(t) & \cdots & \phi_n^{n-1}(t) \end{bmatrix} \tag{9}$$

where $\phi_j^i(t) \equiv (d^i/dt^i)\phi_j$ $(1 \le j \le n,\ 0 \le i \le n-1$ with $\phi_j^0(t) \equiv (d^0/dt^0)\phi_j(t) \equiv \phi_j(t))$ (for a proof see any textbook on Differential Equations).

In this case, it is easy to verify that the set of all solutions $\phi_i(t)$ $(1 \le i \le n)$ of $\dot{x} = Ax$ forms an n-vector space. For example, if $x(t)$ and $z(t)$ are two solutions and $s,t \in R$, then $d/dt(sx+tz) = s\dot{x} + t\dot{z} = sAx + tAz = A(sx+tz)$. This brings us to the concept of Fundamental Matrices.

A.2.2 Fundamental Matrix

The set of n linearly independent solutions of $\dot{x} = Ax$ forms a basis $\{x^1(t), x^2(t), \ldots, x^n(t)\}$ of R^n and any solution can be expressed as a linear combination of these. This basis could be chosen such that $x^i(t_0) = e_i$ where e_i = unit vector $(0,0,\ldots,1,0\ldots0)$.

The matrix the columns of which are n linearly independent solutions of $\dot{x} = Ax$ is called the fundamental or transition matrix $\Phi(t,t_0)$, i.e.

$$\Phi(t,t_0) \equiv [x^1(t)\ x^2(t)\ \ldots\ x^n(t)]$$

$$\equiv \begin{bmatrix} x_1^1(t) & x_1^2(t) & \cdots & x_1^n(t) \\ x_2^1(t) & & & \\ \cdot & & & \cdot \\ \cdot & & & \cdot \\ \cdot & & & \cdot \\ x_n^1(t) & x_n^2(t) & \cdots & x_n^n(t) \end{bmatrix} \tag{10}$$

$$\Phi(t_0,t_0) = \begin{bmatrix} 1 & 0 & . & . & . & 0 \\ 0 & 1 & & & & \\ . & & . & & & . \\ . & & & . & & . \\ . & & & & . & . \\ 0 & & . & . & . & 1 \end{bmatrix} = I \ . \tag{11}$$

Clearly, $\Phi(t,t_0)$ satisfies the system $\frac{d}{dt}\,\Phi(t,t_0) = A\Phi(t,t_0)$. The solution to $\dot{x}(t) = A(t)x(t)$, $x(t_0) = x^0$, is given by

$$x(t) = \phi(t,t_0)x^0 \ . \tag{12}$$

This is easy to verify since

$$x(t_0) = \Phi(t_0,t_0)x^0 = Ix^0 = x^0$$

$$\dot{x}(t) = \frac{d}{dt}\,\Phi(t,t_0)x^0 = A(t)\Phi(t,t_0)x^0 = A(t)x(t) \ .$$

Similarly, it is easy to verify that the fundamental matrix Φ of the time invariant coefficient system $\dot{x}(t) = Ax(t)$ is

$$\Phi(t,t_0) = e^{A(t-t_0)} \ . \tag{13}$$

Note that (12) simply maps the initial state $x(t_0)$ at time t_0 into the state $x(t)$ at time t. Hence Φ is also called the state transition matrix.

The unique solution $x(t)$ observed at t, of state x^0 at t_0 is written as $\Phi(t,x^0,t_0)$. Using this notation, we represent trajectories as

$$x^3 = \Phi(t_3,x^2,t_0) = \Phi(t_3,t_2)x^2 = \Phi(t_3,t_2)\Phi(t_2,t_0)x^0 \ .$$

Hence

$$\Phi(t_3,t_2)\Phi(t_2,t_0) = \Phi(t_3,t_0)$$

$$\Phi(t_0,t)\Phi(t,t_0) = I \ .$$

Note that if $\Phi(t)$ is a fundamental matrix of $\dot{x} = Ax$, so is $\Phi(t)B$ where B is a non-singular constant matrix. This is easy to verify: $\frac{d}{dt}\Phi(t)B = \dot{\Phi}B = (A\Phi)B = A(\Phi B)$.

For example, given the system $\dot{x} = Ax$ where $A = \begin{bmatrix} 0 & 1 \\ 0 & -2 \end{bmatrix}$. One fundamental matrix is

$$\Phi(t) = \begin{bmatrix} 1 & e^{-2t} \\ 0 & -2e^{-2t} \end{bmatrix}$$

another one is

$$G(t) = \begin{bmatrix} 1 & 1/2(1-e^{-2t}) \\ 0 & e^{-2t} \end{bmatrix}$$

i.e. $G(t) = \Phi(t)B$ where $B = \begin{bmatrix} 1 & 1 \\ 0 & -1 \end{bmatrix}$.

When A in $\dot{x} = Ax$ is a constant matrix, we have

$$\Phi(t+\tau) = \Phi(t)\Phi(\tau) = \Phi(\tau)\Phi(t)$$

from which

$$\Phi(t) = \Phi(\tau)\Phi(t-\tau) .$$

For example consider $G(t)$ above

$$G(t-\tau) = \begin{bmatrix} 1 & 1/2(1-e^{-2(t-\tau)}) \\ 0 & e^{-2(t-\tau)} \end{bmatrix}$$

and

$$\begin{bmatrix} 1 & 1/2(1-e^{-2t}) \\ 0 & e^{-2t} \end{bmatrix} = \begin{bmatrix} 1 & 1/2(1-e^{-2\tau}) \\ 0 & e^{-2\tau} \end{bmatrix} \begin{bmatrix} 1 & 1/2(1-e^{-2(t-\tau)}) \\ 0 & e^{-2(t-\tau)} \end{bmatrix}$$

i.e.

$$G(t) = G(\tau) \ G(t-\tau) .$$

A.2.3 The nth Order Linear Differential Equation

Let us first consider the nth order homogeneous system

$$L(D)x \equiv (D^n + a_1 D^{n-1} + \ldots + a_n D^0)x \tag{14}$$

where

$$D^i \equiv d^i/dt^i \quad (0 \le i \le n) \text{ with } D^0 x = x \; .$$

By a redefinition of variables as $x_i \equiv \dot{x}_{i+1}$ $(1 \le i \le n-1)$ where $\dot{x}_i \equiv Dx_i \equiv dx_i/dt$, i.e. $\dot{x}_1 = x_2$, $\dot{x}_2 = x_3$ etc.. $L(D)x$ could be written as

$$\dot{x}(t) = Ax(t) \text{ with } x(t_0) = x^0 \tag{15}$$

where

$x(t)$ is an n-vector and $\dot{x}(t)$ its derivative vector

and

$$A \equiv \begin{bmatrix} 0 & 1 & 0 & \cdots & & 0 \\ 0 & 0 & 1 & 0 \cdots & & 0 \\ \cdot & \cdot & \cdot & \cdot & & \\ \cdot & \cdot & \cdot & & \cdot & \\ \cdot & \cdot & \cdot & & & \cdot \\ 0 & 0 & 0 & & 0 & 1 \\ -a_n & -a_{n-1} & -a_{n-2} & \cdot & \cdot & \cdot \; -a_1 \end{bmatrix} \tag{16}$$

A is called the companion matrix of $L(D)x$.

With this, an nth order linear differential equation has been converted into a system of first order linear differential equations discussed in the last section.

A.2.4 Non-homogeneous First Order Differential Equation Systems

Consider first the following constant coefficient differential equation system

$$\dot{x}(t) = Ax(t) + Bu(t), \quad x(o) = x^0 \tag{17}$$

where x is an n-vector, u is an r-vector, A is an $n.n$ constant matrix and B an $n.r$ constant matrix. Writing (17) as $\dot{x}(t) - Ax(t) = Bu(t)$ and multiplying both sides by the integrating factor e^{-At} gives

$$e^{-At}[\dot{x}(t) - Ax(t)] \equiv \frac{d}{dt}[e^{-At}x(t)] = e^{-At}Bu(t) \tag{18}$$

i.e. the solution of (17) is, for $t_0 = o$,

$$x(t) = e^{At}x^0 + \int_0^t e^{A(t-\tau)}Bu(\tau)d\tau \;. \tag{19}$$

In terms of the fundamental matrix, (19) can be written as

$$x(t) = \Phi(t)x^0 + \Phi(t)\int_0^t \Phi^{-1}(\tau)Bu(\tau)d\tau \tag{20}$$

where $\Phi(t) = e^{At}$. Since $\Phi(t)\Phi^{-1}(\tau) = \Phi(t-\tau) = e^{A(t-\tau)}$, (20) can be written more simply as

$$\begin{aligned} x(t) &= \Phi(t)x^0 + \int_0^t \Phi(t-\tau)Bu(\tau)d\tau \\ &= e^{At}x^0 + \int_0^t e^{A(t-\tau)}Bu(\tau)d\tau \end{aligned} \tag{21}$$

with $x^0 = x(t_0)$.

If $u(t) = u$, time invariant, and A is non-singular, (21) is simply

$$x(t) = e^{At}(x^0 + A^{-1}Bu) - A^{-1}Bu \;. \tag{22}$$

To see this, define $y(t) \equiv x(t) + A^{-1}Bu$ and $\dot{y}(t) = \dot{x}(t)$ and write (17) as

$$\dot{x}(t) = A[x(t) + A^{-1}Bu],$$

or

$$\dot{y}(t) = Ay(t)$$

the solution of which, by (6) is

$$y(t) = e^{At}y^0$$

or

$$x(t) = e^{At}(x^0 + A^{-1}Bu) - A^{-1}Bu \ . \tag{23}$$

Note that $\int_0^t e^{A(t-\tau)}Bu(\tau)d\tau$ in (21)

or $[e^{At}-I]A^{-1}Bu$ in (22) is the particular solution of (17).

For the general case of time variant $A(t)$, $B(t)$, the solution of

$$\dot{x}(t) = A(t)x(t) + B(t)u(t), \ x(t_0) = x^0 \tag{24}$$

is

$$x(t) = \Phi(t,t_0)[x^0 + \int_{t_0}^{t} \Phi(t_0,t)B(\tau)u(\tau)d\tau \ . \tag{25}$$

This can be shown by using Lagrange's method of variation of parameters. Let $x(t) \equiv X(t)w(t)$ where $X(t)$ is the unique non-singular $n\times n$ matrix satisfying $\dot{X}(t) = A(t)X(t)$, $X(o) = I$. (For a proof of the existence and uniqueness of $X(t)$, see, for example, Bellman 1970, p. 167) Differentiation and substitution give

$$\dot{x} = \dot{X}w + X\dot{w}$$

$$= AXw + Bu \ .$$

Hence

$$X\dot{w} = Bu$$

$$\dot{w} = X^{-1}Bu$$

which gives, on integration

$$w(t) = w(t_0) + \int_{t_0}^{t} X^{-1}(\tau)B(\tau)u(\tau)d\tau \ .$$

Since $w(t_0) = X(o)w(o)$, this yields for x the formula

$$x(t) = X(t)w(o) + \int_{t_0}^{t} X(t)X^{-1}(\tau)B(\tau)u(\tau)d\tau$$

$$\equiv \Phi(t,t_0)[x^0 + \int_{t_0}^{t} \Phi(t_0,\tau)B(\tau)u(\tau)d\tau] \;.$$

A.3 Difference Equations

Differential equations deal with continuous changes $d^i x/dt^i$ $(1 \leq i \leq n)$. When these changes are discrete, or when observations are only made at finite intervals, $\dot{\Delta}x(t) \equiv x(t+1) - x(t)$, or more generally $\Delta^n x(t) \equiv \Delta^{n-1}x(t+1) - \Delta^{n-1}x(t) \equiv x(t+n) - nx(t+n-1) + \ldots + (-1)^n x(t)$ difference equations will be used. In view of the similarities between differential and difference equations, this section will be brief. More details can be found in Kenkel (1974), Goldberg (1958), Charlton (1965), Levy and Lesman (1959), for example.

A relationship of the form

$$F(x(t),\Delta x(t),\ldots,\Delta^n x(t),t) = o \tag{26}$$

or equivalently

$$F(x(t),x(t+1),\ldots,x(t+n),t) = o \tag{27}$$

is a difference equation of the nth order, the order being given by the highest order difference.

E.g. $ax(t+2) + bx(t+1) + cx(t) = o$. (28)

A function $x^*(t)$ is said to be a solution of (26) or (27) when its insertion into (26) or (27) makes these hold as identities.

Again, as in differential equations, there is a particular

solution and a general solution and a complete solution is obtained as a combination of these.

Note that an n-order difference has n (not necessarily distinct) solutions and once initial conditions are given, these are unique.

As higher order equations are always reducible to first order systemswe shall briefly deal with the latter.

Consider the following system:

$$\underset{(n.n)(n.1)}{x(t+1) = A(t)x(t)} + \underset{(n.r)(r.1)}{B(t)u(t)} \tag{29}$$

or in scalar notation,

$$x_i(t+1) = \sum_{j=1}^{n} a_{ij}(t)x_j(t) + \sum_{j=1}^{r} b_{ij}(t)u_j(t) \qquad (i = 1,2,\ldots,n)$$

with initial conditions $x_i(o) = x_{io}$ $(1 \le i \le n)$.

The solution to the homogeneous part is obtained by putting $u(t) = o$ in (29) and noting that, just like the case of differential equations, there exist a fundamental matrix X satisfying

$$X(t+1) = A(t)X(t) \ . \tag{30}$$

Note that the n columns of a fundamental matrix constitute a basis of the n-dimensional vector space. There are several fundamental matrices which differ from one another by a multiplicative constant matrix, but once initial conditions are taken into account, the fundamental matrix is unique. Consider the unique matrix $\Phi(t,h)$ which satisfies

$$\Phi(t+1,h) = A(t)\Phi(t,h), \ \Phi(h,h) = I \ . \tag{31}$$

Any fundamental matrix $X(t)$ can be written as $\Phi(t,h)C$ where C is any $n.n$ non-singular matrix. The fundamental matrix $\Phi(t,h)$ has the property that

$$\Phi(h,h) = I$$
$$\Phi(h+1,h) = A(h)$$
$$\Phi(h+2,h) = A(h+1)Ah$$
$$\vdots$$
$$\Phi(t,h) = A(t-1)(t-2) \ldots A(h+1)A(h)$$

the solution to the homogeneous part of (29) is then

$$x(t) = \Phi(t,h)x(h) = \Phi(t,h)x_o \ . \tag{32}$$

If $A(t)$ is a constant matrix, then by choosing $h = o$ and writing $\Phi(t,h)$ as $\Phi(t)$, we obtain

$$\Phi(t) = A^t$$

where $\Phi(t)$ is the unique fundamental matrix satisfying

$$\Phi(t+1) = A\Phi(t), \ \Phi(o) = I$$

i.e. the solution to the homogeneous part of (29), i.e.

$$x(t+1) = Ax(t); \ x(o) = x_o$$

is simply

$$x(t) = A^t x_o$$

which could be obtained by the recursive relation

$$x(1) = Ax(o) = Ax_0$$

$$x(2) = Ax(1) = A^2x_0$$

$$\cdot$$
$$\cdot$$
$$\cdot$$

$$x(t) = A^t x_0 = P\,\Lambda^t P^{-1} x_0 \text{ by (6)} \tag{33}$$

Now the solution to the non-homogeneous system (29) is

$$x(t) = \Phi(t,h)x_0 + \Phi(t,h) \sum_{j=H+1}^{t} \Phi^{-1}(j,h)B(j-1)u(j-1) \tag{34}$$

where $\Phi(t,h)$ is the unique fundamental matrix satisfying

$$\Phi(t+1,h) = A(t)\Phi(t,h),\ \Phi(h,h) = I \tag{35}$$

$$(t = h, h+1, h+2, \dots).$$

We can verify (30) by direct substitution: for $t > h$, (30) gives

$$x(t+1) = \Phi(t+1,h)x_0 + \Phi(t+1,h) \sum_{j=h+1}^{t+1} \Phi^{-1}(j,h)B(j-1)u(j-1)\ . \tag{36}$$

Noting that

$$\Phi(t,h) = A(t-1)A(t-2)\ \dots\ A(h)$$

(32) gives

$$x(t+1) = A(t)\Phi(t,h)x_0 + A(t)\Phi(t,h) \sum_{j=h+1}^{t} \Phi^{-1}(j,h)B(j-1)u(j-1)$$
$$+ \Phi(t+1,h)\Phi^{-1}(t+1,h)B(t)u(t)$$

$$= A(t)x(t) + B(t)u(t) \tag{37}$$

i.e. (34) is the solution of (29). Note that the first part on the RHS of (34) is the homogeneous solution and the remainder is a particular solution. Considerable notational simplification could be made by putting $h = o$ in the above.

For the constant coefficient case, i.e. A and B are constant matrices,

$$x(t+1) = Ax(t) + Bu(t)$$

(34) is simplified to

$$x(t) = A^t x_0 + A^t \sum_{j=1}^{t} A^{-j} Bu(j-1)$$

$$= A^t x_0 + \sum_{j=1}^{t} A^{t-j} Bu(j-1) \; . \qquad (38)$$

A.4 Stability of Differential and Difference equations

Since the stability conditions of differential and Difference equations have been exhaustively investigated (see, for example, Samuelson 1947, Metzler 1945, Hicks 1939, Arrow and Hurwicz 1958, 1959, Newman 1959, Takayama 1974) we shall be brief.

For a differential equation system $\dot{x} = Ax$, A is defined as a stable matrix if its eigen values all have negative real part, and for a difference equation system, A is defined to be a stable matrix if and only if $|\lambda_i| < 1$. With these definitions, it is almost tautological to say that differential and difference equations systems are stable if and only if their coefficient matrix is stable.

For a differential equation system $\dot{x} = Ax$, it can be seen from the solution (8) $x(t) = e^{At}x_0 = Pe^{\Lambda t}P^{-1}x_0 \equiv c_1 e^{\lambda_1 t} v_1 + \ldots + c_n e^{\lambda_n t} v_n$

where v_i = eigen vectors associated with λ_i and $P^{-1}x_0 = c \equiv (c_1,\ldots,c_n)'$

that if each eigen value λ_i has negative real part then as $t\to\infty$ the solution $x(t)$ tends to zero either monotonically (in the case of real λ_i) or periodically (in the case of complex λ_i).

For the case of symmetric A with real distinct eigen values, the condition $Re(\lambda_i) < 0$ implies

(i) A is negative definite, i.e. $x'Ax < 0$, $\forall\, x \neq 0$.

This can be seen by pre-multiplying $Ax = \lambda_i x$ by x, i.e. $x'Ax = \lambda_i x'x$ implies $\lambda_i = x'Ax/x'x$. Clearly $x'Ax < 0$ implies $\lambda_i < 0$ for all i.

(ii) The principal minors of A alternate in signs, i.e. sgn $A_i = (-1)^i$ where A_i = minors of order i ($i = 1,2,3,\ldots$) of A. This can be seen by noting that A is similar to Λ, i.e. $P^{-1}AP = \Lambda$ and $\det A = \det \Lambda = \Pi\lambda_i$ and the order of A is arbitrary. Thus $\det A_i = \det \Lambda_i = \lambda_1 \lambda_2 \ldots \lambda_i \gtrless 0$ depending on whether the number of eigen values, i, is odd or even.

(iii) A is a dominant or quasi-dominant matrix (see Newman 1959 or McKenzie 1960).

For example, the system

$$\begin{bmatrix} \dot{x}_1 \\ \dot{x}_2 \end{bmatrix} = \begin{bmatrix} -1 & 1 \\ 1 & -3 \end{bmatrix} \begin{bmatrix} x_1 \\ x_2 \end{bmatrix}$$

is stable by all the above criteria. Clearly $\det (A - \lambda I) = \lambda^2 + 4\lambda + 1 = 0$, both eigen values are negative, the principal minors of A alternate in signs, i.e. $-1 < 0$, $-3 < 0$, $\det A = 3-1 > 0$ and A is diagonal dominant.

The matrix $A = \begin{bmatrix} 1 & 2 \\ 2 & 1 \end{bmatrix}$ in the example of A.2.1 above, on the other hand, is unstable by all the above three criteria.

For the difference equation system $x(t + 1) = Ax(t)$, the condition $|\lambda_i| < 1$ implies $|\mathrm{tr}\, A| < n$ and $|\det A| < 1$.

It is easy to see that the solution (33) $x(t) = A^t x_0 = P\Lambda^t P^{-1} x_0 \equiv \sum_1^n c_i \lambda_i^t v_i$ where v_i are eigen vectors and $c \equiv P^{-1} x_0$ is stable if and only if $|\lambda_i| < 1$. Clearly $|\lambda_i| < 1 \;\forall\, i \Rightarrow \Pi|\lambda_i| < 1$ and $\sum_1^n |\lambda_i| < n$. But $A = P \Lambda P^{-1}$, $\det A = \det (P \Lambda P^{-1}) = \det \Lambda = \prod_1^n \lambda_i$, and $tr\ A = tr\ (P\ \Lambda\ P^{-1}) = tr\ (P\ \Lambda)P^{-1} = tr\ (P^{-1}P)\ \Lambda = tr\ \Lambda = \sum_1^n \lambda_i$.

Hence $|tr\ A| = \sum_1^n |\lambda_i| < n$ and $|det\ A| = \prod_i |\lambda_i| < 1$.

Note that $\|A\| < 1$ where $\|A\|$ is any norm of A (see, for example, Lancaster 1969) is sufficient for the above difference equation system to be stable. To see this, note that the maximum eigen value, called the spectral radius of A, defined as $\lambda_M(A) = \max_i |\lambda_i|$ is no greater than any norm $\|A\|$, i.e. $\lambda_M(A) \leq \|A\|$. Taking the norm of $Ax = \lambda_i x$, we have $\|Ax\| = \|\lambda_i x\|$, $|\lambda_i|\ \|x\| \leq \|A\|\ \|x\|$ giving for all λ_i including λ_M,

$$|\lambda_i| \leq \lambda_M \leq \|A\|\|x\|/\|x\| \quad = \quad \|A\|$$

This provides a quick stability check (Conlisk 1973).

A.5 Phase Diagrams and Non-Linear Differential Equations

So far, differential equations have been solved analytically. In many problems, however, quantitative solutions are either unobtainable or uninteresting, for example, when the exact form of the equation is not known, or when the system is non-linear or when the only interest is in stability, not in explicit solution. In these cases, a qualitative approach is used. This consists of analysing the motion of the system in the $\dot{x}.x$ plane or in the case of more than one equation, in the $x_i.x_j$ plane, called the phase diagram of differential equations.

For example, the explicit solution of the scalar function $\dot{x}(t) = ax(t)$, $x(o) = x_o$, is $x(t) = x_o e^{at}$ which is stable (unstable) if $a < o$ ($a > o$). An alternative way is to consider the slope of the line $\dot{x} = ax$ in the $\dot{x}.x$ plane (see Fig. A.1). Clearly, above the x-axis,

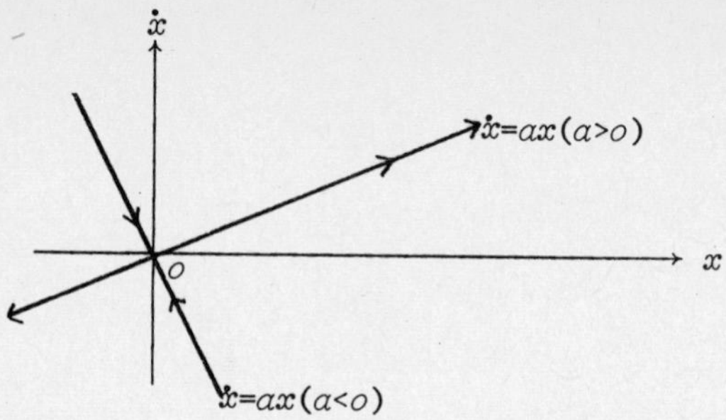

Fig. A.1 Phase diagram of $\dot{x} = ax$

$\dot{x} > 0$, i.e. x increases over time and below it, $\dot{x} < 0$. The law of motion is thus from left to right above the x-axis and from right to left below it (see arrows in Fig. A.1).

The above can also be represented on a single line

Fig. A.2 Phase diagram of $\dot{x} = ax$

Clearly only in the case $a < 0$ is the system stable in the sense it moves to the equilibrium point $\dot{x} = 0 = x$ over time, regardless of the position of the initial value of x_0.

For a system of two autonomous linear differential equations, the phase diagram represents the two variables of the system. We shall start with the linear time invariant system

$$\dot{x}(t) = Ax(t) \ .$$

By the change of variables $x = Py$ (giving $y = P^{-1}x$ and $\dot{y} = P^{-1}\dot{x}$ where P is the non-singular modal matrix discussed earlier), A is diagonalised as $P^{-1}AP = B = \mathrm{diag}(\lambda_1,\lambda_2)$, i.e. $\dot{x} = Ax$ is now changed to

$$\dot{y}(t) = By(t) \tag{39}$$

$$\equiv \begin{bmatrix} \lambda_1 & 0 \\ 0 & \lambda_2 \end{bmatrix} \begin{bmatrix} y_1(t) \\ y_2(t) \end{bmatrix} .$$

Clearly $\dot{y} = 0$, implying $y_1 = 0 = y_2$, i.e. the origin is the equilibrium of the system. The laws of motion are given by the signs of the eigenvalues λ_1 and λ_2. If the real parts of both eigenvalues, $\mathrm{Re}(\lambda_i)$ ($i = 1,2$) are negative, the system is stable, i.e. will move to the origin, if they are both positive, the system will move away from the origin, and if $\mathrm{Re}(\lambda) = 0$, solutions are periodic. If λ_1 and λ_2 are both real andof opposite signs, one variable will move to the origin and the other, away from it over time. This is called a saddle point equilibrium. The relevant phase diagrams with their various names are as follows (see Fig. A.3):

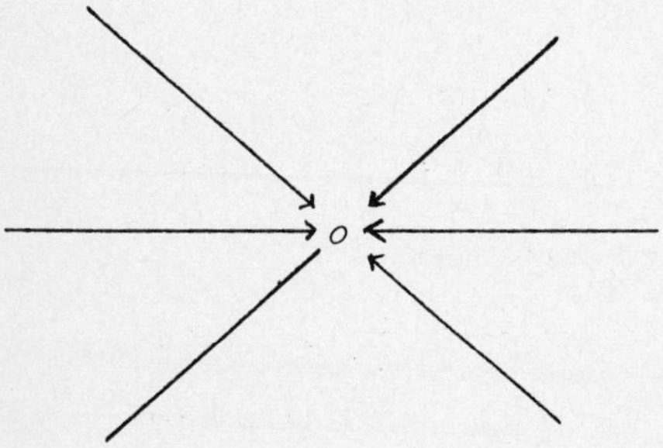

Fig. A.3(i). Focus: equal negative eigenvalues $\lambda < 0$

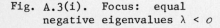

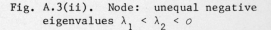

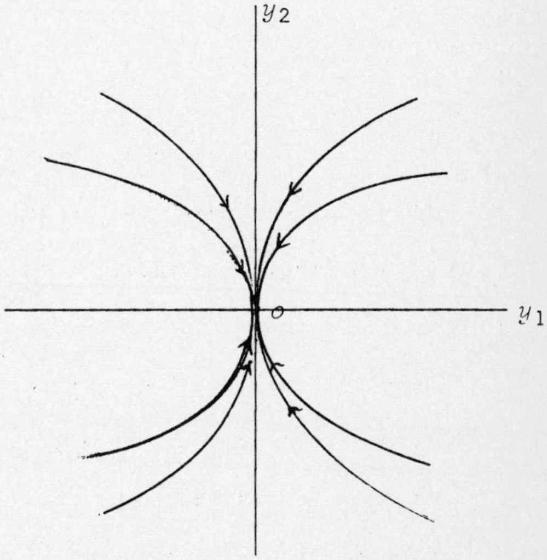

Fig. A.3(ii). Node: unequal negative eigenvalues $\lambda_1 < \lambda_2 < 0$

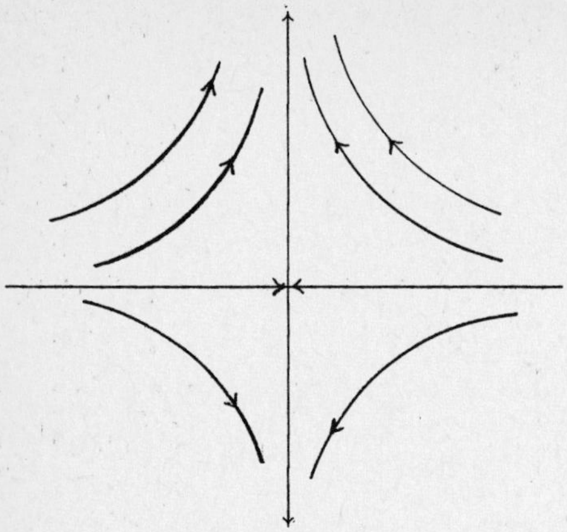

Fig. A.3(iii). Saddle: real λ_i of opposite signs for ex. $\lambda_1 < o < \lambda_2$

Fig. A.3(iv). Spiral sink: $B \begin{bmatrix} u & -v \\ v & u \end{bmatrix}$ $u < o < v$: counterclockwise spiral sink $u, v < o$: clockwise spiral sink(where $\lambda = u \pm iv$ and $i = \sqrt{-1}$)

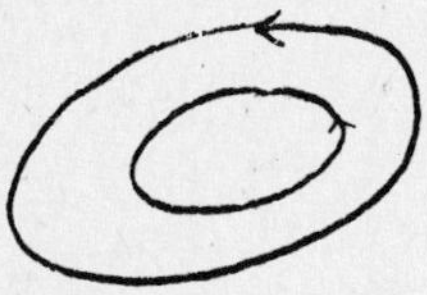

Fig. A.3(v). Centre: $B = \begin{bmatrix} o & -v \\ v & o \end{bmatrix}$ with $u = o < v$

If all eigenvalues have positive real parts, the critical point is called a source. The relevant phase diagrams are like (i), (ii) and (iv) with the arrows pointing in the opposite direction.

Note that all but the last case are structurally stable. The centre is unstable in that a small perturbation of a_{ij} will transform the centre into a stable or unstable focus.

The relationships between λ_i and the elements a_{ij} of matrix A can be seen by the characteristic polynomial $\alpha(\lambda)$ where

$$\alpha(\lambda) \equiv \det(A-\lambda I) = (a_{11}-\lambda)(a_{22}-\lambda) - a_{12}a_{21}$$

$$= \lambda^2 - (\mathrm{tr}\ A)\lambda + \det A = o$$

giving

$$\lambda = 1/2(\mathrm{tr}\ A \pm\sqrt{\Delta}) \tag{40}$$

where $\Delta \equiv (\mathrm{tr}\ A)^2 - 4 \det A$.

It is clear that if $\mathrm{Re}(\lambda) = 1/2\ \mathrm{tr}\ A < o$ ($> o$), the system is stable (unstable). If $\Delta > o$ ($\Delta < o$), both eigenvalues are real (imaginary). If $\mathrm{Re}(\lambda) = o$, i.e. $\mathrm{tr}\ A = o$ and $\Delta < o$, the critical point is a centre. All this is summarised in the following diagram (see Hirsch and Smale 1974 p. 96):

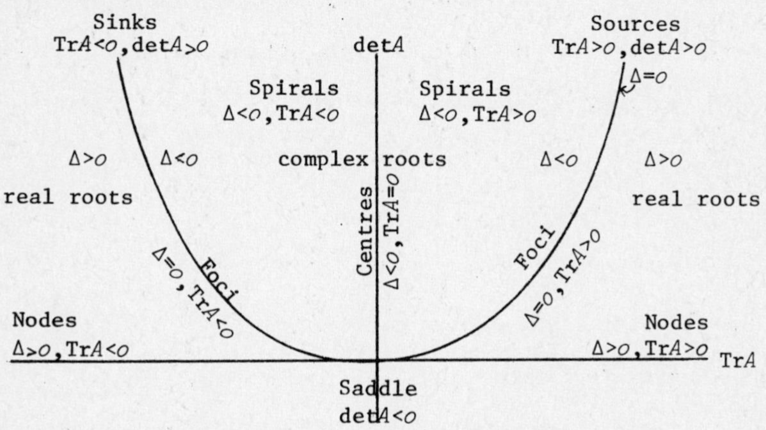

Fig. A.4 Relationships between λ and a_{ij}

A.6 Isoclines

Given the differential equation $\dot{x} = f(x,t)$, a curve along which the slope of f is constant is called an isocline of $\dot{x} = f(x,t)$, i.e. a line or a curve along which all paths have a fixed inclination. For example, the isoclines of $\dot{x} = f(x,t) \equiv x+t$ are $f = c$ (c = constant), i.e. $x = -t+c$ which is a straight line with slope -1 and intercept c. (See Fig. 2.9)

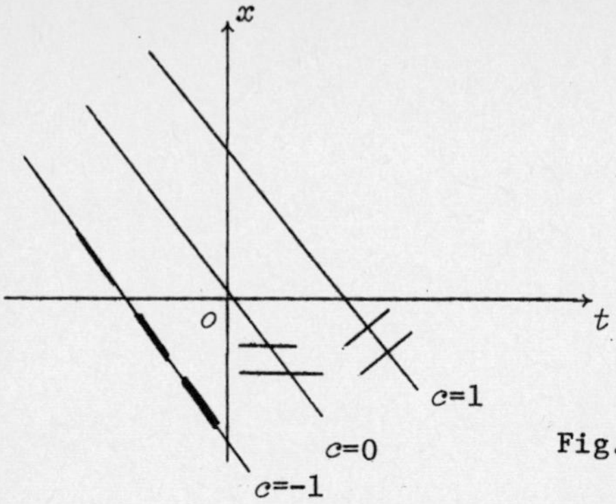

Fig. A.5 Isoclines of $\dot{x} = x+t$

In Fig. A.5 the slopes of the solution $x(t)$ to $\dot{x} = x+t$ are 1, 0, -1 respectively on the isocline, with $c = (1,0,-1)$. Along the isocline $x = -t+o$ ($c = o$) for example $\dot{x} = o$, i.e. $x(t)$ is in equilibrium on the straight line through the origin with slope -1. This is indicated by the horizontal dashes.

In a 2×2 system,

$$\dot{x}_1 = f_1(x_1,x_2)$$
$$\dot{x}_2 = f_2(x_1,x_2)$$

the particular isoclines $f_1 = o$ gives the equilibrium of x_1: any trajectory must cross f_1 vertically (indicating $\dot{x}_1 = o$, i.e. x_1 is in equilibrium. Similarly $f_2 = o$ gives the equilibrium of x_2 and any path must cross it horizontally. These form the vertical and horizontal manifolds. The intersection of f_1 and f_2 (which exists in a non-degenerate system) gives $\dot{x}_1 = o = \dot{x}_2$, i.e. the equilibrium of the whole system. Thus they divide the plane into four isosectors.

As an example, consider the linear case

$$\dot{x}(t) = Ax(t)$$

where $A = \begin{bmatrix} 1 & 2 \\ 2 & 1 \end{bmatrix}$. This has the eigenvalues $(\lambda_1, \lambda_2) = (3,-1)$ and the corresponding eigenvectors (1,1) and (1,-1). The modal matrix P and its inverse are

$$P = \begin{bmatrix} 1 & 1 \\ 1 & -1 \end{bmatrix} \; ; \; P^{-1} = 1/2 \begin{bmatrix} 1 & 1 \\ 1 & -1 \end{bmatrix} .$$

By similarity transformation, using $x = Py$, $P^{-1}AP = B = \text{diag}(3,-1)$ and $\dot{y} = By$, i.e.

$$\dot{y}_1 = 3y_1$$
$$\dot{y}_2 = -y_2 \; .$$

Equilibrium is given by $\dot{y} = o \Rightarrow y_1 = o = y_2$, i.e. $\dot{y}_1 = o$ on the whole y_2-axis and $\dot{y}_2 = o$ on the whole y_1-axis. These isoclines serve as separatrices: $\dot{y}_1 = 3y_1 > o$ $(< o) \Rightarrow y_1 > o$ $(< o)$ on the right (left) of the y_2-axis, i.e. on the right of the y_2-axis, y_1 rises and on the left of it, y_1 falls over time. On the y_2-axis, y_1 is unchanged, as indicated by the arrow on the y_2-axis and pointing to the origin. Similarly, $\dot{y}_2 = -y_2 = o$ on the whole y_1-axis and $\dot{y}_2 = -y_2 > o$ for $y_2 < o$, i.e. below the horizontal axis and $\dot{y}_2 = -y_2 < o$ for $y_2 > o$, i.e. above it. Clearly this is a saddle point equilibrium where the stable arm is the vertical axis itself (see Fig. A.6).

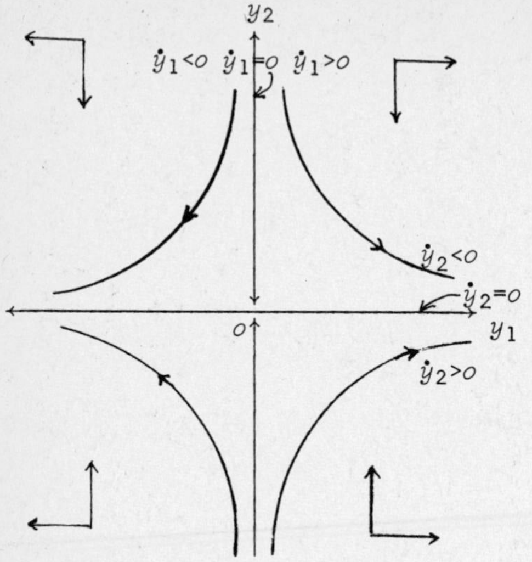

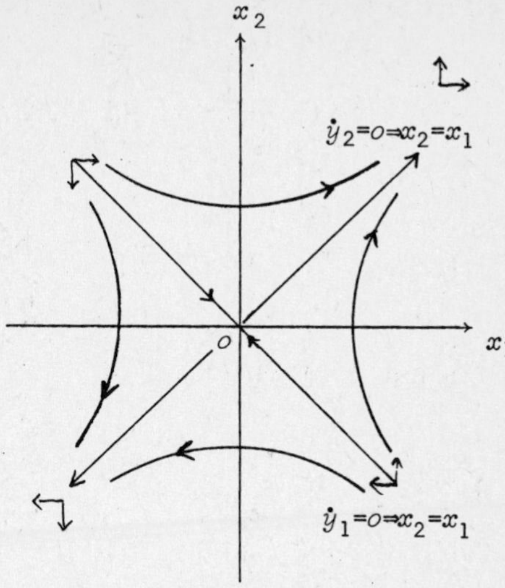

Fig. A.6 Saddle equilibrium in y_1y_2 plane

Fig. A.7 Saddle equilibrium in x_1x_2 plane

In the x_1x_2 plane, with the transformation $x = Py$, these separatrices are now the lines $x_2 = x_1$ and $x_2 = -x_1$. They divide the x_1x_2 plane into four isosectors: only the trajectory with the initial values lying on the curve $\dot{y}_1 = o$, i.e. $x_{20} = -x_{10}$ will bring the system to its equilibrium at the origin (0,0): all trajectories starting with any other initial values will veer away from equilibrium (0,0). (See Figs. A.6 and A.7.)

A.7 Non-Linear Differential Equations

When the system is non-linear, the usual way is to linearize, by using Taylor's expansion, about a particular point. In this case, the above analysis applies.

Note, however, that the solution obtained is a local one which may or may not give the true global picture. An example is the limit cycle case of Van de Pol's equation $\ddot{x} + \mu(x^2-1)\dot{x} + x = 0$ where μ is a parameter. If the initial values $x(o)$, $\dot{x}(o)$ lie inside the limit cycle (i.e. near enough to the origin (0,0), the system spirals out to the limit cycle and if they lie outside the limit cycle (i.e. far enough from the origin), the system will spiral inwards toward the limit cycle. Thus opposite conclusions are obtained depending on how far from the origin the linearisation is carried out.

An alternative way is to use the phase diagram technique just developed. Let us illustrate it with two examples.

Example 1. Optimal Growth Model

Consider the non-linear differential equation system in Optimal Growth Economics examined in Chapter 11.

$$\dot{k}(t) = f(k(t)) - \lambda k(t)$$

$$\dot{c}(t) = [f'(k) - a]c/\sigma(c)$$

where $f \in C^2$, $f(o) = o$, $f''(k) < o < f'(k)$ and $\lim_{k\to o} f'(k) = \infty$ $\lim_{k\to\infty} f'(k) = o$; α, λ are positive constant, $\sigma(c) > o$; $k = k(t)$ = capital and $c(t)$ = consumption per worker. Its phase diagram is (see Fig. A.8):

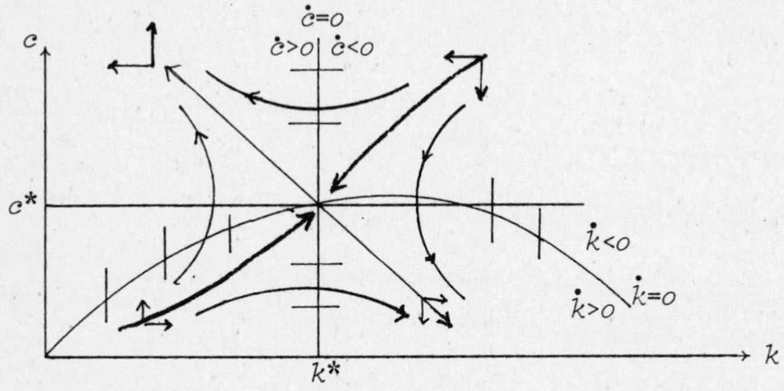

Fig. A.8 Growth Economics phase diagram

Since $f(k)$ is increasing and concave, and $\lambda k(t)$ linear, $f(k) - \lambda k$ is concave and $\dot{k} = o \Rightarrow c = f(k) - \lambda k$, a concave function above which $\dot{k} < o$ and below which $\dot{k} > o$. Similarly, $\dot{c} = o \Rightarrow f'(k) = a$. In view of the concavity of $f(k)$, only one value of k, say k^*, does this, i.e. $\dot{c} = o$ is a vertical line through k^*. On the right (left) of this line, $\dot{c} < o$ ($\dot{c} > o$). Thus $\dot{c} = o$ and $\dot{k} = o$ are the two isoclines of the system with the horizontal manifold on $\dot{c} = o$ and vertical manifold on $\dot{k} = o$. These curves ($\dot{k} = o = \dot{c}$) divide the $k.c$ plane into four iso-sectors and the solution path follows the relevant arrows in the phase diagram. It can be seen that this is a saddle point equilibrium: corresponding to each initial $k(o) = k_o$, a unique corresponding value of $c(o)$ must be chosen on the stable branches: any other choice would send the economy away from the equilibrium (k^*, c^*).

Example 2.

Next, consider the system

$$\dot{x} = 3y^2 + x$$

$$\dot{y} = -(3x^2 + y) \ .$$

These are zero on the two parabolae, i.e. $x = -3y^2$ (for $\dot{x} = o$) and $y = -3x^2$ (for $\dot{y} = o$). They form two isoclines with the vertical manifold on $\dot{x} = o$ and horizontal manifold on $\dot{y} = o$: the contours must cross the curve $\dot{x} = o$ vertically and the curve $\dot{y} = o$ horizontally. Inside the parabola $\dot{x} = o$ (i.e. $x = -3y^2$), $\dot{x} < o$ and outside it, $\dot{x} > o$. Similarly, $\dot{y} > o$ inside the parabola $\dot{y} = o$ (i.e. $y = -3x^2$) and $\dot{y} < o$ outside it. The gradient of the level lines are $\dot{y}/\dot{x} = dy/dx = -1$ for $x = y$ and in general $dy/dx = -(3x^2+y)/(3y^2+x)$. The trajectories follow the arrows

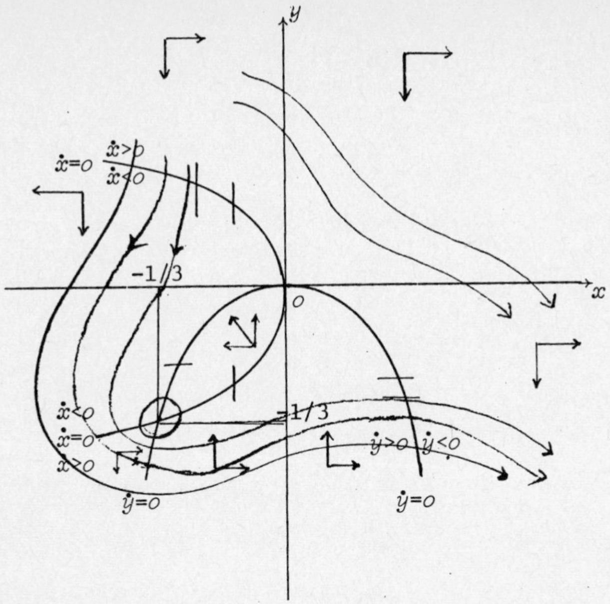

Fig. A.9 Phase diagram of Example 2

indicated. It can be seen that the system has two equilibria, at the intersections of the two parabolae, namely at (0,0) and (-1/3,-1/3).

A.8 Phase Diagrams of Difference Equations

The phase diagram technique is equally applicable to difference equations.

Consider the first order equation

$$x_{t+1} = ax_t + b \ , \quad x(0) = x_o \ .$$

The solution is

$$x_t = Ka^t + b/(1-a) \qquad (a \neq 1)$$

$$= x_o + bt \qquad (\text{if } a = 1)$$

where $\qquad K \equiv x_o - b/(1-a) \ .$

Stability requires that $|a| < 1$.

The phase diagram now gives this condition as the slope is flatter than the 45° line, i.e. $|y_{t+1}| = |y_t|$. This line is the locus of equilibrium points.

Consider for example the familiar dynamic multiplier model of the Keynesian system

$$Y_t = cY_{t-1} + b = f(Y_{t-1})$$

where Y_t is national income, c is the marginal propensity to consume and b is the intercept reflecting autonomous investment (I) and subsistence consumption (C_o) corresponding to zero income. At $t = o$, $Y_t = Y_o$. Economic theory tells us that $0 < c < 1$. This, in fact, turns out to be the stability condition of the system (see Fig. A.10):

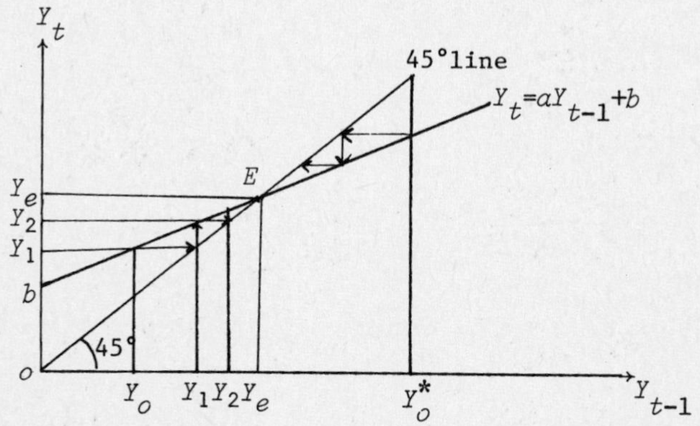

Fig. A.10 Phase diagram of a first order difference equation

If initially income is Y_o, on the horizontal axis, the next period, it will be mapped to Y_1 on the vertical axis, by the line function f, i.e. by the relation $Y_t = aY_{t-1} + b$. The 45° line where

$Y_t = Y_{t-1}$ now allows Y_1 to be translated to Y_1 on the horizontal axis. But Y_1 now generates Y_2 on the vertical axis in the next period, and so on. The path, indicated by the arrows, can be seen to converge to the equilibrium E (Y_e, Y_e). Similarly, if $Y_o > Y_e$, say Y_o^* in Fig. A.10, the system will move back to E. The system is stable. If $a > 1$, the equation line $Y_t = aY_{t-1} + b$ will be steeper than the 45° line and Y_t will move away from the equilibrium level E as time goes on: the system will be unstable.

A.9 Liapunov's Second (or Direct) Method

An alternative to the phase diagram technique is Liapunov's second method.

The idea of Liapunov's method is as follows. Suppose the system has an equilibrium, say at the origin, Liapunov's function $L(x)$ measures the distance or deviation from equilibrium. Obviously $L(x) > o$ whenever $x \neq o$ and $L(o) = o$. If $L(x(t)) \to o$ as $t \to \infty$ the system is stable. Consider, for example, the system $\dot{x} = Ax$ where A is a constant $n.n$ matrix. A Liapunov function $L(x) = x'Vx$ where V is a real symmetric and constant matrix, indicating the various weights associated with the elements x_i of x.

$$\begin{aligned} dL(x)/dt &= \dot{x}'Vx + x'V\dot{x} \\ &= x'A'Vx + x'VAx \\ &= x'(A'V + VA)x \\ &\equiv x'(-W)x \end{aligned} \tag{41}$$

where $-W \equiv A'V + VA$

clearly $dL/dt < o$ if and only if W is positive definite. It is well known (see any text for ex. [Hahn (1963) or Lancaster (1969)] that given W, there exists a unique solution V if and only if A and $-A'$ have no eigenvalues in common. Then there exists a one-to-one correspondence between V and W. Liapunov's contribution lies in pointing out that, given a definite form $x'(-W)x$, the stability of A can be assured by the existence of a definite solution matrix V for $A'V + VA = -W$. Thus $\dot{x} = Ax$ is stable if A is a stable matrix, and this is the case if and only if $\dot{L} < o$, i.e. an $n.n$ matrix A is stable if and only if W defined by $A'V + VA = -W$ is positive (non-negative) definite. For a proof, see, for example, Lancaster (1969).

Similarly, in the discrete case $x(t+1) = Ax(t)$, a Liapunov function L may be defined as $L = x'Vx$, giving

$$\begin{aligned}\Delta L &= L(t+1) - L(t) = x'(t+1)Vx(t+1) - x'(t)Vx(t) \\ &= x'(t)[A'VA - V]x(t) \equiv -x'(t)Wx(t). \qquad (42)\end{aligned}$$

The system $x(t+1) = Ax(t)$ is stable if and only if for any symmetric positive definite matrix W there exists a symmetric positive definite matrix V which is the unique solution of the system

$$A'VA - V = -W \ .$$

For a proof, see, for example, Hahn (1963) or any advanced text book on Differential Equations.

Let us summarize the various steps of Liapunov's method. To ascertain whether the system $\dot{x} = Ax$ is stable, set up $A'V + VA = -W$ and solve this for V. If V is positive definite, the system is stable.

This is both a necessary and sufficient condition for stability. Since W may be any symmetric positive definite matrix, we can choose it to be I such that $A'V = VA = -I$.

Example 1:

$$\dot{x} = Ax \text{ where } A = \begin{bmatrix} -1 & 3 \\ 0 & -2 \end{bmatrix} \text{ setting } -W = -I,\ A'V + VA = -I \text{ gives}$$

$$\begin{bmatrix} -1 & 0 \\ 3 & -2 \end{bmatrix} \begin{bmatrix} v_{11} & v_{12} \\ v_{12} & v_{22} \end{bmatrix} + \begin{bmatrix} v_{11} & v_{12} \\ v_{12} & v_{22} \end{bmatrix} \begin{bmatrix} -1 & 3 \\ 0 & -2 \end{bmatrix} = \begin{bmatrix} -1 & 0 \\ 0 & -1 \end{bmatrix}.$$

Since is symmetric, this is reduced to the following system:

$$\begin{bmatrix} -2 & 0 & 0 \\ 3 & -3 & 0 \\ 0 & 6 & -4 \end{bmatrix} \begin{bmatrix} v_{11} \\ v_{12} \\ v_{22} \end{bmatrix} = \begin{bmatrix} -1 \\ 0 \\ -1 \end{bmatrix}$$

whose solution is $(v_{11}\ v_{12}\ v_{22}) = (1/2\ 1/2\ 1)$, i.e.

$$V = \begin{bmatrix} 1/2 & 1/2 \\ 1/2 & 1 \end{bmatrix}.$$

A Liapunov function is $L(x) = x'Vx = 1/2\ x_1^2 + x_1x_2 + x_2^2$ and $\dot{L}(x) = \nabla L(x)\dot{x} = -x_1^2 - x_2^2 < o$: the system is stable. In this simple example, stability is obvious from the simple observation that the eigenvalues of A are -1 and -2.

Example 2:

Consider the excess demand $E(p) = b/p - a$ and the market dynamic law $\dot{p} = kE(p)$, $(k > o)$. A simplest Liapunov function is $L(p) = (p - p_e)^2$

(i.e. $V = 1$), and the equilibrium price is $p_e = b/a$

$$\frac{dL(p)}{dt} = 2(p-p_e)\dot{p} = -\frac{2k(ap-b)^2}{ap} < 0 \text{ since } k,\ a > 0, \text{ i.e.}$$

the market is stable.

Example 3. The stability of optimal economic growth

Consider the Hamiltonian $H(p, x)$ which is concave in p and convex in x. Samuelson (1972) has shown that the equilibrium is stable. His Lyapunov function is $V \equiv (x - x^*)'(p - p^*)$. Differentiating V, making use of the fact that at (x^*, p^*), the gradient of H vanishes, we have

$$\dot{V} = -\Sigma(x_i - x_i^*)(H_{x_i} - 0) + \Sigma(p_i - p_i^*)(H_{p_i} - 0) < 0$$

by virtue of the posited curvature of H.

By further substitutions and making use of the various properties and assumptions of economic growth model, Cass & Shell (1976), Brock and Scheinkman (1976) and Magill (1977) have shown that the derivative of the Liapunov function V above gives, for all non-zero vector ζ

$$\dot{V} = \zeta' A\, \zeta < 0$$

where $A \equiv \begin{vmatrix} -H^\circ_{xx} & (\delta/2)I \\ (\delta/2)I & H^\circ_{pp} \end{vmatrix}$ is negative definite. Thus the equilibrium is stable, i.e. as time goes on, $p \to p^*$ and $x \to x^*$.

REFERENCES

Adams, F. G. and E. Burmeister. "Economic Models," *IEEE Transactions on Systems, Man and Cybernetics*, SMC-3 No. 1, January 1973.

Allen, R. G. D. *Mathematical Analysis for Economists*, Macmillan, London 1938.

Allen, R. G. D. and J. R. Hicks. "A Reconsideration of the Theory of Value," Parts I-II, *Economica*, NS 1: 52-76; 196-219, February, May, 1934.

Alonso, W. *Location and Land Use*, Harvard University Press, Cambridge Mass. 1964.

Aoiki, M. *Optimal Control and System Theory in Dynamic Economic Analysis*, North Holland, N.Y., 1976.

Arnold, V. I. *Ordinary Difference Equations*, MIT Press, 1978.

Arrow, K.J. and L. Hurwicz. "On the Stability of the Competitive Equilibrium: I & II," *Econometrica*, 26: 522-552, October 1958; *Econometrica*, 27: 82-109, January, 1959.

Arrow, K. J. and M. Kurz. *Public Investment, The Rate of Return and Optimal Fiscal Policy*, John Hopkins Press, Baltimore, Md., 1970.

Arthur, W. B. and G. McNicoll. "Optimal Time Paths with Age-Dependence: A Theory of Population Policy," *Review of Economic Studies*, XLIV(1): 111-123, February 1977.

Athans, M. and P. L. Falb. *Optimal Control*, McGraw-Hill, New York 1966.

Bardhan, P. K. "Optimal Accumulation and International Trade," *Review of Economic Studies*, 32: 241-244, 1965.

Bardhan, P. K. "Optimum Foreign Borrowing" in Shell, K. (ed) *Essays on the Theory of Optimal Economic Growth*, MIT Press, Mass. 1967.

Barnett, S. *Introduction to Mathematical Control Theory*, Clarendon Press, Oxford, 1975.

Bellman, R. *Dynamic Programming*, Princeton University Press, Princeton, N.J. 1957.

Bellman, R. *Introduction to Matrix Analysis*, (2nd edn.), McGraw-Hill, NY 1960.

Benavie, A. *Mathematical Techniques for Economic Analysis*, Prentice-Hall, N.J. 1972.

Ben-Porath, Y. "The Production of Human Capital and the Life Cycle of Earnings," *Journal of Polictical Economy*, 75: 352-365, August, 1967.

Bensoussan A., Hurst, E. G. and B. Naslund. *Management Applications of Modern Control Theory*, North Holland, Amsterdam 1974.

Bensoussan A., P. R. Kleindorfer, and C.S. Tapiero (eds.), *Studies in Management Sciences*, Vol. 9, 1978, North Holland 1978.

Berkovitz, L. D. "Variational Methods in Problems of Control and Programming," *Journal of Mathematical Analysis Applications*, 3, 145-169, 1961.

Berkovitz, L. D. "An Optimum Thrust Control Problem," *Journal of Mathematical Analysis and Application*, 3: 122-132, August 1961.

Bliss, G. A. *Calculus of Variations*, Mathematical Association of America, The Open Court Publishing Co., LaSalle, Ill. 1925.

Bode, W. H. *Network Analysis and Feedback Amplifier Design*, D. Van Nostrand Co., N.Y. 1945.

Bolza, O. *Lectures on the Calculus of Variations*, University of Chicago Press, Chicago 1904.

Brito, D. L. "A Dynamic Model of an Armaments Race" *International Economic Review*, 13(2): 359-375, June 1972.

Brock, W. A. and J. A. Scheinkman. "Global Asymptotic Stability of Optimal Control Systems with Applications to the Theory of Economic Growth," in Cass, D. and Shell, K. (eds.), (1976).

Bryson, A. E., W. F. Denham and S. E. Dreyfus. "Optimal Programming Problems with Inequality Constraints I: Necessary Conditions for Extremal Solutions," *AIAA Journal*, 1, No. 11: 2544-2550, November, 1963.

Bryson, A. E. and Y. C. Ho. *Applied Optimal Control*, Ginn & Co., Waltham, Mass., 1969.

Burmeister, E. and A. R. Dobell. *Mathematical Theories of Economic Growth*, Macmillan, London, 1970.

Calvo, G. A. "Devaluation: Levels vs. Rates," *Journal of International Economics*, Vol. 11: 165-172, 1981.

Cass, D. "Optimal Growth in an Aggregate Model of Capital Accumulation: A Turnpike Theorem, *Econometrica* 34, October, 1966.

Cass, D. and K. Shell (eds.). *The Hamilton Approach to Dynamic Economics*, Academic Press, New York, 1976.

Cass, D. and K. Shell. "The Structure and Stability of Competitive Dynamical Systems," in Cass, D. and K. Shell (eds.), (1976).

Chakravarty, S. and A. S. Manne. "Optimal Growth when the Instantaneous Utility Function depends upon the Rate of Change in Consumption," *American Economic Review* Vol. 58(5): 1351-1354, December 1968.

Chiang, A. *Fundamental Methods of Mathematical Economics*, McGraw-Hill 1974.

Chorlton, F. *Ordinary Differential and Difference Equations*, D. Van Nostrand Co., Princeton, N.J. 1965.

Chow, G. C. "Problems of Economic Policy from the Viewpoint of Optimal Control," *American Economic Review*, 43(5): 825-837, December 1973.

Chow, G. C. *Analysis and Control of Dynamic Economic Systems*, Wiley, N.Y., 1975.

Clark, C. W. *Mathematical Bioeconomics*, J. Wiley & Sons, N.Y. 1976.

Clark, C. and G. R. Munro. "The Economics of Fishing and Modern Capital Theory," in Mirman, L. J. and D. F. Spulber, (eds.), *Essays in the Economics of Renewable Resources*, North Holland, Amsterdam 1982.

Coddington, E. A. and N. Levinson. *Theory of Ordinary Differential Equations*, McGraw-Hill, N.Y. 1955.

Conlisk, J. "Quick Stability Checks and Matrix Norms," *Economica*, NS 40: 402-409, November 1973.

Cremer, J. "On Hotelling's Formula and the Use of Permanent Equipment in the Extraction on Natural Resources," *International Economic Review*, Vol. 20(2): 317-324, June 1979.

Dasgupta, P. S. "On the Concept of Optimal Population," *Review of Economic Studies*, 36(3): 295-318, July 1969.

Datta-Chaudhuri, M. "Optimum Allocation of Investments and Transportation in a Two-Region Economy," in Shell, K. (ed.): *Essays on the Theory of Optimal Growth*, MIT, Mass. 1967.

Deger, S. and S. sen. "Optimal Control and Differential Game Models of Military Expenditure in Less Developed Countries," *Birkbeck College Discussion Paper* No. 104, London 1981.

Dixit, A. "The Optimal Factory Town," *Bell Journal of Economics*, 4: 637-651, Autumn 1973.

Dobell, A. R. and Y. C. Ho. "Optimal Investment Policy: An Example of a Control Problem in Economic Theory," *I.E.E.E. Transactions on Automatic Control*, AC 12, No. 1: 3-14, February, 1967.

Domar, E. D. "Capital Expansion, Rate of Growth and Employment," *Econometrica*, 14: 137-147, April 1946.

Dorfman, R., P. A. Samuelson and R. M. Solow. *Linear Programming and Economic Analysis*, McGraw-Hill, New York, 1958.

Dorfman, R. "An Economic Interpretation of Optimal Control Theory," *American Economic Review*, 59(5): 817-831, December, 1969.

Dreyfus, S. "Variational Problems with Inequality Constraints," *Journal of Mathematical Analysis and Applications*, 4, No. 2: 297-308, April 1962.

Dreyfus, S. E. *Dynamic Programming and the Calculus of Variations*, Academic Press, N.Y. 1965.

Ekman, E. V. "A Dynamic Financial Model of a Managerial Firm," in Feichtinger, G. (ed.) *Optimal Control Theory and Economic Analysis*, North Holland, Amsterdam 1982.

Elsgolc, L. E. *Calculus of Variations*, Addison-Wesley, Reading, Mass. 1962.

Evans, G. C. "The Dynamics of Monopoly," *American Mathematical Monthly*, Vol 31(2): 77-83, February 1924.

Evans, G. C. *Mathematical Introduction to Economics*, McGraw-Hill, N.Y. 1930.

Fan, L. T. and C. S. Wang. *The Discrete Maximum Principle*, Wiley & Sons, N.Y. 1964.

Feichtinger, G. (ed.) *Optimal Control Theory and Economic Analysis*, North Holland, Amsterdam, 1982.

Feichtinger, G. "Optimal Policies for Two Firms in a Non-cooperative Research Project," in Feichtinger, G. (ed.) *Optimal Control Theory and Economic Analysis*, North Holland, Amsterdam 1982.

Forster, B. A. "On a One-State Variable Optimal Control Problem: Consumption-Pollution Trade-offs," in Pitchford, J. D. and S. J. Turnovsky (eds.), (1977).

Forsyth, A. R. *Calculus of Variations*, Cambridge University Press, England, 1927.

Fox, C. *An Introduction to the Calculus of Variations*, Oxford University Press, New York, 1950.

Garfinkel, B. "Inequalities in Variational Problem," in G. Leitmann (ed.) *Topic in Optimization*, Ch. 1: 3-25, Academic Press, N.Y. 1967.

Gelfand, I. M. and C. V. Fomin, *Calculus of Variations*, Prentice-Hall, Englewood Cliffs, N.J., 1963.

Gillespie, R. P. *Partial Differentiations*, Oliver & Boyd Ltd., 2nd edn., London 1954.

Goldberg, S. *Introduction to Difference Equations*, J. Wiley & Sons, N.Y., 1958.

Gordon, H. S. "Economic Theory of a Common-property Resource: the Fishery," *Journal of Political Economy*, 62: 124-142, April 1954.

Gordon, M. J. *The Investment, Financing and Valuation of the Corporation*, R. D. Irwin Inc., Homewood, Ill. 1962.

Gordon, R. L. "A Reinterpretation of the Pure Theory of Exhaustion," *Journal of Political Economy*, 75(3): 274-286, June 1967.

Gould, J. P. "Diffusion Processes and Optimal Advertising Policy," in Phelps, E. S. (ed.) *Microeconomic Foundations of Employment and Inflation Theory*, Norton Co., N.Y. 1970.

Grandville, O. de la, "Capital Theory, Optimal Growth and Efficiency Conditions with Exhaustible Resources," *Econometrica*, 48(7): 1763-1776, November 1980.

Hadley, G. and M. C. Kemp. *Variational Methods in Economics*, North Holland, Amsterdam 1971.

Hahn, W. *Theory and Application of Liapunov's Direct Method*, Prentice-Hall, Englewood Cliffs, N.J. 1963.

Halkin, H. "Necessary Conditions for Optimal Control Problems with Infinite Horizons," *Econometrica*, Vol. 42: 267-272, March 1974.

Hamada, K. "Economic Growth and Long Term International Capital Movements," *Yale Economic Essays*, Vol. VI, Spring 1966.

Hamada, K. "Optimal Capital Accumulation by an Economy Facing an International Capital Market," *Journal of Political Economy*, 77(4, Pt. II): 684-697, July/August 1969.

Harrod, R. F. "An Essay in Dynamic Theory," *Economic Journal*, vol. XLIX: 14-33, March 1939.

Henderson, D. W. and T. J. Sargent. "Monetary and Fiscal Policy in a Two-Sector Aggregate Model," *American Economic Review*, 63(3): 345-365, 1973.

Hestenes, M. R. "A General Problem in the Calculus of Variations with Applications to Paths of Least Time," *RAND Corp.*, RM-100, February 1949.

Hestenes, M. R. "On Variational Theory and Optimal Control Theory," *Journal of SIAM Series A*, Control 3: 23-48, 1965.

Hestenes, M. R. *Calculus of Variations and Optimal Control Theory*, Wiley & Sons, New York, 1966.

Hicks, J. R. *Value and Capital*, Oxford University Press, London 1939.

Hicks, J. R. *The Theory of Wages*, MacMillan, London 1932.

Hicks, J. R. "The Foundations of Welfare Economics," *Economic Journal*, 49: 697-712, December 1939.

Hicks, J. R. *A Contribution to the Theory of Trade Cycle*, Oxford University Press, Oxford 1950.

Hirsch, M. W. and S. Smale. *Differential Equations, Dynamic Systems and Linear Algebra*, Academic Press, NY 1974.

Holt, C, F. Modigliani, J. Muth, and H. Simon. *Planning Production Inventories and Work Force*, Prentice-Hall, Englewood Cliffs, N.J. 1969.

Holtzman, J. M. "Convexity and the Maximum Principle for Discrete Systems," *IEEE Transactions Automatic Control* AC-11, January 1966.

Hotelling, H. "The Economics of Exhaustible Resources," *Journal of Political Economy*, 39(2): 137-175, April 1931.

Hurewicz, W. *Lectures on Ordinary Differential Equations*, MIT, Cambridge, 1958.

Hwang, C. L., L. T. Fan, and L. E. Erickson. "Optimal Production Planning by the Maximum Principle," *Management Science*, 13: 750-755, 1967.

Intrilligator, M. D. *Mathematical Optimization and Economic Theory*, Prentice-Hall 1971.

Isaacs, R. *Differential Games*, Wiley, N.Y. 1965.

Jacquemin, A. P. "Optimal Control and Advertising Policy," *Metroeconomica*, 25: 200-209, May- August 1973.

Jammernegg, W. "Conditions for Optimality of Isotone Policies in Production-Inventory Systems," in Feichtinger, G. (ed.), *Optimal Control Theory and Economic Analysis*, North Holland, Amsterdam 1982.

Johnson, C. D. "Singular Solutions in Problems of Optimal Control," in C. T. Leondes, (ed.): *Advances in Control Systems: Theory and Applications*, Vol. 2, Ch. 4, Acandemic Press, N. Y. 1965.

Johnson, C. D. and J. E. Gibson. "Singular Solutions in Problems of Optimal Control," *IEEE Transactions on Automatic Control*, Vol AC-8 No. 1, January 1963.

Jorgenson, D.W. "The Theory of Investment Behavior," in *Determinants of Investment Behavior*, NBER, N.Y. 1967.

Kalman, R. E. "Contributions to the Theory of Optimal Control," *Sociedad Matematica Mexicana Boletin*, 5: 102-119, 1960.

Kalman, R. E. "Mathematical Description of Linear Dynamical Systems," *Journal of SIAM Control, Series A*, 1: 152-192, 1963.

Kalman, R. E. "The Theory of Optimal Control and the Calculus of Variations," in R. Bellman (ed.) *Mathematical Optimization Techniques*, University of California, Berkeley, California, 1963.

Kamien, M. I. and N. L. Schwartz. *Dynamic Optimization: The Calculus of Variations and Optimal Control in Economics and Management*, Elsevier North Holland Inc., New York 1981.

Kamien, M. I. and N. Schwartz. "Sufficient Conditions in Optimal Control Theory," *Journal of Economic Theory*, 3(2): 207-214, June 1971.

Kareken, J. A., T. Muench, and N. Wallace. "Optimal Open Market Strategy: The Use of Information Variables," *American Economic Review*, LXIII(1): 156-172, 1973.

Keeler, E., M. Spence, and Zeckhauser, R., "The Optimal Control of Pollution," *Journal of Economic Theory*, 4(1): 19-34, February 1972.

Kemp, M. C. and N. V. Long (eds.). *Exhaustible Resources, Optimality and Trade*, North Holland, Amsterdam 1980.

Kemp, M. C. and N. V. Long. "Toward a More General Theory of the Mining Firm," in Kemp, M. C. and N. V. Long (eds.), 1980.

Kendrick, D. and L. Taylor. "Numerical Methods and Nonlinear Optimizing Models for Economic Planning," in Chenery, H. B. (ed.) *Studies in Development Planning*, Harvard University Press, Cambridge 1971.

Kendrick, D. "Applications of Control Theory to Macroeconomics," in M. D. Intrilligator, (ed.), *Frontiers of Quantitative Economics*, Vol. III A, Ch. 7: 239-261, North Holland, Amsterdam 1977.

Kendel, J. L. *Dynamic Linear Economic Models*, Gordon & Breach Science, New York, 1974.

Kirk, D. E. *Optimal Control Theory: An Introduction*, Prentice-Hall, Englewood Cliffs, N.J. 1970.

Kline, M. *Mathematics: A Cultural Approach*, Addison-Wesley, Reading, Mass. 1962.

Knowles, G. *An Introduction to Applied Optimal Control*, Academic Press, N.Y. 1981.

Koizumi, T. and K. J. Kopecky. "Foreign Direct Investment, Technology Transfer and Domestic Employment Effects," *Journal of International Economics* vol. 10: 11-20, 1980.

Koo, D. *Elements of Optimization with Applications in Economics and Business*, Springer-Verlag, N.Y. 1977.

Krouse, C. G. "Optimal Financing and Capital Structure Program for the Firm," *The Journal of Finance*, Vol. XXVII(5): 1057-1071.

Krouse, C. G. and W. Y. Lee. "Optimal Equity Financing of the Corporation," *Journal of Finance and Quantitative Analysis*, Vol. 8: 539-563, September 1973.

Lancester, P. *Theory of Matrices*, Academic Press, New York 1969.

Lane, J. S. "A Synthesis of the Ramsey-Meade Problems when Population Change is Endogenous," *Review of Economic Studies*, XLII(1): 57-66, January, 1975.

Lee, E. B. and L. Markus. *Foundations of Optimal Control Theory*, McGraw-Hill, New York 1967.

Leitman, G. *Introduction to Optimal Control*, McGraw-Hill, N.Y. 1966.

Levhari, D. and N. Liviatan. "Notes on Hotelling's Economics of Exhaustible Resources," *Canadian Journal of Economics*, 10: 177-192, May 1977.

Levine, J. and J. Thepot, "Open Loop and Closed Loop Equilibria in a Dynamic Duopoly," in Feichtinger, G. (ed.), 1982.

Levy, H. and F. Lesman. *Finite Difference Equations*, Isaac Pitman & Sons, London 1959.

Lindorff, D. P. "Sensitivity in Sampled Data Systems," *I.R.E. Transactions on Automatic Control*, A.C.-8: 120-124, April 1963.

Linter, J. "The Coil of Capital and Optimal Financing of Corporate Growth," *Journal of Finance*, 23: 292-310, 1963.

van Loon, P. "A Dynamic Theory of the Firm," in Feichtinger, G. (ed.) 1982.

Lucas, R. E. "Adjustment Costs and the Theory of Supply," *Journal of Political Economy*, 75: August 1967.

Luenberger, D. G. "A Nonlinear Economic Control Problem With a Linear Feedback Solution," *IEEE Transactions on Automatic Control*, AC-20(2): 184-191, April 1975.

Luptacik, M. and U. Schubert. "Optimal Investment Policy in Productive Capacity and Pollution Abatement Processes in a Growing Economy," in Feichtinger, G. (ed.), 1982.

Magill, M. J. P. "On a General Economic Theory of Motion," *Lecture Notes in Operation Research and Mathematical Systems*, No. 36, 1970.

Magill, M. J. P. "Some New Results on the Local Stability of the Process of Capital Accumulation," *Journal of Economic Theory*, 15(1): 174-210, June 1977.

Mangasarian, O. L. "Sufficient Conditions for the Optimal Control of Non-linear Systems," *Journal of SIAM Control*, Vol. 4, February, 1966.

Manning, R. "Issues in Optimal Educational Policy in a Context of Balanced Growth," *Journal of Economic Theory*, 13: 380-395, December 1976.

Manning, R. "Optimal Wage Differentials in Balanced Growth in an Aggregative Model of Education with Higher Degrees," *Australian Economic Papers* 17: 81-90, June 1978.

Manning, R. "Two Theorems Concerning Optimal Educational Policy in Balanced Growth," *Journal of Economic Theory*, 21(3): 465-472, December 1979.

Marshall, A. *Principles of Economics*, MacMillan, London 1890.

Masse, P. *Optimal Investment Decisions*, Prentice-Hall, Englewood Cliffs, N.J. 1962.

McKenzie, L. W. "Matrices With Dominant Diagonals and Economic Theory," in K. Arrow, S. Karlin, and P. Suppes, (eds.) *Mathematical Methods in the Social Sciences*, Stanford University Press, California 1960.

McShane, E. J. "On Multipliers for Lagrange Problems," *American Journal of Mathematics*, 61: 809-819, 1939.

Meade, J. E. "Trade and Welfare," in *The Theory of International Economic Policy*, Oxford University Press 1966.

Meade, J. E. "The Growing Economy," in *The Principles of Political Economy*, Allen & Unwin, London 1968.

Menger, K. "What is Calculus of Variations and What are its Applications," in J. R. Newman (ed.): *The World of Mathematics*, Vol. 2, Simon & Schuster, N.Y. 1956.

Merton, R. C. "Optimal Consumption and Portfolio Rules in a Continuous Time Model," *Journal of Economic Theory*, Vol. 3(4): 373-413, 1971.

Metzler, L. A. "Stability of Multiple Markets: The Hicks Conditions," *Econometrica*, 13(4): 277-292, October 1945.

Miller, M. H. and F. Modigliani. "Dividend Policy, Growth and Valuation of Shares," *The Journal of Business*, 34: 411-433, 1961.

Miller, R. E. *Dynamic Optimization and Economic Applications*, McGraw-Hill New York 1979.

Mills, E. S. "An Aggregate Model of Resource Allocation in a Metropolitan Area," *American Economic Review*, 57: 197-210, May 1967.

Mills, E. S. and J. MacKinnon. "Notes on the New Urban Economics," *Bell Journal of Economics*, 4: 593-601, Autumn 1973.

Mirman, L. J. and Spulber, D. F. (eds.) *Essays in the Economics of Renewable Resources*, North Holland, Amsterdam, 1982.

Mirrlees, J. A. "The Optimal Town," *Swedish Journal of Economics*, 74: 114-135, 1972.

Muth, R. F. *Cities and Housing*, University of Chicago Press, Chicago 1969.

Muth, F. R. "Recent Developments in the Theory of Urban Spatial Structure," in M. D. Intrilligator (ed.) *Frontiers of Quantitative Economics*, IIIB, North Holland Publishing Co., Amsterdam, 1977.

Myoken, H. "Optimal Stabilization Policies for Decentralized Macroeconomic Systems with Conflicting Targets," in Liu, P. T. (ed.) *Dynamic Optimization and Mathematical Economics*, Plenum Press, N.Y. 1980.

Neck, R. and U. Posch. "On the Optimality of Macro-economic Policies: An Application to Austria," in Feichtinger, G. (ed.) (1982).

Nerlove, M. and K. J. Arrow. "Optimal Advertising Policy under Dynamic Conditions," *Economica* 39: 129-142, May 1962.

Newman, P. K. "Some Notes On Stability Conditions," *Review of Economic Studies*, 27:1-9, 1959-1960.

Oniki, H. "Comparative Dynamics in Optimal Control Theory," *Journal of Economic Theory*, 6: 265-283, June 1973.

Oron, Y., D. Pines and Sheshinski, E. "Optimum vs. Equilibrium Land Use Patterns and Congestion Toll," *Bell Journal of Economics*, 4: 619-636, 1973.

Perrakis, S. and I. Salun. "Resource Allocation and Scale Opterations in a Monopoly Firm: A Dynamic Analysis," *International Economic Review*, 13: 399-407, June 1972.

Peston, M. "Econometrics and Control: Some General Comments," *IFAC/IFORS International Conference on Dynamic Modelling and Control of National Economics*, (pp. 15-30), University of Warwick 1973 Proceedings.

Peterson, D. W. and Lerner, E. M. "Optimal Control And Monetary Policy," *International Economic Review*, 12(2): 186-195, June 1971.

Petrov, I. P. *Variational Methods in Optimal Control Theory*, (Translated by M. D. Friedman), Academic Press, N.Y. 1968.

Pindyck, R. S. "An Application of the Linear Quadratic Tracking Problem to Economic Stabilization Policy," *IEEE Transactions on Automatic Control*, AC-17(3): 287-300, June 1972.

Pindyck, R. S. *Optimal Planning for Economic Stabilization*, North Holland Publishing Co., Amsterdam, 1973.

Pindyck, R. S. "The Optimal Phasing of Phased Deregulation," *Journal of Economic Dynamics and Control*, 4(3): 281-294, August 1982.

Pitchford, J. D. *Population in Economic Growth*, North Holland, Amsterdam 1974.

Pitchford, J. D. and S. J. Turnovsky, (eds.) *Applications of Control Theory to Economic Analysis*, North Holland, Amsterdam 1977.

Plourde, C. G. "A Model of Waste Accumulation and Disposal," *Canadian Journal of Economics*, 5(1): 119-125, February, 1972.

Pontryagin, L. S. *Ordinary Differential Equations*, Addison-Wesley 1962.

Pontryagin, L. S., V. G. Boltyanskii, R. V. GamKrelidze and E. R. Mishchenko. *The Mathematical Theory of Optimal Processes*, Interscience, New York, 1962.

Preston, A. J. "A Paradox in the Theory of Optimal Stabilization," *Review of Economic Studies*, 120: 423-432, October 1972.

Rahman, M. A. "Regional Allocation of Investment," *Quarterly Journal of Economics*, LXXII: 26-39, February 1963.

Ramsey, F. P. "A Mathematical Theory of Savings," *Economic Journal*, 38(152): 543-549, December 1928.

Rapp, B. *Models for Optimal Investment and Maintainance Decisions*, Halsted Press, Stockholm 1974.

Richardson, L. F. *Arms and Insecurity*, Boxwood Press, Pittsbugh 1960.

Riley, J. G. "Gammaville: An Optimal Town," *Journal of Economic Theory*, 6(5): 471-482, October 1973.

Roberts, B. And D. L. Schulze. *Modern Mathematics and Economic Analysis*, Norton, N.Y. 1973.

Rose, H. *An Asymptotic Necessary Condition for Optimum Programs When Time is Continuous and the Horizon is Infinite*, John Hopkins University Working Papers in Economics, No. 31, June, 1977.

Rothschild, M. "On the Cost of Adjustment," *Quarterly Journal of Economics*, 85: 605-622, 1971.

Rozonoer, . T. "L. S. Pontryagin's Maximum Function Principle in its Application to the Theory of Optimum Systems," I, II, III, translated in *Automation and Remote Control*, 20 (1959): 1288-1302, 1405-1421 and 1517-1532.

Rozonoer, L. I. "L. S. Pontryagin's Maximum Principle in Optimal Control Theory," *Automation and Remote Control*, Vol. No. 20, October, November, and December 1959.

Ryder, H. E. Jr. "Optimal Accumulation and Trade in an Open Economy of Moderate Size," in Shell, K. (ed.): *Essays on the Theory of Optimal Economic Growth*, MIT, Mass. 1967.

Ryder, H. E. Jr. "Optimal Accumulation in a Two-Sector Neo-Classical Economy with Non-Shiftable Capital," *Journal of Political Economy*, 77(4, Pt. II): 665-683, July/August 1969.

Sage, A. P. *Optimal Systems Control*, Prentice-Hall, Englewood Cliffs, N.J. 1968.

Samuelson, P. A. "Some Aspects of the Pure Theory of Capital," *Quarterly Journal of Economics*, 51(2): 469-496, May 1937.

Samuelson P. A. "Interactions Between the Multiplier Analysis and the Principles of Acceleration," *Review of Economic Statistics*, 21: 75=78, May 1939.

Samuelson, P. A., *Foundations of Economic Analysis*, Harvard University Press, Cambridge 1947.

Samuelson, P. A. "A Catenary Turnpike Involving Consumption and the Golden Rule," *American Economic Review*, LV: 486-496, June 1965.

Samuelson, P. A. "The General Saddle-Point Property of Optimal Control Motions," *Journal of Economic Theory*, 5: 102-120, August 1972.

Samuelson, P.A. and R. M. Solow, "A Complete Capital Model Involving Heterogeneous Capital Goods," *Quarterly Journal of Economics*, LXX: 537-562, November 1956.

Samuelson, P. A. "Efficient Paths of Capital Accumulation in Terms of the Calculus of Variations," in Arrow, K. J., S. Karlin, and P. Suppes (eds.) *Mathematical Methods in the Social Sciences*, 1959 Proceedings of the First Stanford Symposium, Stanford 1960.

Sargent, T. J. *Macroeconomic Theory*, Academic Press, N.Y. 1979.

Sato, R. And Davis, E. G. "Optimal Savings Policy When Labour Grows Endogenously," *Econometrica*, 39(6), November 1971.

Scott, A. D. "The Fishery: The Objectives of Sole Ownership," *Journal of Political Economy*, 63:116-124, April 1955.

Seirstad, A. and K. Sydsaeter. "Sufficient Conditions in Optimal Control Theory," *International Economic Review*, 18(2): 367-391, June 1977.

Sethi, S. P. "Dynamic Optimal Control Models in Advertising A Survey," *SIAM Review*, 19(4): 685-725, October 1977.

Sethi, S. P. "Optimal Equity and Financing Model of Krouse and Lee: Corrections and Extensions," *Journal of Financial and Quantitative Analysis*, 13(3): 487-505, September 1978.

Sethi, S. P. "A Survey of Management Science Applications of the Deterministic Maximum Principle," *TIMS Studies in Management Sciences*, 9 (1978): 33-67.

Sethi, S. P. and Thompson, G. L. *Optimal Control Theory: Management Science Applications*, Martinus Nijhoff, Boston 1981.

Shell, K. "Applications of Pontryagin's Maximum Principle to Economics," in Szego, G. P. and H. W. Kuhn (eds.) *Proceedings of the Varenna Summer School on Mathematical Systems, Theory and Economics*, Springer-Verlag, New York 1969.

Shell, K. (ed.) *Essays on the Theory of Optimal Economic Growth*, MIT, 1967.

Shupp, F. R. "Optimal Policy Rules for a Temporary Incomes Policy," *Review of Economic Studies*, pp. 249-259.

Simaan, M. and Cruz, J. B., "Formulation of Richardson's Model of Arms Race from a Differential Game Viewpoint," *Review of Economic Studies*, XLII(1): 67-77, January, 1975.

Simaan, M. A. and T. Takayama, "Optimum Monoposlist Control in a Dynamic Market," *IEEE Transactions on Systems, Man and Cybernetics*, SMC 6(12): 799-807, December 1976.

Simon, H. A. "On the Application of Servomechanism Theory in the Study of Production Control," *Econometrica* 20: 247-268, 1952.

Slutsky, E. E. "On the Theory of The Budget of the Consumer," *Gionale degli Economisti*, 51: 1-26, July 1915.

Smith, D. R. *Variational Methods in Optimization*, Prentice-Hall, N.J. 1974.

Smith, V. L. "Economics of Production from Natural Resources," *American Economic Rewiew*, 58: 409-431, June 1968.

Smith, V. L. "On Models of Commercial Fishing," *Journal of Political Economy*, 77: 181-198, March, April 1969.

Smith, V. L. "Dynamics of Waste Accumulation: Disposal vs. Recycling," *Quarterly Journal of Economics*, 86: 600-616, November, 1972.

Soderstrom, H. T. "Production and Investment Under Costs of Adjustment: A Survey," *Zeitschrift fur National okonomie*, 36(1976): 369-388.

Solow, R. M. "A Contribution to the Theory of Economic Growth," *Quarterly Journal of Economics*, 70: 65-94, February 1956.

Solow, R. M. "Congestion Cost and the Use of Land for Streets," *Bell Journal of Economics*, 4: 406-618, Autumn 1973.

Solow, R. M. "Intergenerational Equity and Exhaustible Resources," *Review of Economic Studies*, Symposium 1974.

Takayama, A. *Mathematical Economics*, Dryden Press, Ill. 1974.

Takayama, A. "Regional Allocation of Investment: A Further Analysis," *Quarterly Journal of Economics*, LXXI: 330-337, May 1967.

Terborgh, G. *Dynamic Equipment Policy*, McGraw-Hill, N.Y. 1949.

Thom, R. *Stabilite Structurelle et Morphogenese*, Benjamin Inc. Reading, Mass. 1972.

Thompson, G. L. "Optimal Maintenance Policy and Sale Date of a Machine," *Management Science*, 14: 543-550, 1968.

Tomovic, R. *Sensitivity Analysis of Dynamic Systems*, McGraw-Hill, N.Y. 1963.

Tomovic, R. and M. Vukobratovic. *General Sensitivity Theory*, American Elsevier, N.Y. 1972.

Treadway, A. B., "On Rational Entrepreneurial Behaviour and the Demand for Investment," *Review of Economic Studies*, 36: 227-239, 1969.

Tsurumi, H. and Y. Tsurumi."Simultaneous Determination of the Market Share and Advertising Expenditure Under Dynamic Conditions," *The Economic Studies Quarterly*, 22: 1-23, 1971.

Tu, P. N. V. "Optimal Educational Investment Program in an Economic Planning Model," *Canadian Journal of Economics*, 2: 52-64, February 1969.

Turnovsky, S. J. *Macroeconomic Analysis and Stabilization Policy*, Cambridge University Press, 1977; reprinted 1981.

Uzawa, H. "On a Two-sector Model of Economic Growth," *Review of Economic Studies*, 29: 40-47, October 1961; and 30: 105-118, June, 1963.

Uzawa, H. "Optimal Growth in a Two Sector Model of Capital Accumulation," *Review of Economic Studies*, 31: 1-24, January, 1964.

Uzawa, H. "Optimal Techical Change in an Aggregative Model of Economic Growth," *International Economic Reniew*, 6: 18-31, January 1965.

Uzawa, H. "An Optimal Fiscal Policy in an Aggregative Model of Economic Growth," in Adelman, I. and E. Thorbecke (eds.) *The Theory and Design of Economic Development*, John Hopkins Press, Md. 1966.

Valentine, F. A. *The Problem of Lagrange with Differential Inequalities as Added Side Conditions, Contributions to the Calculus of Variations, 1933-1937*, University of Chicago Press, Chicago, Ill. 1937.

Varaiya, P. "On the Design of Rent Control," *IEEE Transactions on Automatic Control*, AC-21(3): 316-319, June 1976.

Wong, R. E. "Profit Maximization and Alternative Theory: A Dynamic Reconciliation," *American Economic Review*, 65(4): 689-694, September 1975,

Yamane, T. *Mathematics for Economists*, Prentice-Hall, N.J. 1968.

Zeeman, E. C. *Catastrophe Theory: Selected Papers 1972-1977*, Addison-Wesley, Mass. 1977.